Science for Agriculture
A Long-Term Perspective

Second Edition

Science for Agriculture
A Long-Term Perspective

Second Edition

W.E. Huffman and R.E. Evenson

Wallace E. Huffman is C.F. Curtis Distinguished Professor of Agriculture and Professor of Economics and Agricultural Economics at Iowa State University, Ames. He received his Ph.D. from the University of Chicago.

Robert E. Evenson is Professor of Economics at Yale University, New Haven, CT. He received his Ph.D. from the University of Chicago.

©2006 Blackwell Publishing

Blackwell Publishing Professional
2121 State Avenue, Ames, Iowa 50014, USA

Orders:	1-800-862-6657
Office:	1-515-292-0140
Fax:	1-515-292-3348
Web site:	www.blackwellprofessional.com

Blackwell Publishing Ltd
9600 Garsington Road, Oxford OX4 2DQ, UK
Tel.: +44 (0)1865 776868

Blackwell Publishing Asia
550 Swanston Street, Carlton, Victoria 3053, Australia
Tel.: +61 (0)3 8359 1011

Authorization to photocopy items for internal or personal use, or the internal or personal use of specific clients, is granted by Blackwell Publishing, provided that the base fee of $.10 per copy is paid directly to the Copyright Clearance Center, 222 Rosewood Drive, Danvers, MA 01923. For those organizations that have been granted a photocopy license by CCC, a separate system of payments has been arranged. The fee code for users of the Transactional Reporting Service is ISBN:13: 978-0-8138-0688-4; 10: 0-8138-0688-7/2006 $.10.

First edition, (c)1993 Iowa State University Press
Second edition, 2006

Library of Congress Cataloging-in-Publication Data

Huffman, Wallace E.
Science for agriculture : a long-term perspective / W.E. Huffman and R.E. Evenson.—2nd ed.
　　p. cm. —
　　Includes index.
　　ISBN-13: 978-0-8138-0688-4 (alk. paper)
　　ISBN-10: 0-8138-0688-7 (alk. paper)
　　1. Agriculture—Research—United States. 2. Agriculture—Research—Economic aspects—United States. I. Evenson, Robert E. (Robert Eugene), 1934– II. Title.

S541.H84 2006
630'.72'073—dc22

2005011946

The last digit is the print number: 9 8 7 6 5 4 3 2 1

Contents

Foreword

In little more than a decade since the publication of the first edition of *Science for Agriculture* the book has become a classic in its field. The book celebrated over a century of contributions by the United States Department of Agriculture—State Agricultural Experimentation research system to the growth of production and productivity of American agriculture.

This second edition is much more than an updating of their earlier work. It raises a series of questions about the future intellectual, economic, and political viability of the system in a world in which the structure of both the agricultural industry and the agricultural-research system has experienced dramatic change.

Crop production has become increasingly concentrated. Livestock production has become increasingly industrialized. And the traditional farm-level linkages between crop and livestock production have been severed.

Private-sector agricultural research has expanded to exceed public-sector agricultural research by several multiples. The federal-state agricultural research system is now primarily a producer of intermediate inputs—of scientific and technical knowledge that reaches the farmer in the form of materials produced by the private sector.

In a final chapter Huffman and Evenson raise a series of challenging issues that bear on the future of the system: the system's response to the biotechnology revolution; the changing structure of the industry; private support for agricultural research; the method of funding public agricultural research; and future productivity growth.

The book should be required reading for every U.S. Department of Agriculture official and by the presidents, deans, and experiment-station directors of state land grant universities. It should be placed on the desks of state governors and of state and federal legislators who are concerned about the future of the agricultural industry in the areas they represent.

Vernon W. Ruttan
University of Minnesota
vruttan@apec.umn.edu

Preface

Since the first edition of *Science for Agriculture* was published 12 years ago several events have markedly altered the agricultural landscape in the United States. Perhaps most prominent has been the continued, even accelerated, pace of structural change in U.S. agriculture. Structural change is reflected in the growing size of the farm and in the "industrialization" of livestock production. A reinforcing factor in structural change has been the introduction in 1993 of Bovine Somatotrophin Hormone (BST) for dairy cows and, in 1996, of several genetically modified (GM) crop products. The introduction of GM products for U.S. farmers has been undertaken by private-sector firms, most of which were agricultural chemical firms. The public-sector USDA-SAES system has not, to date, introduced successful GM products for U.S. farmers.

The SAES system has long relied on state political support for its public budget. SAES units have long been able to point to the public development of crop varieties and related technologies as indicators of service to farmers in the state. The failure to produce new GM products, ten years after private firms introduced GM products, is a source of concern and has the potential to undermine future public support of SAES research.

The biotechnology revolution also has implications for the teaching of science in land grant universities that are probably more important than the production of new GM products. In Chapter 9 of this edition, we review this issue and conclude that PhD students in College of Agriculture programs are receiving "modern science" training.

As new GM products have been made available to farmers, structural change has accelerated. The herbicide-tolerant GM products enable more land to be cultivated, as herbicides can be used to control weeds. Livestock production has effectively been industrialized, and with improved markets, the benefits from feed production on livestock-producing farms have been reduced. Many modern systems of poultry, pork, and dairy production are now characterized by contractual arrangements with grain dealers and meat processors.

Livestock productivity has improved; milk per cow, pigs per sow, and eggs per hen have increased, and days-to-slaughter weight for beef, pork, and poultry has decreased over recent decades.

The bulk of U.S. agriculture has effectively become large scale. The 2002 agricultural census data indicate that the 73,885 farms selling more than 500,000 dollars in products now produce 62.7 percent of all farm products in the United States. The associated changes in food retailing through supermarkets and increases in the "fixed costs" of farm production have accelerated these changes.

Traditional rural values appear to have remained intact, but they are under stress from the industrialization of agriculture. Land-grant colleges of agriculture have kept abreast of developments in science, but they too are under stress. We suggest in the final chapter of the revised edition that the pressures on the SAES system to introduce more competitive research grants will have regional implications. It is very likely that we will see more specialization in the small SAES units.

There is little question that changes in the biological sciences and in the structure of U.S. agriculture have been profound. The USDA-SAES research program and the land-grant university teaching programs have responded quite effectively to scientific changes. Structural changes in U.S. agriculture, however, will very likely lead to further changes in the conduct of public-sector agricultural research in the United States. Hence, this second edition is written because of the dramatic changes in both the structure of agriculture and in the advances in scientific opportunities for agriculture over the past decade or so. ESCOP (Experiment Station Committee on Organization and Policy) provided us with a small grant to examine the productivity of federal formula-funded SAES research, the so-called counter-factual study, and to re-examine the public-choice aspects of states' SAES funding. Xing Fan provided excellent research assistance for updating tables and figures and for making new ones. This project provided new empirical results that were shared and discussed with the state agricultural experiment-station directors at their fall meeting in 2002 and 2004 and are reported in Chapters 7 and 8. Also, a few thousand dollars of these funds were used to facilitate the final production of this second edition. For this funding we owe a debt to the late David MacKinzie, Executive Experiment Station Director for the Northeast Region, and his predecessor Tom Fretz. We also shared our conclusions in Chapter 9 with Cathy Woteki, Dean, College of Agriculture, Iowa State University, for early scrutiny by an SAES administrator. We thank these individuals for their input.

Introduction

This study presents a long-term perspective on a successful science system, one for agriculture in the United States. We piece together a story covering considerably more than a century. The primary objectives of this book are to (1) describe the evolution, development, and financing of the U.S. agricultural research system, (2) report some of its major long-term accomplishments or economic impacts, and (3) address new challenges facing the state agricultural experiment station (SAES)-U.S. Department of Agriculture (USDA) system in the 21st century. These objectives will be accomplished in the following nine chapters. See Ruttan (1982) for an early treatment of similar issues.

Chapter 1 examines the factors that led to the establishment and original design of U.S. institutions for research, teaching, and extension serving agriculture. Key federal legislative acts were ones establishing the USDA, 1862; the Morrill or Land-Grant College Act, 1862; the Hatch Act, 1887; the Smith-Lever Act, 1914; and various other acts providing for agricultural research and extension support. We show that these institutions were not borrowed directly from Europe nor were they the "divine inspiration" of legislators. There were precursors that led up to and helped shape new institutions to the needs of America. The original design of the U.S. system was responsive to political, economic, and scientific conditions prevailing at the time.

In Chapter 2, the transition from a small, embryonic research-teaching-extension system to the current complex system is examined. We show that the system developed by responding to changes in scientific and economic conditions. Science for agriculture first developed at the Yale Scientific School starting in the 1860s and later spread to the land-grant colleges. We show that agricultural science came of age during the 1920s by developing ties with existing upstream or pre-invention sciences. As the land-grant colleges matured into full-fledged universities, they were major contributors to the development and advances in pre-invention science.

The scientists, the source of intellectual discoveries and inventions, are the center of attention in Chapter 3. This chapter examines the education, nature of work, and training of U.S. agricultural scientists over the long term. We conclude that Samuel W. Johnson, professor of analytical chemistry at Yale University, 1856–1896, and director of the Connecticut Agricultural Experiment Station, 1877–1900, was the first modern U.S. agricultural scientist. He and his students had a major impact through the turn of the century. Furthermore, the early research and growth of the U.S. public agricultural research system was severely constrained by a shortage of scientists. Scientists located in the state universities, e.g., at Wisconsin and Cornell, did not emerge as major forces in agricultural research and graduate education until after 1900. Since the 1920s, the land-grant colleges and universities and the agricultural experiment station system have been training a large share, but not all, of the new agricultural scientists. Over 1980–2000, a 20 percentage point increase in the share of doctorates awarded in applied agricultural sciences were to women relative to the 1960–1979 period.

Chapter 4 presents evidence about the long-term trends and compositional characteristics of agricultural research and extension expenditures and in sources of funds for public agricultural research.

Over the 100-year period 1900–2000, the rate of growth of U.S. real total public agricultural research expenditures was 4.0 percent per annum. This is a very high rate of growth, but we estimate that the average rate of growth of real private agricultural R&D was a larger 4.5 percent per annum. We conclude that up to 1906, U.S. private agricultural R&D expenditures most likely exceeded U.S. public expenditures, but for the next 44 years, total real expenditures of the public sector were larger. Using our computations, we conclude that since 1950, U.S. private agricultural R&D expenditures have exceeded total U.S. public expenditures. The overtake date is, however, affected by decisions about the breadth and depth of private sector R&D. Since about 1980, real private R&D expenditures have been growing much more rapidly than public expenditures on agricultural research. In the U.S., little public funding is allocated to support private agricultural research. Within the USDA-SAES agricultural research system, the dominant partner in terms of share of research expenditures has shifted several times over the past century, but since 1950, the size of the SAES system has exceeded the size of the USDA system by about 50 percent.

The USDA obtains almost all of its research funds from the federal government, but the SAESs have diverse sources and the composition has been changing over time. The federal government provided almost all of the funds for the agricultural experiment stations in the 1880s, but 100 years later it accounted for less than 30 percent and "regular federal support" for only 17 percent. State governments have provided slightly over half of total funds for agricultural experiment stations for the past half century. Much of public agricultural research is not general discovery but has a commodity focus. With the exception of research on cotton, fruits, and vegetables, public agricultural research has concentrated most heavily on enhancing the biological efficiency of crops and livestock. For other research commodities, a large share of the research resources has been for protection-maintenance research. A very small share of public agricultural research funds has been allocated to agricultural mechanization research.

In Chapter 5, the contributions of the private sector to R&D for agriculture, especially to agricultural biotechnology, are examined. Although trade secrets have been a long-term form of intellectual property protection, a new intellectual property rights (IPRs) system, in the form of a patent, was established by the U.S. Constitution. Over time IPRs have been strengthened and broadened by new legislation and court decisions. IPRs are a critical factor in private sector R&D decisions because they have been used to establish limited exclusions on the use of inventions. These exclusions are the route to potential private sector profits from R&D. IPRs for plants were strengthened in the 1970s with Plant Variety Protection Certificate Legislation. In the 1980s, patents were extended to living material—first to microorganisms, then plants, and finally to animals. The SAES-USDA system has been impacted by the Bayh-Dole Act of 1980, and some land grant universities are leaders in university patenting. The public system is now facing key science policy decisions as to whether to continue to heavily support the development of the "Intellectual Commons," which is vital to the long run discovery process of the private sector, or to pursue inventions for more near-term profit-licensing income and start-up company creation.

The subject of Chapter 6 is the international dimension of the U.S. agricultural research and higher education system. It addresses the role of technology capital in the economic development process, and the role of U.S. land-grant universities in creating technological capital. We document that the U.S. land grant system trained a high proportion of the agricultural scientists who created the Green Revolution of the 1960s and 1970s and helped develop new plant germplasm that was made freely available to the national agricultural research system of developing countries. In contrast, the Gene Revolution, starting in the late 1980s, is being led by the private sector—currently by 6 large companies with research budgets that are large relative to aggregate public agricultural research budgets. However, given that Gene Revolution products must undergo significant screening, Green Revolution techniques are still required for progress.

We also document that the U.S. university system played a major role in research capacity building in the developing countries in the 1960s through the 1980s. But resources for international capacity building withered in the 1990s. The U.S. universities continue to train foreign students; in fact these students have filled a void left by domestic students. But these recent foreign graduates are less inclined to return to their home countries. Hence, international capacity building has stalled—at least temporarily, and developing countries are increasingly being challenged to find the research capacity needed to adapt new technologies to local conditions. This is especially problematic for biotechnology research, which uses the new science of molecular and microbiology biology rather than the science of population genetics.

In Chapter 7, we examine the economics of public provision of agricultural research. The SAES systems and USDA research agencies are the main beneficiaries of these activities. One aspect of funding is institutional factors enacted by a long sequence of new and amended federal legislation. Another factor is the economic incentives that drive the legislative process of state and federal governments and administration of public agricultural research. These incentives are best viewed from the perspective of economic theories of provision of publicly provided goods, and several models are examined in this chapter.

We document that state government support of SAES research weakened in the 1990s. We show that the response of state governments has been driven through price and income elasticities of demand for research. We also provide new statistical evidence showing how agricultural experiment stations' funding shares from federal grants and contracts, federal formula funds, state government appropriations, and private sector grants and contracts respond to local research quality, type of agriculture, state demographics, and the overall size of the SAES budget. Overall, we conclude that the competitive interest-group model receives the strongest empirical support as a model of state-level decisions on public agricultural research. Finally, we show how the principle of fiscal equivalence might be used to provide insights about the matching of jurisdictional authority for public funding to the geographical range of benefits from research.

U.S. agriculture has an exceptional record of economic growth and productivity change during the 100-year period between 1900–2000. In Chapter 8, we show that U.S. real aggregate agricultural output grew at an average annual rate of 1.61 percent per annum over 1900–2000 and 2.08 percent over 1970–1999. We also highlight some of the important changes in technology that have occurred over the past century—from the invention of hybrid corn varieties at the beginning the century to herbicide and insect resistant transgenic traits for field crop varieties at the end of the century. As a result of rapid and sustained technical change, multifactor productivity for U.S. agriculture grew at an average of 1.62 percent per annum over the last century and 2.09 percent over the past 30 years. We also present econometric evidence covering more than 100 years, and over 1970–1999, showing that investments in public and private R&D have been major contributors to U.S. agricultural multifactor productivity growth. For the latter period, we also show that composition of SAES funding sources affects the impact of public agricultural research on agricultural productivity.

In Chapter 9, six major challenges facing the USDA-SAES system during the 21st century are identified and addressed. They are: (1) the response to the agricultural biotechnology revolution, (2) the response to the changing structure of U.S. agriculture, (3) the increasingly complex issues arising from growing private support of SAES research, (4) the system's productivity, (5) the appropriate balance between federal formula and competitive grant funding of SAES research, and (6) the research commodity allocation record. We end this second edition with the judgment that the SAES-USDA system remains viable and productive.

Reference

Ruttan, V.W. 1982. *Agricultural Science Policy*. Minneapolis, MN: University of Minnesota Press.

Science for Agriculture
A Long-Term Perspective

Second Edition

1
The Evolution of the U.S. System

The present-day set of research-teaching-extension institutions serving U.S. agriculture have enjoyed relatively long lives as institutions. This is in part due to the nature of their design at the time of establishment, and in part due to their capacity to evolve and change over time in response to new economic, political, and scientific conditions. This first chapter examines the factors that led to the establishment and original design of these institutions.

The key federal legislation founding the USDA-SAES research-teaching-extension system included the act establishing the USDA, 1862; the Morrill or Land-Grant College Act, 1862; the Hatch Act, providing for state agricultural experiment station research support, 1887; the Smith-Lever Act, 1914; and various other acts providing for agricultural extension support.[1] These legislative acts and the institutions developed and supported by them were major institutional innovations. The institutions as embodied in legislative acts were not simply the product of exceptional "inspiration" or of the "divine creativity" of legislators and policymakers of the day. By the time each of these major pieces of legislation was passed, considerable institutional development and experience with earlier institutions had been realized.

It is convenient to discuss these developments in terms of historical periods. In the first section of this chapter, three sets of institutions are identified as being precursors to the 1862 legislation. The U.S. Patent Office is discussed as a precursor to the establishment of the USDA. State colleges of agriculture served as precursors to land-grant colleges. Agricultural societies were also important in providing the political support at the state level for the early colleges of agriculture that served as models for the Land-Grant College Act.

In the second section, the developments leading up to the Hatch Act, which established the state agricultural experiment station system in 1887, are considered. We argue that the "experiment station model" established in Europe in the mid-19th century had an important impact on the development of research programs in the USDA and on the creation of state experiment stations in a number of states that served as models for the Hatch Act. In addition, the early land-grant colleges created a demand for research to enhance their teaching activities. The land settlement of western areas facilitated a strong sense of regional or state competitiveness and enhanced the power of state institutions, a factor leading to the federal-state decentralization of the public research system.

In the third section, we show how the early research and teaching institutions led to the development of the extension system. Finally, we close this chapter with a discussion of implications for the further development of the system.

Precursors to the 1862 Legislation

The earliest relevant development to precede the 1862 legislation was the establishment of the U.S. patent system. It provided stimulation for invention by private individuals and firms. The Patent

Office, however, eventually recognized the *limitations* of the patent system for providing economic incentives to encourage certain types of inventions (particularly biological inventions). So it engaged in public research and related activities. Some of these activities were later transferred to the U.S. Department of Agriculture when it came into existence in 1862.

Early Patent Law

The principle of "protection for new means of manufacture" was first enunciated in 1623 in England during the reign of James I.[2] A change in the Statute of Monopolies abolished a wide range of monopolies then in existence except for monopolies dealing with inventions. The basic principle underlying this statute was that protection (i.e., the power to prevent others from using an invention) encouraged invention and technical progress. There was an explicit recognition that a trade-off or bargain was being made. The society would accept certain costs or losses associated with the granting of a monopoly in return for the gains from the encouragement of invention and for making inventions public.

In the early 18th century, changes in European law refined the definition of intellectual property rights and established guidelines for determining which inventions might qualify for protection (i.e., novelty requirements). The patent systems in England and France established in the 18th century thus served as prototypes for the U.S. system.

The founding fathers incorporated intellectual property protection into the U.S. Constitution, which contains the oft-quoted clause (U.S. Constitution, Art. 1, Sec. B), "Congress shall have the power . . . to promote the progress of science and useful arts, by securing for limited times for authors and inventors the exclusive right to their respective writings and discoveries."

Table 1.1 provides a summary of the federal legislation that defined and expanded the role of the Patent Office during 1787–1842. The legislation reveals the groping but steady efforts of the government to encourage private invention via "protecting the rights of men of genius in the fruits of intellectual labor."

The cardinal incentive of a patent law was the monopoly power granted to the inventor. This incentive depended upon establishing the novelty of the invention. Because the invention was new, the patentee had the legitimate right to exclude others from using it during the life of the patent. Novelty under the first patent act was certified by two cabinet members and the Attorney General. Agricultural inventions were among the first granted under this law.

The effective functioning of a patent system ultimately required that an invention meet three standards of patentability: (1) novelty—the invention must be new; (2) usefulness—the invention must have potential usefulness; and (3) an inventive step—the invention must be "unobvious to one skilled in the art." In addition, a full disclosure of the invention was required so that the invention could be used by others.

After 1793, the failure to require novelty in newly patented inventions left resolution of conflicting claims to privately initiated lawsuits.[3] The record shows that, on both foreign and domestic fronts, Congress was sensitive to the enforcement of inventors' rights. The citizenship requirement was relaxed in 1800 and again in 1832 to permit the more rapid introduction of foreign-origin technology into the American economy. Patent disputes were treated similarly to other property disputes by the Act of 1819, which gave litigants access to the appellate courts.[4]

During the 1830s it became apparent that the system, overburdened with complaints, favored imitators at the expense of "men of genius." In 1836, Congress made the first comprehensive

Table 1.1 Legislative acts affecting the Patent Office, 1787–1862

Year	Significant Provisions	Patent Standards				Legal Instruments					Judicial Protection				Citizenship Requirements				Patent Office Duties		
		Novelty	Disclosure	Utility	Priority	Assignments	Renewals	Extensions	Reissues	Designs	Standing to Sue	Appellate Courts	Internal Appeals	Infringement Penalties	Citizenship	Residency	Intent	Working	Patent Fund	Annual Report	Agricultural Statistics
1787	Constitutional authority granted for patent law																				
1790	First Patent Act	X																			
1791	Enabling disclosure; assignments instituted; U.S. citizenship required		X			X									X						
1794	Revival of suits that lapsed after 1793 law		X			X					X				X						
1800	Aliens with 2 years' residency eligible		X			X					X					X					
1819	Circuit courts granted jurisdiction, with appeal to Supreme Court										X	X				X					
1832	Renewals, extensions, and reissues sanctioned; aliens need only intend to become citizens; their patents void if not worked in 1 year		X			X	X	X	X		X	X					X	X			
1836	Patent Office established; Patent Fund created from revenue; novelty exam required	X	X	X	X	X	X	X	X		X	X							X		
1837	Annual reports required; non-judicial appeals of conflicting claims allowed	X	X	X	X	X	X	X	X		X	X						X	X	X	
1839	Collection of agricultural statistics mandated	X	X	X	X	X	X	X	X		X	X	X				X	X	X	X	X
1842	Design patents; infringement penalties specified	X	X	X	X	X	X	X	X	X	X	X	X	X			X	X	X	X	X
1862	USDA created	X	X	X	X	X	X	X	X	X	X	X	X	X			X	X	X	X	X

Source: U.S. Congress 1790, 1791, 1794, 1800, 1819, 1832, 1836, 1837, 1839, 1842, and 1862.
Note: X denotes provisions in force.

attempt to restrict patents to true inventions. If the invention were known or used in the country, or published anywhere in the world, the inventor was barred from obtaining a patent. Furthermore, the inventor was required to furnish a description of the invention that was sufficiently exact to permit an ordinary person skilled in the art to make and use it (the beginning of the "enabling disclosure" requirement). The inventor could not protect what was not disclosed. Then in the Supplementary Act of 1837, Congress further specified that a patent claim could not be "too broad,"[5] which extended the doctrine regarding "reduction to practice."

It is relevant to note that the early patent laws did not protect living organisms and that in particular they did not provide incentives for the invention of improved plant varieties or animals. There were three reasons for this. First, and foremost, it was difficult to establish novelty for plant and animal inventions. Second, the patent laws, even in their earliest form, excluded certain products of nature, such as genetic resources, from protection. In addition, plant and particularly animal breeding was seen to be the activity involving skills of the ordinary farmer.

Early Agricultural Inventions in the U.S. Economy

In keeping with agriculture's large share of GNP and labor force, agriculture led all other fields in the share of patents granted during 1790–1849. (Table 1.2 reports agricultural inventions in 22 technology fields for 5-year periods from 1790 to 1849.) It was followed by textiles, stoves, metallurgy, and chemicals. Several other technology fields had agriculturally oriented patents: textiles (flax and hemp breakers, cotton gins); presses (cotton and hay balers, cider presses); mills; leather (horse collars and harnesses). A large share of the early agricultural patents were in agricultural tools and machinery. Although interest ran strong in fertilizers beginning in the 1840s, no true fertilizers or other agricultural chemical patents had been granted by 1849. Frontier agriculture required extensive land clearing. Thus stump-pullers and other devices for making land suitable for cultivation received some attention from inventors. A handful of irrigation-related patents had also been obtained by the end of this period.

The westward expansion had a major effect on invention. Table 1.3 shows the number of patents awarded, by state, during the period 1790–1849 for these 22 different agricultural technology fields. By 1850, New England soils suffered from nutrient depletion. Their crop-yielding potential had never been as high as in the Hudson Valley or the Midwest. New England had already begun the shift towards an industrial base for its economy. Only 4.6 percent of the patents of Massachusetts residents were in the agricultural technology field (1790–1849). The relative strengths of Massachusetts inventors were in textiles (ranked first, and just slightly behind New York's textile total), metallurgy (second), calorific devices (third), and lumber and leather manufacture. By 1850, Pennsylvania inventors had taken out nearly three times as many agricultural patents as inventors from Massachusetts. Patents granted to foreigners were of limited importance during 1790 to 1849.

Virtually all early inventions were mechanical (Table 1.4). Chemical and electrical inventions were not produced in significant numbers until after 1850. These early inventions were not produced by trained engineers or chemists but were the product of people engaged in economic activity. Farmers made a number of them. (Also see discussion of this point by Marcus 1987, pp. 16–17.) Biological inventions were by and large not protected.

Table 1.2 U.S. patents by agricultural technology field, 1790–1849

Technology Field	1790 –1794	1795 –1799	1800 –1804	1805 –1809	1810 –1814	1815 –1819	1820 –1824	1825 –1829	1830 –1834	1835 –1839	1840 –1844	1845 –1849
Soil Working												
Plows	0	1	2	5	14	14	40	33	63	57	51	67
Cultivators and Harrows	0	1	0	0	1	2	4	13	13	15	10	22
Planters	0	1	0	2	1	0	2	9	8	21	28	29
Animal and Farm Equipment												
Milking Equipment	0	0	0	0	0	0	0	3	1	1	0	1
Harnesses and Yokes	0	0	1	1	1	0	1	1	0	5	10	35
Fencing	0	0	1	1	1	1	2	5	2	1	2	11
Cutting and Reaping												
Cutting Grass and Grain	0	0	0	6	7	8	10	7	38	49	39	37
Horse Rakes	0	0	0	0	0	1	3	3	1	7	3	16
Hay Handling (no horses)	0	0	0	1	2	0	4	4	4	9	3	2
Reapers and Harvesters	0	0	0	2	0	0	1	3	2	2	3	24
Cotton & Tobacco Cutters	0	1	1	0	0	0	0	1	1	1	0	1
Threshing and Shelling												
Grain Threshers ($)	5	6	20	37	46	29	65	79	235	127	112	92
Hemp Breaks	0	0	2	0	3	3	10	16	25	26	15	21
Corn Shellers & Huskers												
Post-Harvest Technology												
Distilleries	4	1	25	18	53	21	15	31	27	6	4	4
PHT: General Food	1	3	11	9	17	13	12	32	20	27	14	21
PHT: Dairy	0	1	4	42	20	8	4	28	60	44	18	31
PHT: Grains	1	0	5	6	12	2	7	39	46	32	27	30
PHT: Meats	0	0	0	0	2	1	1	2	13	5	6	9
PHT: Fruits & Vegetables	0	0	1	2	4	2	2	11	10	9	7	5
PHT: Cotton (gins, etc.)	1	3	10	3	6	4	13	32	24	37	49	35
PHT: Tobacco	0	1	0	2	3	2	2	4	13	8	6	3
Beehives	0	0	0	0	3	0	2	2	7	11	34	22
Total	12	19	83	137	194	110	197	352	611	494	429	471

Source: U.S. Patent Office, Reports.

Table 1.3   Total number of patents and agricultural patents awarded by state of residence of inventor, 1790–1849

Residence of Inventor		Total No. (22 fields)	Agriculture	
			No.	% of Total
ME	(1820)[1]	494	113	22.9
NH	(1789)	366	44	12.0
VT	(1791)	353	52	14.7
MA	(1789)	2221	103	4.6
RI	(1790)	257	9	3.5
CT	(1789)	1193	115	9.6
(New England)		(4884)	(436)	(8.9)
NY	(1789)	4904	565	11.5
NJ	(1789)	480	56	11.7
PA	(1789)	2222	272	12.2
(Atlantic)		(7606)	(893)	(11.7)
DE	(1789)	71	12	16.9
MD	(1789)	678	84	12.4
DC	–	227	14	6.2
VA	(1789)	532	134	25.2
NC	(1789)	140	30	21.4
SC	(1789)	131	39	29.8
GA	(1789)	78	12	15.4
FL	(1847)	1	0	0.0
(South)		(1858)	(325)	(17.5)
OH	(1802)	775	146	18.8
IN	(1816)	117	27	23.1
IL	(1818)	69	22	31.9
MI	(1836)	53	18	34.0
WI	(1848)	10	4	40.0
IA	(1846)	2	0	0.0
MO	(1821)	49	7	14.3
(North Central)		(1075)	(224)	(20.8)
KY	(1792)	197	28	14.2
TN	(1796)	132	29	22.0
AL	(1820)	71	7	9.9
MS	(1817)	37	8	21.6
AR	(1836)	1	0	0.0
LA	(1812)	80	4	5.0
TX	(1845)	4	0	0.0
(South Central)		(522)	(76)	(14.6)
Foreign		192	2	1.0
TOTAL (1790–1849)		16,137	1956	12.1

Source: U.S. Patent Office, *Reports*.
[1]Year state admitted to the Union.

Table 1.4   Early U.S. agricultural inventions: First three inventions by field

Invention	First State		Second State		Third State	
Soil-Working Inventions						
Improvement in ploughs	1797	NJ	1800	—[a]	1804	—
Machine for planting (corn or beans)	1799	—	1809	PA	1809	NY
Machine for cultivating (corn)	1799	—	1823	VT	1829	VT
Harrow	1807	CT	1824	PA	1824	NY
A planting plough	1809	NY	1826	ME	1827	SC
Cotton planter	1825	VA	1826	NA	1826	MS
Animal and Farm Inventions						
Machine for watering cattle	1804	—	1824	PA	1826	NY
Improvement in the method of feeding horses	1828	NY				
Improvement in manufacturing horse shoes	1809	—	1828	PA	1830	NY
Improvement in the horse shoe	1822	MA	1824	NY	1831	MA
Boring machine for posts for fencing	1801	—	1811	VT	1826	NY
Improvements in fencing	1819	PA	1823	PA	1832	NY
Improvement in beehives	1810	MA	1811	CT	1820	ME
Destruction of insects	1810	VA	1812	MA	1816	MA
Improvement in vaccination	1822	MD				
Cutting and Reaping Invention						
Improvement in manufacturing scythes	1813	PA	1813	MS	1816	NH
Improved machine for cutting straw and hay	1804	—	1806	VA	1807	PA,NH
Improvement in cutting tobacco	1797	—	1800	NY	1831	NY
Improvement in cutting grass and grain	1803	—	1811	PA	1811	VA
Mowing machine	1812	PA	1814	NY	1822	PA
Machine for picking oakum	1806	MA	1826	ME		
Improvement in the horse rake	1818	NJ	1822	PA	1824	DE
Machine for reaping grain	1805	—	1825	NJ	1828	ME
Harvesting machine	1836	MI				
Threshing Inventions						
Improvement in machine for breaking sugar	1822	PA	1829	CT		
Breaking hemp and flax	1791	—	1795	—	1799	—
Threshing grain	1791	—	1795	—	1794	—
Scouring rice and other grains	1796	—	1797	—	1808	SC
Machine for hulling rice	1803	—	1801	VA	1809	NC
Clover huller	1804	—	1804	—	1805	PA
Smut fanning mill	1804	NY	1805	NY	1808	OH
Machine for shelling corn	1803	—	1804	—	1810	MD
Post-Harvest Inventions						
Mode of making salt	1794	—	1796	—	1799	—
Improvement in distilling	1791	—	1791	—	1794	—
Grain elevators	1812	—	1812	PA	1812	OH
A kiln for drying grain	1800	—	1801	—	1805	DE,NJ
Improvement in grist mills	1800	—	1802	—	1805	DE
Preserving flour from becoming sour	1824	OH	1825	KY		
Sugar mills	1817	NY	1818	LA	1827	NC
Improvement in making and refining sugar	1808	PA	1816	DC	1828	NY
Clarifying cane juice and making sugar	1828	GA	1829	MA	1830	NY
Post-Harvest Inventions (continued)						
A machine for churning	1802	—	1803	—	1806	NY
Preserving milk (chemical composition)	1836	NY				
Improvement in the cheese press	1807	NY	1808	MA	1808	CT
Machine for paring or coring apples	1803	—	1809	MA	1809	CT
Machine for grinding or grating apples	1825	NY	1825	ME	1826	NC

(Continues)

Table 1.4   *(Continued)*

Invention	First State		Second State		Third State	
Cotton and Tobacco Inventions						
Machine for ginning cotton	1794	CT	1796	—	1796	—
Machine for pressing cotton, other bales	1802	—	1808	VA	1810	NY
Extracting oil from cotton or other seeds	1799	—	1822	NJ	1824	NY
Improvement in preventing cotton rot	1824	MA				
Improvement in machine for picking cotton	1828	VA				
Improvement in curing tobacco	1809	VA	1826	VA	1828	VA
Improved machine for manufacturing tobacco	1793	—	1810	NY	1827	VA

Source: U.S. Patent Office, *Reports*.
[a]Residence of inventor is unknown.

The Agricultural Division of the Patent Office

The first Commissioner of Patents, Henry Ellsworth (from Connecticut), was an agricultural enthusiast. He understood the need for better data describing the largest sector of the economy. Largely because of his efforts, Congress not only appropriated funds to begin collecting agricultural data, but the Patent Office also began a concerted effort to collect seeds and other plant materials from many foreign countries, and to introduce and distribute the best varieties to regions that were deemed inadequately endowed by nature. He was among the first public officials actively to encourage treating agriculture like any other industry in efforts to improve productivity (U.S. Patent Office 1845, pp. 6–9).

Agricultural societies were active during the early 1800s in obtaining foreign seeds and distributing them among their members (see below). There was a problem, however, with relying on private incentives for individuals to go abroad to obtain seeds. In general they did not have a way of making these ventures a commercial success. There was no legal protection of plant material, so reproduction and resale were uncontrolled. Thus, the individual could only expect to gain through increased profitability on his own farm, and this could be diminished as other individuals obtained and planted the new seeds.

The Secretary of the Treasury in 1819 (the Patent Office was first located in this department) took the first step for federal government involvement in seed collection (Baker 1963, pp. 4–8; Kloppenburg 1988). He sent requests to the U.S. consuls and naval officers asking them to collect seeds in foreign locations and send them back to the United States. He also gave a rationale for government aid to germplasm collection. The U.S. institutions did not give exclusive advantage to the importer for his or her introductions. Furthermore, it was known that a very small percentage of the new seeds would be successful. Thus, to have agricultural development of the whole U.S. land area, steady massive importation of new seeds from abroad was required to find plants that would produce food and feed. In addition, massive plant introductions were later required to stay ahead of crop pests; diseases and insects frequently became a problem with locally used varieties. This task was too large and costly for individuals to undertake.

A second Treasury request was sent in 1827 encouraging more attention to seed/plant collection work and providing detailed instructions on procedures for preserving and shipping seeds. The U.S. Navy proved particularly cooperative. Between 1836 and 1849 the U.S. Patent Office took special interest in novel plant varieties. In 1839, the Commissioner persuaded Congress to allocate money directly for the collection and distribution of seeds, plants, and agricultural statistics.[6]

Simple mass selection of varieties was the only plant breeding principle available for selecting among the newly imported seeds. At this time only very general things were known about where or why plants were adapted to particular locations. Thus, a critical step in the new plant introduction program was the large-scale distribution of the seeds to farmers for trial in diverse geographical locations and climates. In 1848, 70,000 packages of seeds were distributed (U.S. Patent Office 1849, p. 13). This was the only way of learning whether or not new seeds would grow in the diverse agricultural areas of the United States. Although most did not, there were some outstanding successes, and taking the 19th century as a whole, the program seems to have been immensely successful.

The foreign plant/seed introduction program of the Patent Office was transferred to the U.S. Department of Agriculture after 1862. This was a major activity of the USDA in its early years.

In 1839, Congress appropriated funds for the Patent Office to collect agricultural statistics. The original intention of this effort seems, at least in part, to have been the prevention of artificial monopoly occasioned by false claims of scarcity. Within a short time, however, the Patent Office had begun to gather additional information besides basic production data. Gathering data on exports and imports of the United States and its major trading partners, disaggregated by commodity, and various other international comparisons, became more and more routine.

The Patent Office also commissioned reports on various agricultural subjects beyond the simple collection of data. For example, the report for 1844 contained an examination of the potato blight that had recently devastated the country, and an analysis of the Hessian fly, which had become a "dreadful" pest. Several chemical analyses pertaining to agriculture were initiated as well, among them a comparison of corn with cane sugars (which showed corn syrup to be "equal to the best muscovado sugar"), and a comparison of the oil content of several grains with an eye toward finding those which best fattened livestock (U.S. Patent Office 1845).[7]

While the Patent Office conducted its surveys of domestic and foreign agricultural production and innovation with increasing enthusiasm, its efforts were not universally well received. Despite the relatively comprehensive and innovative format of the 1844 and 1845 reports, Congress declined to appropriate funds for a separate agricultural report in 1846.

The second Commissioner of Patents, Edmund Burke, took the occasion to argue, in his 1845 summary, for the beneficial and cost-effective results of the Patent Office's publications. His illuminating arguments revealed the view, held apparently with some controversy, that government oversight of business or agriculture was a legitimate function and, especially, that this oversight was properly exercised by the department most concerned with technical progress (U.S. Patent Office 1846).[8] Burke cited an address by Washington in 1796 in which he favored the creation of a department of agriculture, as well as two congressional reports, dated 1812 and 1817, that concurred or recommended that Congress create a National Board of Agriculture.[9] The political will did not yet exist, however, to centralize and empower an agency having dominion beyond cataloguing private behavior. Perhaps out of bureaucratic self-interest, or sensitivity to congressional fears of "big government," Burke stopped short of endorsing such a move himself.[10]

Ellsworth and Burke, then, through their expansion of Patent Office duties into agricultural statistics gathering and assembling, research conduct and funding, gathering and diffusing agricultural research results, and collecting and distributing plant materials and seeds, developed programs that became fundamental responsibilities for the U.S. Department of Agriculture when it came into being.[11]

Early Colleges of Agriculture as Precursors to Land-Grant Colleges

The first agricultural colleges were designed to be quite different from other American colleges. The latter had a strong religious or professional-school (law, medicine) orientation. Their curricula and activities were designed primarily to preserve and transmit traditions. Education was mainly recitation from memory (Eddy 1957, p. 4). Few options existed for creative interpretation, there was very little science, and there were no laboratories for experimentation.

During the early 1800s, a few U.S. institutions successfully broke with past traditions and started to teach college-level courses that created "useful" skills. West Point Military Academy established the first college-level department of engineering in 1812. Rensselaer's Institute, Troy, New York, was established in 1824 to apply science to the common purpose of life (True 1929, pp. 42–43). It first focused on agriculture but later shifted to engineering.

Other U.S. institutions, borrowing largely from the German advances in the laboratory-based sciences of chemistry, physics, and biology, established early scientific schools. These taught science using the new laboratory methods and initiated research to discover new and useful knowledge. In 1845, Yale University initiated its new science program with the addition of two professorships. One was in agricultural chemistry and animal and vegetable physiology, and the second was in practical chemistry (True 1929, p. 63). A laboratory was opened to give instruction and initiate research in general and applied sciences. Harvard University established its Lawrence Scientific School in 1847.

During the 1850s several interest groups were supporting the idea of establishing new colleges to teach agriculture and the mechanic arts (or engineering). The classical college curriculum would be modified to emphasize practical courses in agriculture, commerce, and mechanic arts. Jonathan Turner and Justin Smith Morrill were two of the leading spokespersons for institutions to educate common people (Eddy 1957, Chapter 2). This idea was supported by the 1856 U.S. Agricultural Society meeting (Rossiter 1979).

When Morrill's bill first came before Congress in the late 1850s, it failed to pass because of opposition by the southern congressmen. Plantation owners saw education as a threat to cheap farm labor needed for a successful plantation agricultural system (Wright 1987). In anticipation of the land-grant act, state agricultural colleges were established in Michigan (1855), Maryland (1856), Iowa (1858), and Pennsylvania (1862). All but Iowa had agricultural colleges operating before the Morrill Act was passed.

After the South withdrew at the start of the Civil War in 1861, the main obstacle to passage of a land-grant college bill was removed. The Morrill Act was easily passed by Congress and signed by President Abraham Lincoln in 1862. The act provided a one-time land grant to the states for support of at least one college having the main objective of teaching courses in agriculture and the mechanic arts (see Table 2.1). States that were in rebellion against the U.S. government in 1862, and U.S. territories, were excluded from benefits of the original Morrill Act.

The permanent endowment of the new curricula attracted the interest of many groups. Debates and controversies raged in most states about where to assign the income of the Morrill Act (Eddy 1957). One of three policies was initially followed: (1) add agriculture and mechanic arts to the curricula of an existing private (or church-related) college, (2) add them to the curricula of an existing state university, or (3) establish a new college focusing on agriculture and the mechanic arts (Eddy 1957, p. 49). Furthermore, state legislators sometimes changed the institution to which Morrill Act benefits were assigned.

Eight states first assigned all or part of their land-grant benefits to private colleges. All of these states were located in the Northeast except for Oregon. The Ivy League universities of Yale,

Dartmouth, and Brown were recipients of land-grant benefits for a short time. The University of Vermont, Cornell University, and Rutgers University became mixed public-private universities and continue to receive land-grant benefits today. Corvallis College was taken over by Oregon; it and Storrs Agricultural School of Connecticut were converted into state agriculture and mechanic arts colleges. Massachusetts Institute of Technology and Pennsylvania State University remain private but continue to receive Morrill Act benefits.

Eleven states assigned Morrill Act benefits to state universities (or colleges) that were already in existence and were established significantly before the states accepted the conditions of the Morrill Act. Kentucky, North Carolina, South Carolina, and Mississippi later reassigned these benefits to newly created agriculture and mechanic arts colleges. Louisiana later merged an older state college and new agriculture and mechanic arts college.

Political Institutional Support: The Agricultural Societies

Agricultural societies provided early support to teaching and research institutions. Keen interest arose within these societies about the latest techniques, fertilizers, and implements. The climate they generated—a mixture of innovation, competition, and dissemination of results—has served the agricultural community to this day and has formed an integral part of the "client" relationship that exists between farmers, the extension service, and the research institutions themselves.

Although the societies were private institutions, many state and some county societies received appropriations from state governments. The purposes of the societies were to form networks for the interchange of information, from both private and government sources, and to stimulate the kind of informal, unpatentable innovation that was necessary for agricultural progress.

The societies accomplished this by using their appropriation and dues to offer cash prizes at state and county fairs for the best local farm horse, team of oxen, acre of corn, implement, etc. They also built libraries and purchased land for implement trials and other experiments. An 1858 survey by the Patent Office found that Connecticut's state society paid the salary of a chemist to analyze the commercial fertilizers that were offered for sale in the state, and the New York society funded an entomologist and paid him $1,400 in premiums for experimental results (twice the amount of the state appropriation). The societies also obtained and distributed seeds from the Patent Office collection and exchanged seeds and cuttings among themselves. Several societies owned experimental farms.

The state and county fairs provided a unique means of communication and reward among farmers. Several of the prizes offered were substantial for the period, and many were specifically targeted at agricultural implements. The Massachusetts Society for Promoting Agriculture offered $1,000 for the best mowing machine. Illinois offered $5,000 "for a steam-engine that would do all the work of a farm." The New York State Agricultural Society recorded that "many visitors to our Fairs remarked that, had the Society done nothing else than to secure the improvement in implements on exhibition, the state would have been amply compensated for all the outlay which had been made to promote agriculture." Marcus (1987, pp. 16–17) also discusses the role of the agricultural societies and newspapers in promoting research. The early research conducted largely by farmers before the 1870s was simple—largely mechanical, as we have noted in the previous section. But because public monies went to support agricultural newspapers, agricultural societies, and fairs, these public funds aided the early research activities of farmers.

Table 1.5 summarizes the characteristics of state and local agricultural societies in 1858 and 1876. These societies were particularly important in the Northeast, where significant proportions of farmers were members. Many members were not farmers but had strong interests in agriculture and

Table 1.5 Characteristics of state and local agricultural societies, 1858 and 1876

State	Date First Society		No. Societies		Percent w/Fairs	Counties w/Libraries	In 1876		Percent[a] Members
	State	County	1858	1876			No. Members	No. Farms	
Maine	1855	1818	17	62	77.4	48.4	14104	62957	22.4
New Hampshire	1849	1824	7	21	81.0	28.6	9802	31419	31.2
Vermont	1850	1842	16	25	80.0	36.0	4214	35014	12.0
Massachusetts	1792	1811	43	74	85.1	37.8	32876	34834	94.4
Rhode Island	1820	1867	2	6	83.3	33.3	2944	5962	49.4
Connecticut	1852	1803	10	47	61.7	10.6	6956	29071	23.9
New England			95	235	77.4	34.0	70896	199257	35.6
New York	1832	1817	97	153	74.5	27.5	53449	233616	22.9
New Jersey	1866	1852	16	23	52.2	30.4	3061	33211	9.2
Pennsylvania			71	94	72.3	33.0	35357	201685	17.5
Mid-Atlantic			184	270	71.9	29.6	91867	468512	19.6
Delaware	—	1847	4	10	50.0	20.0	604	8409	7.2
Maryland	1869	1819	10	27	44.4	11.1	2981	36462	8.2
Virginia	1854	1857	33	36	52.8	36.1	5228	105117	5.0
West Virginia	—	1848	—	11	36.4	27.3	1516	55805	2.7
North Carolina	—	1852	8	27	37.0	37.0	2210	138396	1.6
South Carolina	1795	1815	13	10	70.0	30.0	962	81271	1.2
Georgia	1846	1853	10	77	41.6	29.9	9169	118025	7.8
Florida	—	1871	—	5	20.0	60.0	160	19479	0.8
South Atlantic			78	203	44.3	29.6	22830	562964	4.1
Ohio	1846	1831	74	138	81.2	17.4	52147	231818	22.5
Indiana	1851	1849	77	99	82.8	21.2	12712	184196	6.9
Illinois	1853	1846	94	133	78.2	29.3	24205	239860	10.1
Michigan	1848	1842	28	70	85.7	40.0	34408	137441	25.0
Wisconsin	1851	1850	35	81	71.6	6.2	12933	124897	10.4
East N. Central			308	521	79.8	22.5	136405	918212	14.9

Minnesota	1859	1856	29	43	79.1	25.6	4678	78620	6.0
Iowa	1854	1844	74	144	73.6	33.3	25661	164633	15.6
Missouri	1859	1852	32	86	58.1	36.0	6856	195401	3.5
Nebraska	1858	1857	11	35	65.7	25.7	2636	48061	5.5
Kansas	1862	1859	1	106	61.3	42.5	8989	108453	8.3
West N. Central			147	414	67.1	34.8	48820	595168	8.2
Kentucky	1876	1836	14	33	78.8	9.1	4332	152044	2.8
Tennessee	1867	1854	20	55	58.2	36.4	3489	151397	2.3
Alabama	1867	1854	7	13	30.8	7.7	1229	115319	1.1
Mississippi	—	1856	16	11	63.6	36.4	1153	91647	1.3
East S. Central			57	112	61.6	25.0	10203	510407	2.0
Arkansas	—	1868	2	15	40.0	60.0	704	80030	0.9
Louisiana	—	1860	4	9	22.2	44.4	960	42349	2.3
Texas	—	1856	24	41	46.3	31.7	3471	140266	2.5
West S. Central			30	65	41.5	40.0	5135	262645	2.0
Montana	—	1870	—	1	100.0	—	150	1319	11.4
Idaho	—	1871	—	4	25.0	75.0	119	1444	8.2
Colorado	1873	1869	—	5	40.0	40.0	404	3676	11.0
Utah	—	1850	1	33	48.5	54.5	2135	8089	26.4
Mountain			1	43	46.5	53.5	2808	14528	19.3
Washington Terrace	—	1863	2	10	50.0	30.0	790	5508	14.3
Oregon	1862	1863	2	7	28.6	28.6	485	13628	3.6
California	1854	1859	8	16	75.0	12.5	4115	32271	12.8
Pacific			12	33	57.6	21.2	5390	51407	10.5
Total			912	1896	68.4	29.8	394354	3583100	11.0

Source: U.S. Patent Office, *Reports*.
[a]Computed as number of members of agricultural societies divided by number of farms $\times$ 100.

its progress. Contemporary farm organizations, while different in orientation, can trace their origins to these early agricultural societies.

These early societies demonstrated the usefulness of research and education in improving agriculture. The societies demonstrated that the sharing of ideas and experiences among farmers and between farmers and innovators was productive. The societies also showed that they were serious about the need for agricultural research by funding research and experiment stations of their own. And they provided a formidable political force supporting the use of public monies for research and education in agriculture. This support was important to the eventual passage of relevant federal legislation beginning in 1862.

Precedents to the Hatch Act—1887

The 1862 legislation, the USDA Act, the Land-Grant College Act, and the Homestead Act, all had important influences on the design of the state agricultural experiment stations created by the Hatch Act in 1887. The USDA in its early years established formal programs of research that guided later research program design. Land-grant colleges created a demand for research at the state rather than the federal level, and this facilitated the establishment of state experiment stations that preceded and importantly influenced the Hatch Act. These early experiment stations were also influenced by the emergence of a new "model" for experiment stations from Europe. Westward expansion facilitated by the Homestead Act encouraged stronger state participation in these activities and a recognition that many of the benefits from these institutions, including research programs, were indeed state-specific. Transportation policy led to a rapid decline in transport costs, and this too led to increased state and regional competitiveness.

Early Animal Breeds in the United Kingdom

The development of animal breeds in Europe, especially in the United Kingdom, and in America during the last half of the 18th and all of the 19th centuries was the forerunner to modern livestock breeding and animal improvement. Before 1700 in Europe, land-holdings were under a feudal system, and livestock mingled together as they grazed the "common pastures." There was little control of mating, and these early animal species were called "common stock." With enclosure of the "commons" in the United Kingdom starting about 1700, the transition to private property, individual holdings, and fences made controlled mating of farm animals possible. Thus, during the early 1700s, farmers took an interest in farm animal improvement, and improved strains began to appear. They were generally adapted to the unusual features of the local geoclimatic conditions; e.g., it is said that during the 18th and 19th centuries a different strain of sheep inhabited every valley of England and Scotland (Acker and Cunningham 1991, pp. 370–371). Farmers have known for a long time that some animal strains or breeds were better adapted to a particular environment than another, i.e., they would save more offspring or fatten faster than another strain or breed.

In 1760 Robert Bakewell, an English farmer, is credited with first establishing the pattern of modern animal breeding. He established an early form of inbreeding, or purification of lines, by "breeding the best to the best, regardless of genetic relationship" (Warwick and Legates 1979, p. 18). He placed an emphasis on selecting for particular traits. This inbreeding led to relatively true breeding strains and became the foundation for "purebred animals." He worked with cattle, sheep, and horses, and his sires were in great demand at the time.

With the establishment of relatively true breeding strains of livestock, the relatively homogenous groups were called "breeds." As a breed grew in number and popularity, it became difficult to know the ancestry of each animal. Breeders then developed a written summary of the pedigree of their animals, frequently summarized in herdbooks (Willham 1985). George Coates, Scotland, established the first official herdbook for a breed, the Shorthorn Cattle Herdbook in 1822. Breed societies were organized to promote particular breeds and to control the herdbooks; e.g., the Shorthorn society was founded in 1874 and other breed associations followed.

The early herdbooks were "open," which meant that new stock could be introduced to the breed when sufficient uniformity of a particular type was established. In America, "closed herdbooks" became the rule, which meant that after some particular date no new stock could be introduced to the breed. All purebred animals of a particular breed had to have ancestors in the breed's herdbook. However, registration in a herdbook did not require meeting any performance criteria, except for a few visual characteristics, e.g., particular coat color, freedom of horns (Warwick and Legates 1979).[12] Thus, the breeds were too generous in their registration criteria, and many "poor" animals were registered.

In America, the inter-war years between the Civil War and World War I were the purebred era. A large number of purebred breed associations and registry books were established and fairs and expositions were held to advertise particular breeds. At this time, however, no scientific methods existed for objectively proving that one breed or animal was superior to another. Most selection was based upon visual characteristics alone.

The Role of the Experiment Station Model from Europe

Much of the invention or technology improvements relevant to agriculture prior to 1840 were the result of activities of private individuals who had little formal research training but who faced practical production problems or were seeking improved methods of production, e.g., innovative farmers, blacksmiths, estate owners. A large share of the advances from this informal system was mechanical rather than biological (see Evenson 1983; Hayami and Ruttan 1971).

Early attempts to formalize agricultural research and to build upon a scientific base occurred during the mid-19th century in Europe. New institutions, established in Scotland, England, and Germany, were built upon the emerging scientific field of agricultural chemistry. The new chemistry of the 19th century was based largely upon soundly designed and executed laboratory experiments.[13]

Justus von Liebig, a German, was the early European agricultural scientific leader. His contributions came through his own research in chemistry, his intellectual stimulation and training of other agricultural chemists, and his book *Organic Chemistry in Its Relation to Agriculture and Physiology,* published in 1840. Liebig's book was outstanding in the sense that it did "bring together and interpret the very considerable mass of chemical and related data pertaining to plants and soils that had accumulated up to that time" (Salman and Hanson 1964, p. 22).

Liebig established a very successful chemistry laboratory in the early 1800s at the University of Giessen for training research students in chemistry. It was the first laboratory of its kind and attracted students from all over the world, including some individuals who were interested in synthetic organic dyestuff and others interested in agricultural chemistry. Among the latter group was the young and aspiring agricultural chemist by the name of Samuel W. Johnson. He worked under Liebig and became the leading U.S. agricultural chemist during the last one-third of the 19th century.

Two institutions to perform agricultural research were established in the United Kingdom during the 1840s. A laboratory was established in Edinburgh, Scotland, in 1842 by the Agricultural

Chemical Association of Scotland, a voluntary society. The laboratory was closed in 1848 because of its inability to respond to the association members' demand for immediate practical results (Ruttan 1982, p. 68). In general, the Scottish farmers failed to appreciate the potential of laboratory work conducted by a researcher who had obtained scientific training, preferring instead farmer-initiated field trials (Knoblauch, Law, and Meyer 1962).

In 1843, an experiment station was established at Rothamsted, England, by Sir John Bennet Lawes on his estate. Lawes, who had been engaged in the manufacture of phosphate fertilizer from bones, also began in 1843 to manufacture superphosphate fertilizer in a nearby village. Sir Henry Gilbert, who had studied under Liebig in Germany, was placed in charge of experimental work. The station was supported by the profits of the Lawes phosphate enterprise until 1889, then it was endowed through the Lawes agricultural trust. As the research program expanded and became more costly, government funds were obtained to finance the work at Rothamsted (Ruttan 1982, p. 68). The Rothamsted Agricultural Experiment Station is the oldest continuously operating agricultural experiment station in the world.

Starting in 1852, German states established 75 publicly supported agricultural experiment stations. They were to seek out methods of applying science, especially chemistry, to agriculture. The first German station, established at Moeckern, Saxony, in 1852, had a single objective of conducting research. Diffusion of results was not to be a major activity. A brilliant young chemist was hired as director and an experienced farmer was hired as superintendent. This mixture of early staff had the merit of representing both the scientific and practical side of agriculture.

The Moeckern station was chartered by the state of Saxony and obtained an annual appropriation from the government to finance its operations. This station and other German agricultural experiment stations were established outside the college or university system and did not have teaching obligations.[14] Control was decentralized, and stations generally received strong support from local farmers' organizations and chambers of agriculture. A move to more-centralized agricultural research started in the early 1900s, and a number of other changes weakened what had been established as the world's leading agricultural research system in the 19th century (Ruttan 1982, p. 75).

The early attempts to apply science to agriculture in the United States drew upon the German example for their model of institutional organization and the education of agricultural scientists. Between 1875 and 1887, agricultural experiment stations were established in 14 states before passage of the Hatch Act.[15]

Formalization of Research in the USDA

During the 1850s, farm groups were lobbying for a separate agriculture department in the federal government (Gaus and Wolcott 1940, p. 4). In 1862, federal legislation was passed establishing the U.S. Department of Agriculture. The USDA, however, did not have cabinet rank and operated with the leadership of a Commissioner rather than a Secretary until 1889. The purpose of the new department was to "acquire and diffuse among the people of the United States useful information on subjects connected with agriculture in the most general and comprehensive sense and to procure, propagate, and distribute among the people new and valuable seeds and plants" (U.S. Congress 1863, p. 387).

Isaac Newton was the last Superintendent of the Agriculture Division of the Patent Office and the first Commissioner of the USDA. He emphasized research organized around disciplines, individuals, and educational activities. The staffing and facilities for USDA research remained small, however, until after the USDA received cabinet status in 1889. James Wilson, who became Secretary of Agriculture in 1887 and continued in that position for 16 years, is credited with reorganizing and building the USDA's research program into a first-rate institution (Moore 1967, p. 16; Office of Technology Assessment 1981; Baker et al. 1963).

Early USDA research focused on four areas: (1) importation of seeds and plants and plant classification, (2) statistics and statistical estimates, (3) chemical analyses, and (4) livestock disease control. The first three research areas had been initiated by the Patent Office. The USDA invested significant resources in new plant/seed identification, importation, and distribution. Plant explorers were sent on international expeditions to search for new plant materials that might grow in the United States or that would be more hardy than varieties that were already growing there. This research included plant classification work by a botanist. Little systematic testing of new plant materials occurred before 1890, but the USDA did reproduce seeds of many plants in Washington (True 1937; Baker et al. 1963, pp. 4–9).[16] Packages of seeds were mailed widely by the USDA to farmers and by Congressmen to their constituents.

Although the USDA's public seed distribution program continued until 1923, importing and testing new plant materials for non-horticultural plants became more selective and systematic after 1898 (Kloppenburg 1988). Some early USDA successes were Brazilian seedless navel oranges in California (Baker et al. 1963, pp. 18–19) and hardier winter wheat varieties from northern Europe in the Great Plains.

The USDA continued the activity initiated by the Patent Office of making annual state and national estimates of crop production. In 1863, it started publishing monthly or bimonthly reports of conditions of crops, based upon voluntary reports of crop correspondents in each county (U.S. Department of Agriculture 1933, p. 3). Regular estimates and reports began in 1866 for acreage, yield per acre, and production of important crops and numbers of livestock. Regular annual estimates of prices of farm products were prepared and published starting in 1867.

To improve their crop estimates, the USDA established a large corps of crop reporters—township reporters—in 1896, and started publishing the *Monthly Crop Reporter* in 1899; then in 1905 the Crop Reporting Board was established to help make further improvements in crop and livestock estimates. Although the statistical tools employed in these early USDA estimates were simple and the estimates were sometimes subject to substantial errors because of sample design problems, they represented a monumental first step in providing national and state market and production information.

In 1862, virtually no information existed on the chemical composition of agricultural products, soils, fertilizers, and agricultural wastes. Furthermore, standard laboratory procedures did not exist for most analyses. This made for slow advances in the chemical-content knowledge base and led to early credibility problems for the chemistry profession (Rossiter 1979; Marcus 1985, pp. 47–48).

In 1862, the USDA initiated studies on the chemical analyses of wine grapes and sorghum, including syrup and sugar. The sorghum research was part of USDA efforts during the Civil War to find substitutes for southern products, e.g., cane sugar. Shortly thereafter, chemical analyses of soils, fertilizer materials, and other agricultural products were initiated. In 1869, the USDA chemists called attention to extensive adulteration of fertilizers and feedstuffs with undesirable materials (Harding 1947).[17] USDA regulatory activities for content and labeling of these materials followed. During the 1890s, the USDA research on alternative methods of chemical analyses of soils and minerals provided the basis for uniform standards or procedures for public and private laboratories doing these analyses.

Research on animal diseases began in the USDA in 1868 with an emphasis on veterinary science. In 1869, studies of tick or Texas cattle fever, fowl and hog cholera, and pleuropneumonia were initiated (Moore 1967). Two early and notable discoveries were the causes of tick fever and hog cholera. USDA researchers showed that cattle ticks were the cause of tick fever in both cattle and humans, a discovery that had great significance to animal and human health (Baker et al. 1963, pp. 32–33). Also, USDA researchers discovered that hog cholera was caused by a virus.

A Division of Chemistry was established in 1862 and a Division of Botany in 1868. In 1878, the USDA had four research-oriented divisions—Botany, Chemistry, Entomology, and Microscopy—and a Division of Statistics and a Seed Division devoted to seed introduction and distribution.

Early State Agricultural Experiment Stations

The Connecticut State Agricultural Experiment Station was established at Wesleyan University, Middleton, Connecticut, in 1875, and was moved to New Haven in 1877. It was the first successful U.S. agricultural experiment station. The station was established with research as its sole function; it had an urban rather than rural location. The founders were convinced of the relative advantage of laboratory experimentation and experimental plots over field experiments, and of the need for access to a library and public utilities (Knoblauch, Law, and Meyer 1962, p. 23). It received an annual appropriation from the state but could also receive funds from other sources.

The management of the New Haven station was assigned to an eight-member board of control. They were to choose a director to be in charge of general management and oversight of experiments and investigations. Samuel W. Johnson, professor of agricultural chemistry at Yale, was appointed the first director. He had been trained at Yale and by Liebig in Germany and had been an enthusiastic proponent of agricultural experiment stations (Browne 1926). Two members of his research staff were also agricultural chemists. Early research in this station focused upon chemical analysis of fertilizer and feedstuffs, seed testing, and scientific relationships between soil and water (True 1937, p. 86).

What the early European and American agricultural experiment stations did was show the usefulness of bringing agricultural scientists together in an organization specifically set up to encourage and conduct research. Their experience also demonstrated the importance of public funding for such work, since most would not have survived without that support. These new stations also helped resolve the struggle over whether farmers or agricultural scientists in agricultural experiment stations and colleges were primarily going to produce new knowledge about agriculture (Marcus 1985, 1987).

U.S. Land Distribution Policy and Advancement of the Frontier

By 1898, the United States had basically taken on the shape and area that it has today, and the frontier had advanced across the country. Immediately after the American Revolution, the governments of the 13 original colonies owned a large amount of land in the interior of America, and between 1784 and 1802, the 13 original states transferred title to lands that they owned to the U.S. government. Between 1802 and 1898, the U.S. government made six major land acquisitions that basically completed the shape and area of the continental United States. This included the two large acquisitions of the Territory of Louisiana (1803) and the Mexican Cession (1848). (See Cochrane 1979, pp. 37–56, 78–88.)

The national government set out immediately to get federally owned lands into the hands of private landowners. They initiated the following policies:

1. The early land distribution ordinances (1784 and 1785) attempted to sell large tracts of land at auction for cash. This was not very successful because of an absence of credit.
2. The next turn was to provide federal credit (1796 act) and to sell in smaller tracts (1800 act). Easy credit seemed to cause unusually optimistic land price speculation, which was followed by a fall in land prices.
3. After 1840, the national government did not provide credit but did give away large quantities of land (a) to canals and railroads (Pre-emptive Act, 1862), (b) to homesteaders (Homestead Act, 1862), and (c) to states for education.

In 1860, the frontier stretched from roughly the middle of Minnesota, then westward and southward down the eastern borders of Nebraska, Kansas, and Oklahoma, and then westward and down the center of Texas. At that time in the west, settlements had occurred only along major river valleys. By 1890, the frontier had essentially advanced across the country. Land acquisition was now largely purchases from private landowners and not settlement of new unoccupied government land. Significant additional agricultural output could not be obtained by bringing more land into production, and land prices could now realistically rise and provide an incentive for development of land substitutes, e.g., fertilizers and higher-yielding crop varieties.

Transportation System Development and Interregional Competition

Before 1815, the means for transporting agricultural products and other goods were very poor. The modes were largely mud roads and rivers. This meant that transportation was frequently impossible and freight rates high. This limited interregional and international trade for commodities produced in the interior of the United States.

Between 1815 and 1840 significant mileage of canals was built to connect major natural waterways. In 1815, there were less than 100 miles of canals in the United States, but by 1840, 3,300 miles of canals were completed. One of the few long-term successful canals was the Erie Canal, which was completed in 1825. It completed a critical link to the interior because it linked the Great Lakes to the Hudson River of New York. This made it possible for the first time to ship grain cheaply from Chicago and the Midwest to New York City and the eastern seaboard. Wagon freight rates were about four times higher than canal rates at this time, and completion of the canal led to dramatically increased competition between farmers of the Midwest and the East Coast. (See Cochrane 1979, pp. 209–225)

Railroads were the major transportation advance during 1840–1890. In 1840, there were about 2,799 total track miles in the United States, but by 1860 this number had grown to 30,283 miles. In 1890, there were 163,597 miles of railroad tracks in the United States. One very critical rail transportation link was the completion of the first transcontinental railroad in 1869. For heavy freight, the railroads had major advantages over other transport modes by providing all-weather routes over varied terrain. By 1900, large trains of about 80 cars were traveling across the United States. (See North 1974, Chapter 9; Fogel 1964, pp. 207–237.)

After 1815, freight rates fell dramatically. In 1815, wagon rates were 30 cents per ton mile, and upstream river and canal rates were 6.7 cents per ton mile. By 1890, freight rates by railroad, upstream rivers, and canals were less than 1 cent per ton mile. This decline in freight rates would be even more spectacular if they were converted to constant prices.

In summary, the rapid disposal of undeveloped federal lands and dramatic drop in interregional transport cost caused farmers to turn to their local state government for assistance. This was a major factor contributing to the development of state support for public land-grant colleges, agricultural experiment stations, and agricultural extension.

The Hatch Act

Passage of the Hatch Act in 1887 was one of the most important legislative steps taken to develop public agricultural research in the United States. Support, however, came gradually. Interest groups' support for a federally subsidized system of state agricultural experiment stations may have started about 1870 (True 1937; Office of Technology Assessment 1981, pp. 33–34). In 1871, representatives of 12 land-grant colleges met to discuss how to accelerate agricultural research. In 1872, the Commissioner of Agriculture called a national agricultural convention in Washington at which time a committee on experiment stations was appointed. In 1882, Hilgard published one of the first proposals for federal funding of experiment stations. He encouraged the use of federal funds in cooperation with land-grant colleges for operation of a station in each state. Other support was building during the 1880s (True 1937, pp. 200–210; Knoblauch, Law, and Meyer 1962).

The experiment station bill was first introduced in Congress in 1882 by Representative Carpenter from Iowa. It was based upon ideas of Seaman Knapp, an Iowa State College agriculture professor, and called for "national experiment stations" at each land-grant college. The bill called for an annual appropriation of $15,000 for each station, and management by the states. The bill was not reported from committee until 1884, and it did not have enough support to become legislation.

A new experiment station bill was reported to Congress by the House Agricultural Committee in early 1886. The chairman of this committee was William Hatch from Missouri. The bill was intensely debated in the Senate during 1886 and 1887. States' rights advocates and some farm interest groups were afraid that federal funding would ultimately lead to federal control of the stations. Revisions in the legislation permitted funds to go to independent (noncollege) stations and left the U.S. Department of Agriculture with the minor role of aiding and assisting the stations. The Hatch Act was signed into law by President Cleveland in 1887, and appropriations were started in 1888 to administer the Hatch Act provisions.

The Hatch Act caused state agricultural experiment stations to be established quickly in all of the states. Under the Hatch Act, the stations were to (1) acquire and spread practical information on subjects connected to agriculture, and (2) perform original science-based research. Each state meeting the provisions of the act was to receive a federal appropriation of $15,000 annually to support investigations and printing and distributing the results (Table 1.6).

When the Hatch Act was enacted, 14 states had 15 stations functioning.[18] By the end of 1888, there were 46 stations. The number increased to 55 in 1893 (Office of Technology Assessment 1981, p. 39). Within the overall numbers, a few stations continued to be wholly state-directed (2 in 1906), and 3 of the territorial stations (Alaska, Hawaii, and Puerto Rico) were sponsored by the USDA. Virtually all the other stations were tied to a land-grant college.[19]

The Hatch Act, as has other legislation that establishes new institutions, left many issues associated with the activities and management of state agricultural experiment stations unspecified. Many of these issues were faced during the first 30 years of SAES system operations (True 1937; Kerr 1987).

Table 1.6 A history of U.S. major legislation affecting federal funding of state agricultural experiment stations, 1887–1998

Year	Legislation	Provision	Funding Mechanism
1887	Hatch Act	Each state could establish an experiment station to conduct original research or verify experiments on subjects bearing directly on the agricultural industry of the United States. Stations were to be established under the direction of the 1862 land-grant colleges, but exceptions were permitted.	Each qualifying state was to receive $15,000 per year.
1906	Adams Act	Each state could receive additional federal funding to pay the necessary expenses of conducting original research and experiments.	Each qualifying state could receive a maximum of $15,000 additional per year. Each state was entitled to an increase of $5,000 for the first year, and to $2,000 over the previous years sum for 5 subsequent years.
1925	Purnell Act	Each state could receive additional federal funding for research to (i) establish and maintain a permanent and efficient agricultural industry and (ii) to develop and improve the rural home and rural life. (Note first emphasis on economics, sociology, and home economics.)	Each qualifying state could receive a maximum of $30,000 per year. Each state was entitled to an increase of $10,000 for the first year, and to annual increases of $5,000 per year for 4 subsequent years.
1935	Bankhead-Jones Act	SAESs and USDA could receive additional funding for research into laws and principles underlying basic problems of agriculture; research relating to improvement of the quality of, and the development of new and improved methods of production of, distribution of, and new and extended uses and markets for agricultural commodities; and research relating to conservation, development, and use of land and water resources for agricultural purposes.	Maximum of $5 million per year with $3 million to SAES. Total increment of $1 million for each of 5 years. Funds were to be distributed to the states on the basis of each state's proportion of the rural population of the U.S. and each state must match the federal contribution with non-federal funding of SAES.
1946	Research and Marketing Act	SAESs and USDA could receive additional funds for marketing and utilization research, and for regional research involving two or more states on a problem of regional significance.	Total increment of $2.5 million in 1947 and 1948; $5 million increase for 1949, 1950, and 1951; and such additional funds as Congress deems necessary for subsequent years. Allocation: 20% of each year's appropriation to be split equally among states; 52% to be allocated by formula (i) 50% according to a state's share of the U.S. rural population and (ii) 50% according to a state's share of the U.S. farm population. These funds must be matched by non-federal SAES funds. Another 25% restricted to regional research; and 3% for Federal Administration of the program.

(Continues)

Table 1.6 (*Continued*)

Year	Legislation	Provision	Funding Mechanism
1955	Amended Hatch Act	SAES funding consolidation. Funding to conduct original and other research, investigations, and experiments bearing directly on and contributing to the establishment and maintenance of a permanent and effective agricultural industry in the U.S., including research basic to the problems of agriculture, and investigations to develop and improve the rural home and rural life and hence the welfare of consumers.	Consolidated federal funding for SAESs into two accounts(formula funds and regional research funds). No set annual amounts were established. Allocation: 20% of each year's appropriation divided equally among stations; 26% according to each state's share of the farm population. These funds must be matched by non-federal SAES funds. Restrictions that 20% must be spent for marketing research. An additional 25% for regional research; and 3% for USDA administration.
1962	McIntire-Stennis Act	Funding available to SAES and schools of forestry for forestry research. It included reforestation, wildlife habitat, wood utilization and other studies for full/effective use of forest resources.	Money to be allocated by formula set up by a committee. $10,000 per year allocated annually to each state and of the remainder, 40% was to be allocated according to a state's share of U.S. total commercial forest land, 40% according to a state's share of value of U.S. timber cut annually, and 20% according to a state's share of non-federal forestry research funding.
1965	Research Facilities Act	Funds available to SAES for construction, acquisition, and remodeling of buildings, laboratories, and other capital facilities.	Allocation: One-third equally to each state; one-third according to a state's share of rural residents; and one-third to states in proportion to their share of the farm population.
1965	Public Law 89-106	Established Special Grants program for selected projects, maximum of 5 years.	Selection of projects to be funded is by Federal Administrators. CSRS was to call annually for proposals in areas singled out by Congress for special attention.
1972	Rural Development Act	SAES and Extension Service could receive funds for rural development and small farm research and education.	Funds were to be distributed 10% for multi-state work, 20% equally distributed among states, 33% to each state based on its percentage of the U.S. rural and farm population, and 4% for federal administration.
1977	National Agricultural Research Extension, and Teaching Policy Act, Title XIV	Continued most previous Hatch programs and initiated a new Competitive Research Grants program for high priority agricultural research to be awarded on the basis of competition among scientific research workers and all colleges and universities. Established mechanism for greater research planning.	Continuation of procedures similar to Amended Hatch Act 1955 and Public Law 89-106. Dropped requirement that 20% amended Hatch funds be spent on agricultural marketing research. Allocation was according to the formula: 20% of each year's appropriation equally among states; 26% by formula according to a state's percentage of the U.S. rural

Year	Legislation	Action	Details
			population; 26% by formula according to a state's share of the U.S. farm population; 25% for cooperative regional research; 3% for federal administration. Permanent funding via the Evans–Allen Research Program, which provides formula funding of the 1890 land-grant institutions.
1981	Amendments to Title XIV, National Agricultural Research, Extension, and Teaching Policy Act	Primarily extended the 1977 Act for 4 years.	Same as 1977 Act, but guaranteed Hatch money at a minimum of 25% of USDA expenditures in cooperative research programs and prohibited the substitution of federal funds in lieu of continued state support.
1985	Amendments to Title XIV of the Food Security Act 1985	Primarily extended the 1981 Act for 4 years.	Competitive Grants Program amended to include emphasis on biotechnology research. A total of $70 million per year authorized for this program.
1990	Farm Bill	Significantly expanded the scope of the competitive grants program, renamed it the National Research Initiative(NRI) Competitive Grants Program	Increased the funding authorization over the 5-year period to a level of $500 million by FY1995.
1996	Farm Bill	Established a new research program called the Fund for Rural America	$100 million per year beginning Jan 1,1997 is allocated to the Fund for Rural America. The funds are split equally among three areas: rural development, research, and an amount to be used at the discretion of the Secretary for research or rural development.
1998	Agricultural Research, Extension, and Education Reform Act	Section 3 of the Hatch Act of 1887 is amended.	No less than 25% shall be allotted to the States for cooperative research employing multidisciplinary approaches in which a State agricultural experiment station, working with another State agricultural experiment station, the Agricultural Research Service, or a college or university, cooperates to solve problems that concern more than one State.
1998	Agricultural Research, Extension, and Education Reform Act	Established in the Treasury of the United States an account to be known as the Initiative for Future Agriculture and Food Systems(IFAFS).	On October 1, 1998, and each October 1 thereafter through October 1, 2002, out of any funds in the Treasury not otherwise appropriated, the Secretary of the Treasury transfer $120,000,000 to the account.

Source: Adapted from True 1937; Knoblauch, Law, and Meyer 1962; and USDA 1986, 2002.

The Agricultural Extension Service

The Cooperative Extension Service was not formally established until 1914 by the Smith-Lever Act. Preceding this legislation, the route to discovery of an effective teaching system for rural adults was a long, searching one. The early agricultural societies sought to collect and distribute information among their members, to improve their member's farming practices, and to affect the farming practices of their neighbors by example (Scott 1970, pp. 26–27). The farm press, local agricultural clubs and societies, county farms, and pioneering instruction in agriculture appeared before 1861. Farmers' institutes and college short courses for farmers were initiated on a broad scale starting in the 1880s. The USDA started to distribute bulletins to farmers in 1889. Other USDA extension activities started shortly after the turn of the century.

Farmers' institutes were labeled initially as schools for farmers (Scott 1970, p. 93). They were an activity where practicing agriculturalists gathered to learn of new techniques and methods from trained scientists and from progressive and successful farmers. The farmers' institutes first emerged in a successful form in Massachusetts about 1859. Institutes were initiated in Connecticut in 1866 and then in other New England states. These institutes represented an expansion of the functions of agricultural societies and state boards of agriculture. In the Midwest, successful farmers' institutes were used to reach local farmers in the 1870s and 1880s. Most were associated with the land-grant colleges, e.g., Michigan, Iowa, Minnesota, Wisconsin. Successful farmers' institutes were established during the 1890s in the West, but they were unsuccessful in the South (Scott 1970, p. 87). The farmers' institute movement grew rapidly during 1901–1914; 8,861 institutes with an attendance of 3 million were held in fiscal 1914 (Scott 1970, p. 105).

College short courses for farmers first emerged in 1867, when the state agricultural society in Michigan urged Michigan Agricultural College to start a program. In 1871, the Agricultural College of Pennsylvania inaugurated a four-day course that was devoted to trials of agricultural equipment and lectures by the agriculture faculty. During 1874–1899, the Illinois Industrial University experimented with special courses for farmers that ranged in length from three months to two years (Scott 1970, p. 153). Wisconsin was the leader in developing short courses in their modern form. The first sessions were offered in 1886 and proved to be a continuing success.

Despite frequent false starts, college short courses were a fixture on most agricultural college campuses by 1914. They varied greatly in length, content, and subject matter, but their purpose was always to reach and instruct rural adults, primarily farmers. By 1907–1910, college short courses were being enthusiastically accepted by farmers.

Traveling college short courses in Illinois (1893–1994) and in other states were initially unsuccessful. Iowa State College's traveling short course started in 1905 and was one of the early successes. Perry G. Holden developed the Iowa program, which was a carefully planned activity—produced on request and with a guaranteed fee (Scott 1970, p. 157). Local groups provided the facilities and advertising. Other states that followed Iowa's lead in developing traveling short courses were Indiana (1907), Ohio (1908), and Virginia (1910). The railroads were a key means of transportation for short course speakers and materials.

In the USDA, extension work with farmers began about 1900. Early work in the South was led by Seaman A. Knapp and in the North and West by William J. Spillman. These two individuals developed very different but successful approaches to extension education. The approaches were tailored to different social structures and information demands (Wright 1987).

The USDA's extension work with farmers in the South was prompted by a serious cotton boll weevil infestation during the early 1900s. In 1889, the Bureau of Plant Industry hired Seaman Knapp and in 1904 assigned him as a special agent to conduct boll weevil control activities in Arkansas, Louisiana, and Texas. His USDA extension work expanded to other states as the weevil spread. He developed a successful control program built around field demonstrations (Baker et al. 1963, p. 43). He sought cooperation with state and local organizations, worked with and through local farmers, and used local demonstration fields to illustrate selection and better management practices.

The General Education Board—a private foundation initiated by John D. Rockefeller to promote education—gave Knapp financial support for demonstration work in states that were not infested by the boll weevil (Scott 1970, p. 223; True 1928, p. 69). He expanded the scope of extension education activities to include the whole family. Boys' corn and calf clubs–later 4-H–girls' canning clubs, and home demonstration work for women were initiated in the South. Knapp's activities relied primarily on USDA information and on local farmers and homemakers for leadership assistance; professional assistance from the faculty members of southern land-grant colleges was not solicited or obtained.

The USDA's extension work in the North and West developed under less-pressing circumstances. In 1901, the Bureau of Plant Industry established a Farm Management Branch and started surveys and studies of farms there to identify practices of successful farmers. In 1902, they hired William Spillman, an agronomist from Washington State College, and he began shortly to conduct farm management research and extension work. Studies were made of farming conditions and practices, especially of successful farms, in various regions. In 1908, studies had been made of business management on the most successful farms, including farm records, farm equipment, livestock feeding systems, and general farm records. These were referred to as early cost-of-production studies. Publications were prepared showing how farmers could improve their management practices—frequently by diversifying their crops.

The first extension bill was introduced in the U.S. Congress in 1909 by James McLaughlin of Michigan, but it did not have enough support to get out of committee. In 1913, Rep. Asbury Lever (S.C.) and Sen. Hoke Smith (Ga.) introduced similar bills that authorized cooperative extension work between the agricultural colleges and the USDA. The agricultural colleges were to establish extension departments (states could designate the institution to administer them) to give instruction and practical demonstration in agriculture and home economics through field demonstrations, publications, and other methods. An annual federal grant to each state of $10,000 was included plus additional federal funds to be allocated to states by a formula based on a state's share of the U.S. rural population (Table 1.7).[20]

The new Smith-Lever bill smoothed over differences between the North and South in organizational philosophies for extension work (Scott 1970, pp. 305–310; Baker et al. 1963, p. 40). Northern agricultural colleges were not attracted to the county agent concept that had developed in the South. They felt that it was not substantive. The farming interests of the South were suspicious of college-trained persons, and farmers' institutes and college short courses had not been successful there. The Smith-Lever Act provided for *cooperative* extension between the land-grant colleges and the USDA. It was passed by the House and Senate and signed by President Woodrow Wilson in 1914.

Table 1.7 History of major U.S. legislation affecting federal funding of Cooperative Extension, 1914–2002

Year	Legislation	Provisions	Funding Mechanism
1914	Smith–Lever Act	Created Cooperative Extension Service to aid in diffusing among the people . . . useful and practical information on subjects relating to agriculture and home economics and to encourage its application.	Provided lump sum grants of $10,000 per state ($480,000 total) and additional formula funding. Formula funds were allocated on the basis of a state's share of the U.S. rural population. Formula funding phased in over 7 years, maximum of $4.1 million. The formula money was to be matched by state funds.
1928	Capper–Ketchmam Act	Provided for expansion of Cooperative Extension Service.	An additional lump sum grant of $20,000 per state ($980,000 total per year) and an additional $500,000 starting in 1929 to be allocated by formula. Required 1/3 of added funds to be matched in 1923 and full matching after 1928.
1935	Bankhead–Jones Act	Provided for expansion of Cooperative Extension Service.	An additional lump sum grant of $20,000 per state ($980,000 total per year) and an additional $8 million to be allocated to states by formula in 1936 and $1 million additional for each of the next 4 years. Formula funds to be allocated by state's share of the U.S. farm population, matching not required.
1945	Bankhead–Flannagan Act	Further expansion of Extension.	Two percent of the federal appropriation was for Federal Adm., 4% was set aside for the Secretary for special need allocation, and 94% distributed by a formula or a state's share in the U.S. farm population.
1953	Amended Smith–Lever Act	Consolidated 9 existing Acts, provided for appropriations for Federal Extension Staff in USDA	Provided that subsequent increases be allocated 4% to special need; 48% based on a state's share of the U.S. rural population and 48% based on a state's share of the U.S. farm population.
1955	Smith–Lever Amendment	Special program system established	Provisions added permitting special nonformula funds.
1961	Amended Smith–Lever Act	Resource and community development extension added.	Provided $700,000 per year for resource and community development work.
1962	Smith–Lever Amendment		Froze distribution of current federal funds to each state. Subsequent increases to be 4% to the Federal Service, and of the remainder 20% in equal proportions to all states, 40% according to a state's share of the U.S. rural population, and 40% according to its share of the U.S. farm population.
1968	Smith–Lever Amendment		Congress abolished special program funding except for $1.6 million for agricultural marketing. These funds were to be allocated by formula.

Year	Act		
1972	Federal Rural Development Act	Title V authorized work in rural communities in agricultural and nonagricultural fields.	Funds were to be distributed 4% for Federal Administration, 10% for multi-state work, 20% equally distributed among states, and 33% each according to a state's share of the U.S. rural and U.S. farm population.
1977	Food and Agriculture Act		Changed the Rural Development Title V formula of 1972 to 19% for farm research programs and 77% for small farm extension programs.
1978	Passage of the Resource Extension Act	Authorized funding for extension forestry and other renewable national resources.	By appropriation.
1981	The Agriculture and Food Act of 1981		Rural development extension funds became part of Smith-Lever formula appropriation.
1998	Agricultural Research, Extension, and Education Act	Section 3 of the Smith-Lever Act is amended by adding terms regarding Multi-state Cooperative Extension Activities.	Of the Federal formula funds that are paid to each State for fiscal year 2000 and each subsequent fiscal year, the State shall expend for the fiscal year for multi-state activities a percentage that is at least equal to the lesser of (1)25%; or (2)twice the percentage for the State determined under subparagragh (a).
2002	Farm Security and Rural Investment Act	Reauthorized and established new agricultural research and extension programs.	Raised the authorized funding level of 1890 extension formula programs from 6 to 15% of funds appropriated under the Smith-Lever Act and from 15% to 25% of funds appropriated under the Hatch Act.

Source: Adapted from True 1928; USDA 1986; and U.S. Congress 1999, 2002.

Conclusion

Public U.S. institutions for agricultural research did not rise full-blown from the minds of their creators. Instead, there was a history of precursors to institutional development and change that led up to and helped shape the major federal legislation starting in 1862.

Our founding fathers provided for the "promotion of progress" in the U.S. Constitution. Patent law encouraged the private individual to be inventive, since the individual was given the exclusive right to the invention for a stated time. At the same time, the rights of the public to share in the benefits of the invention were recognized in that disclosure of the invention in a form that others could reproduce was one of the requirements for obtaining a patent. This requirement made the discovery available to the public. But the patent laws primarily encouraged invention in mechanical and later in chemical inventions because they did not cover plant and animal innovations until the mid-1980s.

The people who directed the entity set up to carry out those laws, the Patent Office, had a much wider impact and goal in mind. This unevenness of impact was clear to them. Because of the efforts of these people, a role for government—public institutions—began to take shape. They began gathering and testing new plant species from around the world, diffusing information useful to agriculture, and funding and conducting research around the world and distributing the results widely across the United States. All of these programs had great impact on the functions these publicly supported agricultural research institutions took on as they began. Also, the public's support of what the Patent Office was doing helped pave the way for the establishment of the USDA and the SAES system.

The early agricultural societies knew the importance of innovation in agriculture and promoted the sharing and recognition of new ideas and practices whenever they could. They used fairs and contests and other means to interest farmers in improving what they were doing. They also financed private research. But they knew that farmers could never finance the needed research and education themselves. So they became a force for the development and passage of the legislation setting up and financing public agricultural research.

The early efforts of educators, too, to develop education relevant to the vast majority who lived and worked in rural America provided models for Congress and others in setting up the land-grant college system. They demonstrated that staff could be assembled to carry out science and laboratory-based research for agriculture in an academic setting. They also led the way in creating appropriate curricular changes to combine classical education with agricultural and mechanic arts educations.

While the European efforts to develop agricultural experiment stations contributed valuable experience to those wanting to establish a similar system in this country, Americans did not follow those models exactly. They saw that the Europeans by and large did not tie researchers to educational institutions, did not expect scientists to take responsibility for diffusing the results of their research to users, and did not provide a way of training future researchers in agriculture. These were all ideas that became part of the U.S. system and contributed to its uniqueness.

These experiences and forces helped shape the legislation and institutions that eventually were set up to help agriculture via publicly supported research, teaching, and extension.

Notes

1. Also, the Land-Grant Act of 1890 established state agricultural and mechanic colleges for blacks in southern states.

2. This section and the following one draw heavily from Evenson and Putnam 1987.

3. Under the first act, novelty was to be determined by a committee consisting of the Secretaries of State and of War, and the Attorney General. By February 11, 1793, when under the second act patent administration fell to the State Department, the United States had granted 56 patents.

4. Although this seems only to be common sense, it was the first concrete recognition in U.S. law of the analogy between the patent and real property. Assignability of patent rights is an important common-law attribute. It also represented a clear efficiency gain, by allowing the inventor to specialize in innovation and others to specialize in production.

5. A claim was too broad "when the patent claims more than that of which the patentee was the original and first inventory" (Sec. 7).

6. Ellsworth (U.S. Patent Office 1838, pp. 57–59) cites the following examples of the benefits of this program:

> A short time since, the most eastern state of our Union was, in a measure, dependent on others for their breadstuffs. That state is now becoming able to supply its own wants, and will soon have a surplus for exportation; and this is effected by the extensive introduction of *spring wheat*. . . .

> From experiments made the last summer, there can be no doubt that the crop of Indian corn may be improved at least one-third, without any extra labor, and this affected by a due regard only to the selection of seeds.

> From the samples transmitted to the Patent Office, especially from the shores of Lake Superior, there is a moral certainty of a good crop of corn in the higher latitudes, if proper attention is paid to the selection of seeds. Inattention to this subject has lost, to the northern portion of our Union, many millions every year.

7. The spread of many agricultural implements to common use by this time can be seen in Ellsworth's summary remarks (U.S. Patent Office 1845, pp. 6–9).

> Among the first inquiries of the political economist is the question, How can the productiveness of the earth be increased? Modern practice answers it easily. Manure and tillage are the instruments employed; either, alone, is comparatively useless. "Grapes will not grow on thorns, nor figs on thistles." Nor will sour land yield sweet food. The nature of the soil must be changed, and this is effected by draining.

> Intimately connected with draining land, is that of sub-soiling; indeed, the last has lately been substituted for the former with good success. The cheapness of subsoil ploughs brings them within the reach of every farmer. . . . (An account of the yields produced by this technique follows.)

> There is much to console the husbandman in the reduction of the cost of the necessities of life which he has occasion to purchase.

> Labor-saving machines are being introduced with still greater success. Mowing and reaping will, it is believed, soon be chiefly performed on smooth land by horse power. Some have regretted that modern improvements made important changes of employment; but the march of the arts and sciences is onward.

8. After describing the extended communications network among the agricultural press and private individuals upon which the Patent Office relied, Burke (U.S. Patent Office 1846, pp. 15–17) defended the special interest the Patent Office had taken in agriculture:

> Assuming the general accuracy of those estimates, it is hardly necessary to speak of their value . . . without a knowledge of the statistics of a nation, which embrace every fact relating to its condition and welfare, moral, or political, it is impossible to legislate wisely for

its interests. And no statistical knowledge is more important than that which exhibits the resources of a nation, as indicated by the products of its labor. An important part of that knowledge the agricultural estimates of the Patent Office were designed to furnish.

9. Congress was not supportive of the first research publications, which were incorporated in later editions of the agricultural report (e.g., see U.S. Patent Office 1849, Appendixes 1 and 2), prior to the founding of the Department of Agriculture in 1862.

10. Burke said that it is not recommended, nor is it desired by the under-signed, that any such department, or national board of agriculture, be instituted by this government. The practices of other enlightened governments are referred to only to show that the exercise of such functions by government is not without precedent, nor without utility.

11. This development of public responsibility for research and for science preceded the more general interest in science by the government.

12. The Red Angus breed was the first to impose a performance requirement on animals to be registered as purebreds. In the late 1950s, the Red Angus Association established a weaning weight standard for purebred Red Angus calves.

13. In Germany, chemistry developed into a science during the 18th century. This was aided by the establishment of the first successful chemical journal by D. Lorenz Crell in 1784. Chemists were then able to have regular access to the research of other researchers and to build upon that research (Hufbauer 1982; Price 1969). Chemistry was seen as having considerable practical value to society through its potential for improving human health and increasing production.

14. Although the modern research university was a German invention, it did not include agricultural sciences. In 1809, Friedrich-Wilhelm University of Berlin was established to aid the development of the new lab-based sciences, e.g., chemistry, physics, biology; but agricultural facilities were not established at leading German universities until 1863–1880.

15. Early agricultural research was also conducted in the United States at the Patent Office, 1836–1862. In 1862, this research activity was transferred to the newly established U.S. Department of Agriculture (True 1937, pp. 41–43).

16. True (1937, p. 47) indicates that in fiscal 1867, 58 percent of the USDA's budget went into the seed distribution activity.

17. However, much earlier, Samuel W. Johnson lectured on the frauds in commercial manures (fertilizers) at the January 7, 1857, meeting of the Connecticut Agricultural Society (Osborne 1913, p. 112).

18. Some states, e.g., New Jersey, maintained two stations for a while—one Hatch-funded and one state-funded.

19. The Hatch Act also aided the establishment of new land-grant colleges in New Hampshire and Rhode Island. In New Hampshire, Dartmouth's loss of land-grant and experiment station funds was also prompted by the receipt, in 1891, by New Hampshire of a large estate near Durham conditional on the establishment of a state agricultural college. See Kerr (1987) for additional details on the establishment of experiment stations under the Hatch Act.

20. No funds from national organizations could be used for matching, i.e., funds from the General Education Board.

References

Acker, Duane and M. Cunningham. 1991. *Animal Science and Industry*. Englewood Cliffs, NJ: Prentice-Hall.

Baker, Gladys, et al. 1963. *Century of Service: The First 100 Years of the United States Department of Agriculture*. Washington, DC: U.S. Government Printing Office.

Browne, C.A. (ed.) 1926. "A Half-Century of Chemistry in America, 1876–1926." *Journal of American Chemical Society*, vol. 48(8a), part 2.

Cochrane, W.W. 1979. *The Development of American Agriculture: A Historical Analysis*. Minneapolis, MN: University of Minnesota Press.

Eddy, Edward D., Jr. 1957. *Colleges for Our Land and Time*. New York: Harper & Brothers Publishing.

Evenson, Robert E. 1983. "Intellectual Property Rights and Agribusiness Research and Development: Implications for the Public Agricultural Research System." *American Journal of Agricultural Economics* 65: 967–975.

Evenson, Robert E. and Jonathan D. Putnam. 1987. "Institutional Changes in Intellectual Property Rights." *American Journal of Agricultural Economics* 69: 403–409.

Fogel, Robert. 1964. *Railroads and American Economic Growth: Essays in Economic History.* Baltimore, MD: Johns Hopkins University Press.

Gaus, John M. and L.O. Wolcott. 1940. *Public Administration and the United States Department of Agriculture.* Chicago, IL: Public Administration Service.

Harding, T. Swann. 1947. *Two Blades of Grass: A History of Scientific Development in the U.S. Department of Agriculture.* Norman, OK: University of Oklahoma Press.

Hayami, Y. and Vernon W. Ruttan. 1971. *Agricultural Development: An International Perspective.* Baltimore: Johns Hopkins University Press.

Hilgard, Eugene W. 1882. "Progress in Agriculture by Education and Government Aid." *Atlantic Monthly* 49:531–541, 651–661.

Hufbauer, Karl. 1982. *The Formation of the German Chemical Community (1720–1795).* Berkeley, CA: University of California Press.

Kerr, Norwood Allen. 1987. *The Legacy: A Centennial History of the State Agricultural Experiment Stations, 1887–1987.* Columbia, MO: Missouri Agricultural Experiment Station, University of Missouri-Columbia.

Kloppenburg, Jack R., Jr. 1988. *First the Seed: The Political Economy of Plant Biotechnology, 1492–2000.* Cambridge, NY: Cambridge University Press.

Knoblauch, H.D., E.M. Law, and W.P. Meyer. 1962. *State Agricultural Experiment Stations: A History of Research Policy and Procedure.* Washington, DC: U.S. Department of Agriculture, Misc. Publ. 904.

Marcus, Alan I. 1985. *Agricultural Science and the Quest for Legitimacy.* Ames, IA: Iowa State University Press.

Marcus, Alan I. 1987. "Constituents and Constituencies: An Overview of the History of Public Agricultural Research Institutions in America." In *Public Policy and Agricultural Technology: Adversity Despite Achievement,* ed. D.F. Hadwiger and W.P. Browne. London, England: Macmillan.

Moore, Ernest G. 1967. *The Agricultural Research Service.* New York, NY: Praeger Publishers.

North, Douglas C. 1974. *Growth and Welfare in the American Past: A New Economic History.* 2nd ed. Englewood Cliffs, NJ: Prentice-Hall.

Office of Technology Assessment. 1981. *An Assessment of the United States Food and Agricultural Research System.* Vol. I. Washington, DC: U.S. Government Printing Office.

Osborne, Elizabeth A. 1913. *From the Letter-Files of S.W. Johnson.* New Haven, CT: Yale University Press.

Price, Derek. 1969. "Citation Measures of the Hard Sciences, Soft Sciences, Technology, and Nonscience." Proceedings of Conference on Communication Among Scientists and Technologists, Johns Hopkins University.

Rossiter, Margaret W. 1979. "The Organization of the Agricultural Sciences." In *The Organization of Knowledge in Modern America (1860–1920),* ed. A. Oleson and J. Voss. Baltimore, MD: Johns Hopkins University Press.

Ruttan, Vernon. 1982. *Agricultural Research Policy.* Minneapolis, MN: University of Minnesota Press.

Salman, S.C. and A.A. Hanson. 1964. *The Principles and Practice of Agricultural Research.* London: Leonard Hill.

Scott, Roy V. 1970. *The Reluctant Farmer—The Rise of Agricultural Extension to 1914.* Urbana, IL: University of Illinois Press.

True, A.C. 1928. *A History of Agricultural Extension Work in the United States, 1785–1923.* Washington, DC: U.S. Department of Agriculture, Misc. Publ. 15.

True, A.C. 1929. *A History of Agricultural Education in the United States, 1785–1925.* Washington, DC: U.S. Department of Agriculture, Misc. Publ. 36.

True, A.C. 1937. *A History of Agricultural Experimentation and Research in the United States, 1607–1925: Including a History of the United States Department of Agriculture.* Washington, DC: U.S. Department of Agriculture, Misc. Publ. 251.

U.S. Congress. 1863. *Statutes at Large* 12:387–388.

U.S. Congress. 1999. "Agricultural Research, Extension, and Education Reform Act of 1998" (Public Law 105-185, June 23, 1998). Washington, D.C.: U.S. Government Printing Office.

U.S. Congress. Various years. Legislative Acts.

U.S. Department of Agriculture. 1933. "The Crop and Livestock Reporting Service of the United States." Washington, DC: U.S. Department of Agriculture, Misc. Publ. No. 171.

U.S. Department of Agriculture. 1986. *Compilation of Statutes Related to Agriculture and Forestry Research and Extension Activities and Related Matters.* Washington, D.C.: U.S. Department of Agriculture, Agricultural Research Service.

U.S. Department of Agriculture, Economic Research Service. 2002. "Farm Bill Policy: Title VII—Research and Related Matters." Available at: http://www.ers.usda.gov/Features/farmbill/Titles/TitileVIIAgriculturalResearch.htm.

U.S. Congress. 2002. "Smith-Lever Act as Amended through Public Law 107-293, Nov. 13, 2002." Washington, D.C.: U.S. Government Printing Office.

U.S. Patent Office. 1838, 1845, 1846, 1849. *Annual Report.* Washington, D.C.

Warwick, Everett J, and J.E. Legates. 1979. *Breeding and Improvement of Farm Animals.* New York, NY: McGraw-Hill.

Willham, Richard L. 1985. "The Legacy of the Stockman." Ames, IA: Iowa State University, Department of Animal Science.

Wright, Gavin. 1987. "The Economic Revolution in the American South." *Journal of Economic Perspectives* 1:161–178.

2

The Development of a System of Agricultural Sciences

In retrospect, the design elements incorporated into the research-teaching-extension institutions of the USDA-SAES land-grant college system when they were formally established have served the system well. Many elements of the original design remain intact and functional today. In Chapter 1, we attempted to show that the original design was responsive to the political, economic, and scientific conditions prevailing at the time in the United States. The special political and economic interests of the states in teaching and later in research and extension were incorporated into the federal-state design. Federal funds were used to support these activities in the states without strong centralized federal control. The experiment-station model was borrowed from European experience, and it influenced the design and conduct of the early research programs. Institutions, even those with good original designs, are subject to changes over time. In the United States over the past century, dramatic changes in the economic organization of agricultural production have taken place. Rural communities have changed in size and population composition. The political organization and interests of the "clientele" have changed. The nature of science and interrelationship between services have evolved. Research and Development (R&D) by private firms has grown in part because incentives for it have been strengthened.

This chapter attempts to interpret the responses of the teaching-research-extension (USDA-SAES land-grant) institutions to these changing conditions as evolutionary changes with selection leading to the complex system that we observe today. The term "evolutionary" is intended to convey the notion that the system responded to change. The slow process that was laden with conflicts, or selection, has produced a more effective and productive system. This process is a continuing one, and the pressures for change during the early 21st century are as great or greater than at any time in the history of the system. (See Chapter 9.)

It is difficult to argue that a system of truly applied agricultural sciences existed in the middle and late 1800s. Such sciences did emerge over time as part of this evolutionary process, and we attempt to describe this emergence by reporting brief histories of these institutions after their establishment and by examining in some detail the knowledge or science system that has evolved.

This chapter is organized in five sections. The first four sections provide historical accounts of the institutional evolution of (1) research in the USDA, (2) research and teaching in the land-grant colleges, (3) research in the state agricultural experiment stations, and (4) extension activities. For each, a summary of legislative history is provided. These discussions attempt to describe responses to and accommodation of several fundamental conflicting interests in the system. These include the conflicting interests of state and federal policies and the inherent centralization-decentralization conflicts. A second fundamental source of conflict is between clientele demand for practical and usable technologies and the demands imposed on research by the organization of science. A third source of conflict (and cooperation) is between the public and private institutions engaged in original research and extension.

The final section of the chapter attempts to formally describe the arrangements that have accommodated these conflicting interests and have evolved into an interrelated system of scientific organization including active scientific or knowledge exchange. It is this larger system of science that has supported the agricultural sciences and maintained their viability and effectiveness in serving agricultural interests. (See Chapter 8 for evidence on this point.)

The Evolution of Research in the USDA

Major structural changes in the USDA have occurred since its establishment in 1862.

Statistics and Economics

An overview of the organizational changes in the institutions of the USDA that deal with statistics and economics research is presented (Figure 2.1). To collect and publish agricultural statistics, a Division of Statistics was created in 1863; it became a bureau in 1903 and was renamed the Bureau of Markets and Crop Estimates in 1913 (Baker et al. 1963, pp. 501–502).

A Farm Management Branch in the Bureau of Plant Industry was organized in 1901, and its name was changed in 1905 to Office of Farm Management as farm management research and extension work advanced in the USDA. In 1919, the name was changed again, this time to Office of Farm Management and Farm Economics, and the research was organized into sections: farm organizations, cost of production, farm labor, farm finance, land economics, agricultural history and geography, and rural life studies (Baker et al. 1963, p. 501).

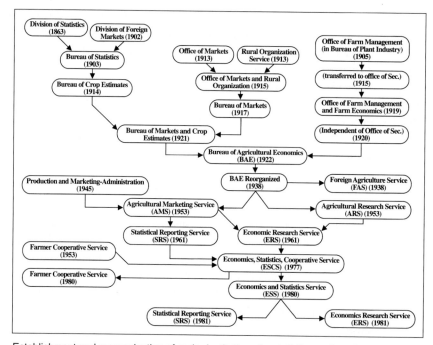

Fig. 2.1. Establishment and reorganization of major institutions for statistics and economics research in the U.S. Department of Agriculture, 1862–1985. (Adapted from Baker et al. 1963, pp. 453–462.)

In 1922, the statistical and agricultural economics research were consolidated into a new Bureau of Agricultural Economics (BAE), formed by combining the Bureau of Markets and Crop Estimates and Office of Farm Management and Farm Economics. During the first years of the bureau, collection of data on production, prices, and markets for farm products was the major activity. Correlation and regression analyses were first applied to crop forecasts in 1929 (U.S. Department of Agriculture 1933, p. 3).

The BAE divisions and staff grew as advances in economics and statistical methods occurred and by 1938 it had 20 program divisions. In that year a reorganization occurred and some of the marketing work, and all of the regulatory research, were transferred to other USDA agencies. In 1945, BAE was given sole responsibility for agricultural economic information and coordination of economic and statistical research in the USDA.

BAE had a relatively long institutional life of 31 years, but in a major USDA reorganization by the Eisenhower administration in 1953, the BAE was abolished. Its market research, agricultural economics (only part), and agricultural estimates were transferred to the newly created Agricultural Marketing Service (AMS). The farm management, land economics, and agricultural finance research were transferred to a newly created Agricultural Research Service (Baker et al. 1963, pp. 498–500; Office of Technology Assessment 1981, p. 120).

The Statistical Reporting Service (SRS) was created in 1961 to direct the former AMS programs for crop and livestock estimates, marketing surveys, and development of statistical standards and techniques. The Economic Research Service (ERS) was created to carry on the major agricultural economics research functions of the USDA.

These institutional developments within the USDA were responsive to growth in the demand for statistical information to inform policymakers. As the role of market interventions and farm income support programs became more important, policy research became both more important and more politically sensitive. The reorganization of the BAE in the 1950s represented a response by USDA bureaucrats to achieve lower-profile and less politically sensitive research projects.

It is relevant to see that the growth of farm programs and the emergence of "supply control" programs designed to curtail supply and achieve higher farm prices introduced an inherent conflict between farm program policymakers and productivity-increasing research after World War II. This conflict was accommodated by the federal-state model of research that evolved over time with the result that the anti-research forces had little net effect on research conducted by the states.

Natural Sciences

In 1878, the USDA had four natural science research-oriented divisions—Botany, Chemistry, Entomology, and Microscopy—and a Division of Statistics and a Seed Division devoted to seed introduction and distribution.[1] A Veterinary Division was established in 1883, and an increase in the origin of livestock diseases during the 1880s (see Chapter 3) resulted in the Veterinary Division being expanded into a new Bureau of Animal Industry (1884). Divisions for Soils, Forestry, and Pomology (fruits) were established before 1897 (Figure 2.2).

During 1901–1905, Secretary Wilson completed a major reorganization of scattered chemical-biological work into seven bureaus: Chemistry, Soils, Plant Industry, Forestry, Entomology, Animal Industry, and Biological Survey.[2]

The Bureau of Agricultural Engineering, which had been established in 1931, was then abolished in 1938, and the work in chemistry and agricultural engineering was combined into a Bureau of

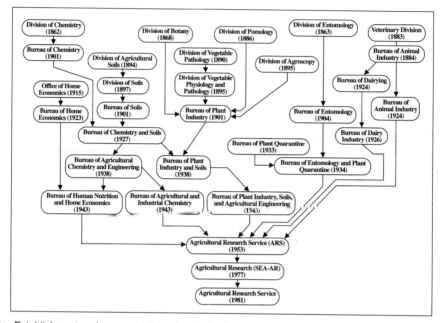

Fig. 2.2. Establishment and reorganization of major institutions for chemical and biological research in the U.S. Department of Agriculture, 1862–1981. (Adapted from Baker et al. 1963, pp. 453–462.)

Chemistry and Agricultural Engineering. This reorganization occurred about 1939 when most of the Department's chemical research was being transferred out of Washington to four regional utilization laboratories.[3]

A Division of Dairy Industry was created in the Bureau of Animal Industry in 1895. The growing economic strength of the dairy industry during the 1920s resulted in a separate Bureau of Dairying being established in 1924. Its name was changed to Bureau of Dairy Industry in 1926, and it later (1953) became part of the Agricultural Research Service.

The Agricultural Research Service (ARS) was established by the Secretary of Agriculture in 1953 as part of a major reorganization of agricultural research in the USDA. It was established to consolidate the chemical, biological, physical, and engineering research of the Department.

The coordination of state agricultural experiment stations was contained in its Experiment Station Division. This activity was also transferred out of ARS in 1961 to a new independent agency, the Cooperative States Research Service (CSRS) (Moore 1967, p. 80).

The Home Economics Division and the Utilization Division were combined in 1957 into a program of Nutrition, Consumer, and Industrial Uses. ARS lost its Production Economics Division to the Economic Research Service in 1961 but gained two AMS marketing research divisions. The research units in 1967 were designated Farm Research; Nutrition, Consumer, and Industrial Use; and Marketing Research (Moore 1967, p. 76).

Field Stations and Laboratories

The expansion of USDA research during the Wilson era extended the location of USDA research beyond the Washington and Beltsville, MD, areas. An increasing amount of the work of the Bureau of Plant Industry was conducted at USDA field stations. By 1913, BPI operated 18 field stations in nine states, eight in cooperation with state experiment stations. In the same year, the Bureau of

Entomology had 35 field laboratories in different parts of the United States (True 1937, pp. 197–198, 203). By 1931, the USDA maintained 51 field stations in 24 states (Waggoner 1976, p. 242). The number of research locations in the states had increased further by 1958, and more than 150 research locations existed in 1982 (U.S. Department of Agriculture 1983).

The Agricultural Adjustment Act of 1938 authorized the USDA to establish four regional-utilization research laboratories—one in each major farm producing region—that were to concentrate on developing new uses and outlets for surplus commodities. The laboratories were located at Philadelphia (Wyndmoor), Pennsylvania; Peoria, Illinois; New Orleans, Louisiana; and Albany, California; and were constructed about 1940. They were placed under the direction of the Bureau of Chemistry and Engineering (later the Bureau of Agricultural and Industrial Chemistry) until 1953. Then they became part of the four Regional Utilization and Development divisions of the Agricultural Research Service. Their first research activities were devoted to developing war material uses of agriculture commodities (Moore 1967, p. 22)—for example, substitutes for petroleum oils and natural rubber.

These organizational changes in the USDA were responses to developments in the states as well as to changes in the structure of scientific research. Experience in the land-grant colleges reflected these factors as well.

The Establishment of Teaching, Research, and Extension in the Land-Grant Colleges

The new land-grant colleges faced several serious problems:

1. When new institutions were needed, the Morrill Act land endowments provided insufficient resources for building and operating new institutions. Restrictions placed on the sale of the land-grant land and farmers' access to "free" Homestead Act land caused the land-grant endowments to have low values (Eddy 1957, p. 51). Furthermore, only current income or interest on the principal could be spent from the land grant, and Morrill Act funds could not be spent for construction, purchase, or repair of buildings.
2. Few youth who had agricultural and mechanical interests had completed a high school diploma. The population of teenagers was small, and only a small share of them completed or attended high school. High college admission standards meant no students, and teaching college preparatory courses became an added drain on limited college resources.
3. A stock of college-trained agricultural or engineering staff did not exist. The traditional U.S. institutions did not offer degrees in these fields.
4. The stock of knowledge on agriculture and engineering subjects was very limited.

The new land-grant colleges also faced intense scrutiny or opposition by some interest groups. The presidents of the affiliated church colleges saw them as a threat to student enrollments and to their moral leadership.[4]

Early Criticisms

Also, during the late 1860s and 1870s, agricultural interest groups were frequently disappointed with and criticized the curricula of the new land-grant colleges (Marcus 1987). The primary issue was relative emphasis—classical and scientific courses versus practical courses in farming and mechanic arts. The leaders of these interest groups could not see a need for mastery of biology, geology, chemistry, mathematics, and English. Farm groups succeeding the agricultural societies, such as the Grange (founded in 1867) and Farmers' Alliance, frequently denounced the agricultural programs of the land-grant colleges and went to state legislatures to register their protests and to demand changes (Kerr 1987; Eddy 1957; Nevins 1962, pp. 53–59; Scott 1970, pp. 52–60). The Grange, however, was in the forefront of the movement during the 1870s for establishing separate state agricultural colleges (Scott 1970, pp. 52–53).[5]

State universities in existence before 1862 and receiving land-grant funds were also frequently criticized.[6] In some states, major conflicts between the land-grant university and farm interests resulted in a compromise. During the early 1880s, the Farmers' Alliance led an assault against the University of Minnesota and its organization of the agriculture program. In 1885, the Alliance charged that the university was ignoring its responsibility for agricultural instruction and was mis-using land-grant funds. It branded the university an utter failure in agricultural programs. In 1886, the Alliance, with support from the Grange, demanded that a separate agricultural college be established. Although the Minnesota authorities rejected these demands, they recognized that changes were needed. A new university campus was established for farm students at St. Anthony Park (the St. Paul campus). In addition, the state launched a farmers' institute program and established a 2-year school of agriculture (Scott 1970, p. 54).[7]

The Illinois Industrial University had difficulty organizing a school of agriculture in the late 1860s (Moores 1970, pp. 15–31). The Chicago and farm press denounced President J.M. Gregory for the university's slow progress. He, however, told the critics that the institution they demanded was impossible. At an introduction to an agricultural short course in 1869, Dr. Gregory declared: "Looking at the crude and disjointed facts which agricultural writers give us, we come to the conclusion that we have no science of agriculture. Botany is a science, chemistry is a science, but agriculture is not a science in any sense. It is simply a mass empiricism." This statement seemed to slow the critics because they sensed its truthfulness (Nevins 1962, pp. 54–57).

The large increase in demand for college teachers and researchers trained in agriculture and engineering had not been anticipated by supporters of the Morrill Act. Illinois, Michigan, Missouri, and other colleges had difficulty finding college-trained faculty for their new agricultural (and engineering) programs (Nevins 1962). The new programs needed faculty who had a portfolio of skills: scientific training, farming experience, and organizational ability.

The demands for useful material to teach in agricultural courses and for information about local practical farm problems were major driving forces behind the initiation of research by the faculty in the land-grant colleges (Hilgard 1882). Because of the very small knowledge base, the tools for discovery frequently needed to be developed. The possibilities for applications of science to agriculture were enormous. Farmers, however, demanded answers to current problems. The establishment of early U.S. state agricultural experiment stations (1875–1887) marked the beginning of organized public agricultural research outside of the USDA and brought improvements in laboratories and facilities required for research.

Expanded Institutions: The Second Morrill Act

A second Morrill Act (1890) provided annual federal appropriations to each of the states to support instruction in land-grant colleges. These funds were much needed by the struggling new colleges. It, however, contained a new provision forbidding racial discrimination in admission to colleges receiving the funds. A state was given the option of a separate institution that was to receive a "just and equitable" share of the money. The southern states took the route of establishing separate black colleges.

Maryland assigned its money to a private black college that subsequently became a state institution. Alabama, Arkansas, Florida, Texas, Kentucky, Virginia, Mississippi, and Missouri gave portions of the funds to existing publicly funded black schools (Kerr 1987). Delaware, Georgia, North Carolina, Oklahoma, South Carolina, Tennessee, and West Virginia created new land-grant schools for their black residents. The "1890 Colleges," as these colleges became known, received very little funding relative to the original "1862 land-grant colleges" or relative to the number of students enrolled.

Table 2.1 provides a summary of the legislation affecting federal funding of the land-grant institutions. It shows that after the 1890 Morrill Act subsequent legislation was intended to support extension and vocational education programs in the land-grant institutions. The development of the departmental structures in the land-grant institutions was strongly influenced by the state experiment stations and is discussed further in the last section of this chapter.

Organized Research under the Hatch Act

The demand for knowledge about relationships in agriculture grew rapidly after the land-grant colleges were established. Agricultural faculty needed information for useful agricultural courses. Also, the farmers of New England were demanding information about fertilizers. These farmers, who had relatively infertile soils, were facing increased competition in grain production from the newly settled agricultural lands of the Midwest and from reduced transport costs (see Chapter 1; Hilgard 1882; Rossiter 1975).

Although agricultural faculty conducted rudimentary research before experiment stations were established, the agricultural experiment stations became a key institutional innovation in the process of making the study of agriculture a science.

The Hatch Act caused state agricultural experiment stations to be established quickly in all of the states. Under the Hatch Act, the stations were to (1) acquire and spread practical information on subjects connected to agriculture and (2) perform original science-based research. Each state meeting the provisions of the act was to receive a federal appropriation of $15,000 annually to support investigations and the printing and distribution of the results. (Table 1.6 reports a summary of legislative acts affecting federal support of state agricultural experiment stations.)

At the time of the Hatch Act, 14 states had 15 stations functioning.[8] By the end of 1888, there were 46 such stations. The number increased to 55 in 1893 (Office of Technology Assessment 1981, p. 39). Within the overall numbers, a few stations continued to be wholly state-directed (2 in 1906), and three of the territorial stations (Alaska, Hawaii, and Puerto Rico) were sponsored by the USDA.[9]

The Hatch Act left many issues associated with the activities and management of state agricultural experiment stations unspecified, as has other legislation that establishes new institutions. Many of these issues were faced during the first 30 years of SAES system operations. Natural phenomena—especially soils, climate, and plants—differ greatly across states. One advantage of having an agricultural experiment station in each state was that these institutions could apply agricultural sciences to develop technology suited to state-specific local conditions. However, natural phenomena frequently differ significantly even across a given state, especially when the state is large or extends over several degrees of latitude, e.g., California, Texas, Minnesota. Thus, arguments for a diffused

Table 2.1 History of major legislation affecting federal funding of land-grant institutions, excluding agricultural experiment stations, 1862–1996

Year	Legislation	Provisions	Funding Mechanism
1862	Morrill Act	Each state may establish and maintain at least one college where the leading object shall be to teach courses related to agriculture and mechanical arts in order to promote the liberal and practical education of the industrial classes, without excluding other scientific and classical studies, and including military tactics.	Each state was to receive 30,000 acres of land for each Senator and Representative in Congress. States where not enough public lands existed to meet this Federal obligation were given land script to public lands in other states. The income from the land or principle was to be used for operating expenses (construction, purchase, or repair of building was excluded).
1890	Morrill Act	Each state could receive additional funds to more completely endow and support land-grant colleges. The receipts were to pay for instruction in agriculture, mechanical arts, the English language and branches of mathematics, physical, natural, and economic sciences related to agriculture and mechanical arts. African Americans were to be admitted to land-grant institutions. States could establish separate land-grant colleges for African Americans.	Proceeds of public land sales or Treasury revenues were to go to each state and territory in the maximum amount of $25,000 per year. Each state and territory was entitled to $15,000 for the first year and an increase of $1,000 per year for a total of 10 years. No expenditures could be made for construction, purchase, or repair of buildings.
1907	Nelson Amendment	Same as Second Morrill Act with the additional specification that a portion of the fund could be used for "providing courses for the special preparation of instructors for teaching the elements of agriculture and mechanic arts."	Double annual appropriation to $50,000.
1914	Smith-Lever Cooperative Extension Act	The land-grant college and U.S. Department of Agriculture were to cooperate in extension work. The extension work was to consist of instruction and practical demonstrations in agriculture and home economics to persons who are not attending the land-grant college. Information was to be supplied through field demonstrations, publications, and other methods.	Ten thousand dollars per year were appropriated to each qualifying state. Starting in 1915, states were to share increases in total funding in proportion to their share of the total U.S. rural population but these increases were to be matched by non-federal funds. The increase for all states was $600,000 in 1915, and the total increased by $500,000 per year for a total of 7 years. The total maximum increment was to be $4.1 million for later years.
1917	Smith-Hughes Vocational Education Act	States were eligible for national grants to stimulate vocational education in agriculture, home economics, and industrial arts. The grants had two forms: (a) for training of teachers by public colleges and (b) for paying for part of the salaries of teachers and directors of these subjects in secondary public schools.	No specific statement.

42

Year	Act	Description
1935	Title II, Section 22 of the Bankhead-Jones Act	Same as Morrill Act of 1862 as "amended and supplemented".	Annual appropriation of $980,000; and "for the fiscal year following the first fiscal year for which an appropriation is made in pursuance of (the $980,000) . . . $500,000, and for each of the two fiscal years thereafter $500,000 more than the amount authorized to be appropriated for the preceding fiscal year and for each fiscal year thereafter $1,500,000."
1960	Amendment to Title II, Section 22 of the Bankhead-Jones Act, Stennis Act	Same as Morrill Act of 1862 as "amended and supplemented".	Annual appropriation of $7,650,000 which is distributed equally among the States and Puerto Rico; $4,300,000 which is allotted based on the proposition of State (Puerto Rico) population to total U.S. and Puerto Rico population.
1977	Food and Agriculture Act	Transfer of the administration of the Bankhead-Jones Act from Office of Education to Department of Agriculture.	
1990	Food, Agriculture, Conservation, and Trade Act	Reauthorized sustainable agriculture research and education program and added new program for training of extension service personnel in sustainable agriculture practices.	Congressional appropriators increased the previous year's allocation of $42.5 to $73 million.
1994	Elementary and Secondary Education Reauthorization Act	Conferred land-grant status to the 29 Native American colleges that compose the American Indian Higher Education Consortium.	The 29 colleges were authorized $23 million endowment over a 5-year period. They would receive interest payments from the endowment each year. Also authorized were funds for the colleges' education and extension programs in agricultural and natural resources.
1994	Equity in Educational Land-Grant Status Act	Authorized land-grant status to the eligible 1994 institutions.	For fiscal year 1996 and for each fiscal year thereafter, there are authorized to be appropriated an amount equal to $50,000 multiplied by the number of 1994 Institutions. Also the Secretary shall make two or more institutional capacity building grants to assist 1994 Institutions, $1,700,000 for each of fiscal years 1996 through 2002 were authorized to the Department of Agriculture to carry out this section.
1996	Federal Agriculture Improvement and Reform Act	Authorized grant program to upgrade agricultural and food sciences facilities at 1890 land-grant colleges.	$15,000,000 for each of fiscal years 1996 and 1997 is appropriated.

Source: Adapted from True 1929, Eddy 1957, and USDA 1980.

system of state agricultural experiment substations could be made at the state level. Organization, control, and economies of staff size seemed to make decentralized research more costly than locating all station staff and facilities in the same location (Ruttan 1982, pp. 161–170).

Substations Established

The first concerted effort to develop outlying state agricultural experiment stations occurred in California. Eugene Hilgard joined the staff of the College of Agriculture in 1874 and became director of the California Agriculture Experiment Station in 1877. He took a strong interest in the agriculture of the state and used his knowledge and skills in agricultural chemistry to build a body of scientific knowledge about California soils, climates, and farming practices in irrigation and cultivation (Stadtman 1970, pp. 141–154). Because the soils and climate at Berkeley, where the university's experimental farm was located, were different from those in the state's principal agricultural regions, Hilgard and the regents decided to spend most of the Hatch Act funds to support outlying experiment stations.

Hilgard divided the state into four main climatic regions and sought land for substations. During 1888–1890, substations were located near Jackson in Amador County, near Paso Robles in San Luis Obispo County, in Tulare in the San Joaquin Valley, and at Pomona (Stadtman 1970, p. 146). These stations flourished for about 10 years. As expected, substations were costly to operate; in addition, foremen were not scientists and were sometimes unreliable in carrying out field experiments. Consequently, in 1903, the substations near Jackson and Paso Robles were closed. Later, the substations at Tulare and Pomona were also closed.

Strong farm interests in the California legislature, however, demanded in 1903 that research on special problems of California agriculture, especially of vegetable and poultry producers, be initiated. This led to the establishment in 1905 of Southern California Pathological Laboratory, at Whittier, with a branch at Riverside. The Riverside branch evolved into the university's citrus experiment station.

At the request of poultry farmers of the Petaluma region, the legislature also appropriated funds to establish a station to carry out research on the causes of disease, the relative value of poultry feeds, and methods of sanitation, and to promote the poultry interests of the state. This work was conducted between 1904 and 1909 and was subsequently moved to the university's farm at Davis. In 1905, the legislature acted upon the interests of California millers and grain dealers to provide financial support for university experiment stations in the Sacramento and San Joaquin valleys. Thus, both experiment station directors and farm interest groups had seen the need for research substations or outlying stations.

In 1900, 13 years after passage of the Hatch Act, nine states had established one or more substations: Minnesota, Louisiana, Texas, Alabama, California, Colorado, Arizona, New York, and Connecticut (Table 2.2).[10] Table 2.2 also presents information on the share of the agricultural experiment station staff assigned to substations. For states that had substations in 1900, the share was largest in 1900 and has trended downward over time. For states that established substations after 1900, there is no trend.

USDA-SAES Collaboration

Cooperation between the USDA and SAES developed early through the USDA's Office of Experiment Stations. W. O. Atwater, who left the directorship of the Storrs Agricultural Experiment

Table 2.2 The number of state agricultural experiment station substations and share of the staff assigned to substations, by state, 20-year intervals, 1900–1980

	1900		1920		1940		1960		1980		No. Geo-Climatic Subregions
	No.	Off-Campus Staff [a] (%)	No.	Off-Campus Staff [a] (%)	No.	Off-Campus Staff [a] (%)	No.	Off-Campus Staff [a] (%)	No.	Off-Campus Staff [a] (%)	
Lake States											
Wisconsin	0	0.0	3	4.0	7	5.0	6	3.9	10	3.8	2
Minnesota	2	25.0	6	18.6	5	14.9	7	2.1	6	7.9	5
Michigan	0	0.0	3	5.3	6	5.3	7	2.3	10	3.0	2
Corn Belt											
Illinois	0	0.0	0	0.0	0	0.0	1	2.9	0	0.0	3
Iowa	0	0.0	0	0.0	0	0.0	0	0.0	8	4.0	3
Indiana	0	0.0	0	0.0	4	3.7	7	4.2	10	10.4	3
Ohio	0	50.0	3	17.6	15	13.5	9	5.7	7	2.0	3
Missouri	0	0.0	1	2.4	0	0.0	4	1.6	5	1.6	4
Delta											
Louisiana	2	76.9	3	22.2	4	11.1	10	21.4	16	10.6	3
Mississippi	0	0.0	5	47.6	7	39.6	11	23.4	10	22.9	3
Arkansas	0	0.0	0	0.0	4	20.9	11	17.9	14	16.7	3
Southern Plains											
Texas	1	12.5	14	41.2	17	31.0	34	42.9	30	31.7	6
Oklahoma	0	0.0	0	0.0	0	0.0	18	8.3	18	13.3	5
Northern Plains											
Kansas	0	0.0	4	12.9	4	10.4	5	8.3	5	8.5	5
Nebraska	0	0.0	3	11.1	5	11.9	5	9.6	7	21.3	3
North Dakota	0	0.0	4	20.8	5	12.5	6	9.2	7	52.0	2
South Dakota	0	0.0	0	0.0	4	11.4	6	4.5	12	14.9	2
Southeast											
Florida	0	0.0	1	12.5	5	28.0	13	54.3	22	39.8	2
Georgia	0	0.0	0	0.0	2	25.8	7	26.3	8	38.0	4

(*Continues*)

45

Table 2.2 (*Continued*)

	1900		1920		1940		1960		1980		No. Geo-Climatic Subregions
	No.	Off-Campus Staff[a] (%)	No.	Off-Campus Staff[a] (%)	No.	Off-Campus Staff[a] (%)	No.	Off-Campus Staff[a] (%)	No.	Off-Campus Staff[a] (%)	
Alabama	1	33.3	1	5.3	5	14.8	10	12.4	11	9.5	4
South Carolina	0	0.0	0	0.0	5	34.8	6	18.5	4	15.0	4
Appalachia											
North Carolina	0	0.0	6	15.8	6	29.7	17	7.4	15	1.8	5
Virginia	0	0.0	8	12.5	9	37.0	14	18.5	12	13.9	4
Tennessee	0	0.0	5	38.5	4	7.0	7	18.6	10	2.5	4
Kentucky	0	0.0	0	0.0	2	4.3	2	3.0	2	0.5	2
West Virginia	0	0.0	2	6.9	0	0.0	4	7.5	6	11.5	1
Pacific											
California	8	b	5	21.9	7	22.7	11	2.3	3	?	6
Oregon	0	0.0	7	13.5	10	17.9	14	13.1	9	9.0	5
Washington	0	0.0	3	22.9	6	29.6	8	39.5	5	29.4	4
Alaska	—	—	—	—	3	b	0	0.0	0	0.0	—
Hawaii	—	—	—	—	4	10.8	5	7.4	5	4.5	—
Mountain											
Colorado	2	16.7	1	2.8	5	6.0	8	7.7	11	10.8	2
Arizona	1	14.3	0	0.0	5	8.5	6	3.9	5	2.1	2
Utah	0	0.0	3	4.8	2	16.2	0	1.1	0	0.0	2
Wyoming	0	0.0	0	0.0	8	22.0	6	8.4	8	14.3	3
Idaho	0	0.0	4	11.8	4	7.8	6	5.8	5	2.6	2
Montana	0	0.0	1	4.2	5	20.0	7	15.9	7	13.3	4
New Mexico	0	0.0	0	0.0	2	6.1	5	13.7	5	9.8	3
Nevada	0	0.0	0	0.0	0	0.0	0	0.0	10	16.9	2

Northeast											
New York (combined)	1	48.0	2	9.8	1	24.7	1	20.8	3	13.0	2
New Jersey	0	0.0	0	0.0	0	0.0	4	2.7	7	3.3	2
Pennsylvania	0	0.0	3	6.2	0	0.0	0	0.0	0	0.0	4
Maryland	0	0.0	1	4.0	0	0.0	0	0.0	0	0.0	2
Massachusetts	0	0.0	2	5.0	2	8.5	0	0.0	2	10.8	1
Connecticut (combined)	2	77.8	1	60.0	1	49.3	2	44.1	2	38.0	1
Vermont	0	0.0	0	0.0	0	0.0	0	0.0	0	0.0	2
New Hampshire	0	0.0	0	0.0	0	0.0	0	0.0	0	0.0	2
Rhode Island	0	0.0	0	0.0	0	0.0	0	0.0	0	0.0	1
Delaware	0	0.0	0	0.0	0	0.0	1	10.0	1	29.9	1
Maine	0	0.0	2	20.0	2	5.6	3	5.3	3	1.9	2

Source: Collected from U.S. Department of Agriculture, Handbook of Professional Workers in Experiment Agricultural *Stations 1901, 1921, 1940, 1950, 1960, 1970, 1980.*

Note: A substation or outlying station is assumed to exist only when professional staff are assigned to a location other than the central station. In New York and Connecticut the stations at Geneva and New Haven are counted as substations. The SAES in Puerto Rico and Guam are excluded. See Appendix 3.A for details about how substation staff were counted.

[a]Computed as the number of professional staff assigned to all substations divided by the number of station research staff holding a Ph.D., M.S., or B.S. degree.

[b]Exceeded 100 percent.

Station, became the new head of this office. Atwater aided experiment stations primarily by providing them with information that might be useful for research. In 1889, he initiated three serial publications: *Bulletins, Circulars*, and the *Experiment Station Record*. The latter publication contained abstracts from domestic and foreign stations' research reports. In 1891, Atwater left the USDA to resume research at Wesleyan University, Middletown, Connecticut (Baker et al. 1963, p. 36).

The Office of Experiment Stations took a much more active role in the operation of SAES under A. C. True, who was its head 1893–1929. Systematic accounting procedures were established, station visits or reviews were conducted for the first time, legitimate station research was more narrowly defined, and expenditure of federal money on substations of SAES was stopped.[11]

Before 1900, cooperative USDA-SAES research was largely a result of individual initiatives, with cooperation based primarily on personal contacts. One of the first efforts to conduct coordinated research programs involving federal and state researchers and cooperating farmers was the work in dryland agriculture in the Great Plains area (Quisenberry 1977). Cooperating units of the Great Plains Cooperative Association were the Bureau of Plant Industry and SAES of North and South Dakota, Kansas, Nebraska, Oklahoma, Texas, and Colorado. The association conducted research at the stations then in existence and also established new stations at Hays, Kansas, in 1901; Nephi, Utah, in 1903; Amarillo, Texas, in 1904; and North Platte, Nebraska, in 1906. By 1910, there were 20 stations operating, and by 1916, there were 29. Eventually 30 stations were involved in this project (Office of Technology Assessment 1981, p. 38). The experimental work was done jointly by state and USDA workers.

Regional research was given a substantial boost by two congressional acts passed during the 1930s that established regional research laboratories. In 1935, the Bankhead-Jones Act authorized the establishment of laboratories in different regions to work on priority problems of each region (Baker et al. 1963, pp. 226–227). Nine were established by 1940: Plant, Soil, and Nutrition (Ithaca, New York); Pasture Research (State College, Pennsylvania); Vegetable Breeding (Charleston, South Carolina); Poultry Research (E. Lansing, Michigan); Soybean Research (Urbana, Illinois); Sheep Research (Boise, Idaho); Salinity (Riverside, California); and Plant-Growth Regulating Substance and Photo-Period and Plant Development (Beltsville) (Office of Technology Assessment 1981, p. 38). These facilities were regarded as federal field laboratories (Office of Technology Assessment 1981).

The U.S. Department of Agricultural has assigned, and state agricultural experiment stations have accepted, collaborators at least since 1900. A USDA collaborator is a scientist located in the state and assigned to work jointly with one or more agricultural experiment station scientists. Frequently, collaborators have been given visiting-professor rank in the land-grant college to which they were assigned. Sometimes they have also taught classes and served on Ph.D. dissertation committees.[12]

Table 2.3 (last row) shows that the relative importance of USDA collaborators assigned to SAES increased during 1920–1940 and declined thereafter. In 1920, the number of USDA collaborators was equal to 2.2 percent of total SAES scientists. The relative importance more than doubled between 1920 and 1930 (4.6 percent in 1930) and doubled again between 1930 and 1940. In 1940, the number of USDA collaborators was equal to 8.8 percent of total SAES scientists. Since 1940, the relative importance of USDA collaborators has steadily decreased, being 8.1 percent in 1950, 6.9 percent in 1960, 4.2 percent in 1970, and 3.3 percent in 1980.

Across each of the 10 ERS production regions, the change in relative importance of USDA collaborators has been similar to changes at the national level. Expressed as a percentage of SAES scientists, collaborators increased until 1940 or 1950 and declined thereafter.

Table 2.3   USDA collaborators as a percentage of SAES research staff, by region, 1920–1980

Regions	1920	1930	1940	1950	1960	1970	1980
Lake States	2.9	5.8	7.6	3.9	3.8	3.6	3.1
Corn Belt	1.7	3.3	10.9	10.0	7.6	4.6	2.8
Delta	1.7	3.7	8.9	7.1	5.6	2.0	3.3
Southern Plains	3.7	4.3	15.8	18.1	12.9	6.8	3.2
Northern Plains	2.9	9.1	12.7	11.3	7.6	4.3	3.2
Southeast	8.0	5.5	15.8	12.5	11.0	9.5	6.7
Appalachian	5.3	4.3	7.5	6.9	2.3	1.3	0.2
Pacific	0.5	2.2	9.4	8.5	11.5	5.3	6.4
Mountain	2.6	12.0	8.7	11.4	8.8	5.6	3.2
Northeast	0.0	1.0	2.4	1.2	0.9	1.3	1.3
Overall	2.2	4.6	8.8	8.1	6.9	4.2	3.3

Source: Collected from U.S. Department of Agriculture, *Handbook of Professional Workers in Agricultural Experiment Stations* 1921, 1931, 1940, 1950, 1960, 1970, 1980.

The rate of assignment of USDA collaborators to SAES has differed across the regions. However, no single region has ranked highest for all seven data points. The Southeast had the highest collaboration rate in 1920, 1940 (tied with the Southern Plains), 1970, and 1980; and the Southern Plains had the highest rate in 1940 (tie), 1950, and 1960.[13]

Cooperative Extension under the Smith-Lever Act

Cooperative extension became a major adult educational activity after passage of the Smith-Lever Act in 1914. In the USDA, extension work was removed from the Bureau of Plant Industry in 1915. A new States Relations Service was established that included the Office of Extension, North-West; Office of Extension, South; Office of Home Economics; and Office of Experiment Stations (True 1928, p. 127; Baker 1939, pp. 503–504). Over time, the regional extension programs were merged. Currently, administration of Cooperative Extension at the national level is by the Cooperative States Research, Education and Extension Service (CSREES) in the U.S. Department of Agriculture.

In the states, the state-level extension organizations, located at the land-grant colleges, have provided administrative leadership and subject matter specialists. At the local level, a county office-county agent system has been established. It has close ties to the state-level extension, but also provides a permanent local contact for farmers and other people seeking information about agriculture, rural areas, and home economics. Although the problems faced by farmers and other rural people have changed since 1914, the current mission of Cooperative Extension is similar to its initial one: to produce and distribute information on applied problems and to train local leaders (Joint USDA-NASULGC Committee on the Future of Cooperative Extension 1983).

A summary of other major federal legislation affecting the funding of Cooperative Extension can be found in Table 1.7. In particular, the amended Smith-Lever of 1953 added a consideration to a state's share of the U.S. farm population in allocating federal funds to the states.

The Development of the Agricultural Sciences: A Formalization

The "demand" for agricultural research results that emerged with the establishment of the USDA, the land-grant colleges, and the experiment stations was real enough. The state of the "supply" of research results, however, was not well developed in the early period. The crop and animal collectors' work and the early work on soils and the use of fertilizer were important (and contributed to agricultural productivity—see Chapter 7). However, the emergence of true agricultural sciences did not take place until well after the agricultural experiment stations were established in the late 19th century.

In this section, we use data from scientific journals to document the evolution of the agricultural sciences. We examine the data on the emergence of new specialized journals and on citations between journals. We also examine data on departmental organization in the land-grant colleges to show more details of the evolutionary process.

We begin this section by considering a structure of a national agricultural research system that could be self-sustaining over time. This scheme for science and invention identifies factors that we believe describe the agricultural sciences and their relationship to other sciences and to the development of technology. We discuss the growth of journals in the relevant categories. We also undertake a citation analysis to demonstrate the usefulness of our classification scheme and to characterize the types of agricultural sciences existent today.

A Self-Sustaining R&D System for Agriculture

A schematic picture of the structure of a research and development system for agriculture is presented in Figure 2.3. It is built on a careful analysis of sources and communication of scientific and technical knowledge (also see Huffman 2001). The system consists of six simultaneously ongoing levels or layers of activity with vertical (upstream and downstream) and horizontal feedback and linkages. Level III contains products from innovation. Level I identifies the final users of new technologies, who are also a source of information about technology needs and problems. Level II is the public and private information system—extension, marketing, and distribution—that links upstream sources to final users and the public and private information systems. Level III refers to the commercialized technologies and knowledge that are the product of applied research. Level IV, technology invention, identifies the engineering and applied science fields that generate new technologies. Level V, pre-invention research, is directed specifically toward producing discoveries that advance the knowledge needed to design new technology and institutions. It is linked upstream to Level VI, to the fundamental or core sciences and downstream to Level IV technology invention. A distinction between Level V and VI is that research in the Level V fields tends to be demand driven and in Level VI research tends to be supply or scientist driven.

Level IV activities are those public and private applied research efforts directed toward the discovery or invention of new technologies. These include mechanical, chemical, biological, managerial, and policy technologies. Indeed, most mechanical, chemical, and biotech inventions are produced by private firms (see Chapter 5). The public sector USDA-SAES research system also engages in Level IV invention activities. In many cases, the public and private sectors are in direct competition (see Chapter 5). Much of public sector invention, however, is in technology fields where intellectual property rights are not marketable and therefore do not stimulate private invention. Innovations in minor and most self-pollinated crops, animal breeding, economic policy analysis, and resource and environmental management methods fall into this category.

The distinction between technology invention (Level IV) and pre-invention (or pre-technology) science fields (Level II) is important because the products of pre-invention research are not generally subject to patent protection. Pre-invention science means research directed specifically toward

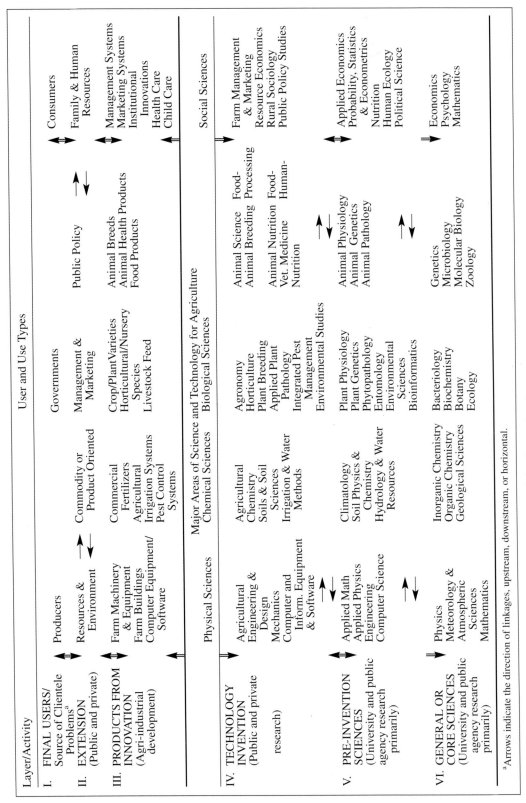

Fig. 2.3. Science and technology in the agricultural research and development system.

producing discoveries that enable and assist technology invention. Pre-invention science is specifically an intermediate product supplied to invention-producing firms, and it is a key input for both public and private invention producers.

The USDA-SAES research engages in significant research designed to produce pre-invention scientific discoveries, e.g., plant/animal genomics, plant/animal physiology (see National Research Council, Committee on Opportunities in Agriculture 2003). However, much pre-invention science relevant to the USDA-SAES system is conducted in universities and departments not directly supported by the USDA-SAES system. It is also largely beyond the range of influence (or scope) of existing intellectual property right laws. Hence, relatively little pre-invention science research is undertaken by the private agribusiness sector (Evenson 1983; Bonnen 1986; National Research Council, Committee on Opportunities in Agriculture 2003).

The Evolution of Knowledge Exchange in the Sciences

Advances in science were very slow during earlier centuries for several reasons, one being a weak communication system. This was closely associated with weak property rights for new discoveries. Scientists of this early era read books to learn about what others were doing (or traveled to visit their laboratories in the case of chemical research). The earliest scientific journals served as newsletters. For example, the *Philosophical Transactions of the Royal Society of London* (established in 1665) had the stated function of digesting new books and the activities of the learned people in Europe (Price 1986, p. 57).

Scientific progress was very slow during earlier eras when scientists had to build their own foundation for investigations. However, communication of results by one individual could lead to someone else claiming priority to ideas, discoveries, or findings. For example, Isaac Newton was reluctant to convey his new discoveries to scientific societies (Committee on the Conduct of Science 1989, p. 9).

Our current system of property rights in scientific discoveries was worked out starting in the 17th century. The Royal Society of London made an institutional innovation that resolved the problem of public communication and priority to scientific discoveries. Henry Oldenburg, secretary, guaranteed rapid publication to contributors of the *Philosophical Transactions of the Royal Society of London* and official support of the society in case the author's priority to ideas or findings was brought into question (Committee on the Conduct of Science 1989, p. 9). This institutional innovation set a precedent in science that continues today: the first to publish a view, discovery, or finding, and not necessarily the first to discover it, gets (almost all the) credit for the discovery. Thus, a strong incentive exists for scientists to release advances in knowledge early to peers and to the professional journals. The incentive for early release can conflict with performing thorough analyses and careful interpretation. These activities consume time, and someone else might publish the idea or finding first. However, when results are released in a scientific mode, other scientists can check for validity of scientific claims.

The Royal Society set a precedent that was needed for the development of modern science in a building block mode. New papers could build on the intellectual foundation of previous papers (and books) and acknowledge their input through citation in the references. Citations serve a number of purposes. They acknowledge the prior work of others, direct the reader to additional related sources of information, acknowledge conflicts with other results, and provide support for views expressed. Furthermore, citations place a paper within its scientific context and assign responsibility if errors should later be discovered (Committee on the Conduct of Science 1989, p. 16). The transformation of journals and scientific papers into their modern state occurred primarily during 1850–1900. Currently, journals, books, and informal arrangements (so-called invisible colleges) are important channels of scientific communication. The relative importance of each channel has changed over time, and this change is partly associated with the size of the total stock of knowledge and the pace of science (Price 1968, 1986; Busch and Lacy 1983, Chapter 4).

To provide additional perspective on the development of agricultural and related sciences, we examined the dates of establishment and rates of birth of new American journals. For journals that have an extended life, the date of birth is important because it shows when the pace of science was sufficient to sustain a professional journal. The birth of new and successful journals in a field indicates in a crude way that the pace of discovery has increased recently.

The journals were grouped into three broad groups: general sciences related to agriculture, pre-invention sciences related to agriculture, and applied agricultural sciences. The applied sciences were further subdivided into field crops, horticulture crops, and forestry; livestock and poultry; post-harvest sciences; and other, including natural resources and wildlife, agricultural and resource economics, rural sociology, and home economics. We tried to follow the structure of R&D laid out in Figure 2.3 in placing journals in groups (see Table 2.4 at end of the chapter).

The earliest general science journal was in the field of chemistry. The *Journal of the American Chemical Society* started in 1878. Before 1900, journals were also established in economics. The *Quarterly Journal of Economics* and the *American Economic Review* began in 1886 and 1887, respectively.

During 1900–1940, the rate of birth of new American journals in the general sciences was slow, except for economics. Two new journals were established in biochemistry, two in biology, and two in bacteriology. In contrast, six new journals were established in economics during this period. The Great Depression seemed to heighten the interest in economic measurement and econometric analyses. *Econometrica* and the *Review of Economics and Statistics* began in 1933.

Very few new American general science journals were born during the 1940s, but the pace picked up during the 1950s and continued to accelerate into the 1960s and 1970s. The 1980s and 1990s also brought new journals as fields of science evolved, e.g., in genetic engineering and biotechnology, information science and technology, and even in economics.

Before 1920, it would have been difficult to differentiate pre-invention sciences from general and applied sciences. We, however, have designated four fields and labeled them with headings of genetics, entomology and environmental sciences, and crop-related and livestock-related pre-technology sciences. Several journals in these fields were established before 1900. The earliest were in entomology; the *Transactions of the American Entomological Society* began in 1869 and *Entomology Americana* in 1885. Three crop-related journals were established before 1900: *Plant World* (1897), *Publication of Botanical Society of America* (1897), and *Bryologist* (1898), which covers a branch of botany dealing with mosses. The establishment of these botany journals in rapid succession indicates that a great burst of botanical/plant research occurred about the turn of the century. Only one livestock-related journal was established at this time, the *American Journal of Physiology* (1898). No journal specializing in genetics was established before the turn of the century.

Between 1900 and 1940, new journals in the pre-invention sciences began to blossom more rapidly across our field designations. The first journals specializing in genetics were established: *Journal of Heredity* (1910), *Genetics* (1916), and *Records of the Genetics Society of America* (1932). In entomology and environmental sciences, six new journals were born. In crop-related sciences, six new journals were also established: *American Journal of Botany* (1914), *Studies from the Plant Physiology Lab* (1923), *Plant Research* (1926), *Plant Physiology* (1926), *Phytologia* (1933), and *Botanical Review* (1935). In livestock-related sciences, the new journals were *Journal of Experimental Zoology* (1904), *Journal of Immunology* (1916), and *Endocrinology* (1917).

During the 1940s and 1950s, few new journals at this level were established, but the pre-invention sciences grew relatively rapidly after 1960 with 26 new journals being established between 1960 and the mid-1980s.

In the applied agricultural sciences, only a few journals began before 1900, but many were established later. The journals in crops and forestry and livestock and poultry fields were grouped into subcategories to reflect their emphasis on production and protection-maintenance. For the crops and forestry fields, the earliest journals dealt with aspects of agricultural engineering and horticulture. *Drainage Journal* and *American Farm Equipment* were established in 1879 and 1898. *Rhodora*, a horticultural journal, started in 1899. Only one journal in plant protection-maintenance was established before 1900, the *Journal of Mycology*, which covers a branch of botany dealing with fungi.

Between 1900 and 1940, a large number of new American journals in the crop and forestry fields were established. In field crop production sciences, the *Agronomy Journal* began in 1907, the *Journal of Agricultural Research* in 1913, *Soil Science* in 1916, *Wheat Studies* in 1924, *Fertilizer Review* in 1926, and *Soil Science Society of America Journal* in 1936. In horticultural production sciences, the *Journal of the American Society of Horticulture Science* was established in 1903, *Citrography* in 1915, *American Horticulturist* in 1922, *Bulletin of the American Horticulture Society* in 1923 (although it was published for only 3 years), and *Journal of Arbiculture* and *Morris Arboretum Bulletin* in 1935.

In forestry production sciences, the *Journal of Forestry* (later renamed the *Forestry Quarterly*) was established in 1902, *Forest Worker* in 1924, and *Tree-Ring Bulletin* and *Timber Statistical Review* in 1934. Taken together, the large (net) number of new American journals established during 1900–1940 in crop and forestry production sciences shows that the pace of scientific discovery and application of science to agricultural production increased rapidly during this period. Four new journals in crop and forestry protection-maintenance sciences were also born during this period: *Mycologia* (1909), *Phytopathology* (1911), *Plant Diseases* (1917), and *Studies in Forest Pathology* (1928).

Few new journals in the crops and forestry fields were established during the 1940s, but the 1950s brought several in the production sciences. And since 1970, there have been eight new journals established in production sciences. Thus, plant science discoveries in the production sciences have been growing at a relatively rapid pace since 1970. The pace of new discoveries in crop and forestry protection-maintenance sciences increased only during the early 1980s.

In the livestock and poultry fields, the short list of production-oriented journals is unusual relative to the longer list of protection-maintenance journals. A second striking feature is that most of the journals, including the *Journal of Animal Science*, were established after 1940. This, however, is not too surprising given that a small number of SAES scientists engaged in the animal research fields before 1940 (see Chapter 4).

Exceptions to the 1940 date are the *Journal of Dairy Science* (1917), *International Review of Poultry Science* (1920), *Poultry Science* (1921), *Animal Veterinary* (later the *Journal of the American Veterinary Medical Association*) (1877), *Cornell Veterinarian* (1911), *Journal of Parasitology* (1914), and *Veterinary Bulletin* (1931). About one-half of these early journals were in protection-maintenance. Thus, the pace of new discoveries in the livestock and poultry fields during this early period was relatively rapid in the protection-maintenance sciences.

Journals in the post-harvest sciences were divided into two groups: food chemistry, technology, and science; and human nutrition. The pace of discovery in the food chemistry, technology, and science fields was slow before 1975. The first journal, *American Food Journal*, was established in 1906, and 18 years later *Cereal Chemistry* was established. For each of the following three decades, two new journals were established. However, starting in 1975, the picture changes abruptly as seven new journals were born during the next eight years, and one journal established in 1967 was discontinued.

Human nutrition research started at Yale during the last half of the 19th century, and the first journal, *Dietetic and Hygienic Gazette*, was established in 1886. The *Journal of American Dietetic Association* and *Journal of Nutrition* followed 41 and 42 years later. New journals in this field have been added occasionally, except after 1976, when six new journals were established in a relative short span of years. The titles of these new journals clearly reflect greater specialization as the pace of discovery in the field has quickened and the field has strengthened its scientific roots.

In the natural resource and wildlife fields, only two journals were established before 1933. They were the *American Naturalist* and the *American Midland Naturalist*. During the remainder of that decade, five new journals were established. The burst in scientific activity was stimulated by the conservation movement, which had President Franklin D. Roosevelt's support. Also, federal money was allocated for the first time to soil erosion control.

The next spurt in journal births occurred during the decade of the 1950s, when four new journals were established. Seven new journals started during the decade of the 1960s (six survived), and nine during the decade of the 1970s. Thus, the pace of scientific discovery in these fields accelerated after 1950 and continued into the early 1980s.

In the agricultural and resource economics fields, the *Journal of Farm Economics* (later renamed the *American Journal of Agricultural Economics*) was established in 1919 and remains one of the main journals in this field. *Land Economics* and *Studies in Land Economics,* forerunners of modern resource economics, were born in 1925 and 1928. After 1930, the titles of new journals show the increased specialization of subject matter and regionality of research. For example, the *Agricultural Finance Review* was established in 1938, *Agricultural Marketing* in 1956, and the *Journal of Agribusiness* in 1983. The regional journals were the *Southern Journal of Agricultural Economics* (1969), *Northeastern Journal of Agricultural Economics* (1972), *Western Journal of Agricultural Economics* (1977), and *North Central Journal of Agricultural Economics* (1979). The *Journal of Environmental Economics and Management* was established in 1974. Since 1960, 10 new journals have been added to these fields, and only one journal terminated; this indicates a quickening pace of discovery.

In contrast, 37 years passed between the establishment of the first and second journals in the field of rural sociology. *Rural Sociology* was born in 1936, but other journals did not appear until the mid-1970s. The conclusion is that this field has moved forward relatively slowly.

The final applied field covers textiles research and social sciences research in home economics. The *Journal of Home Economics* was established in 1912. Starting in 1931 with *Child Development*, other more-specialized journals have gradually been added to cover the expanding scope of research in this area.[14]

Scientific Communication within and between Fields of Science

Low-cost and efficient exchange mechanisms (and economies of size) are the key to specialization in a competitive market-oriented economy. Similar conditions hold for a research economy. In general, the exchange of knowledge among specialists is *inefficient* if it does not meet acceptable standards of credibility. To facilitate this exchange, scientists have developed specialized language and measurement procedures to achieve more exactness and hence credibility. In most sciences, this is the language of statistics, experimental design, and exact measurement.

Researchers who specialize in research problems that are at only one level of science exchange information both *horizontally*, i.e., within their own field of specialization and with other specialists working on similar problems, and *vertically*, i.e., with specialists "upstream" and "downstream" from their specialization. Scientific communications systems have been developed primarily to facilitate horizontal exchange in the general or basic sciences. The journal papers, reference citations, specialized language, and elements of style are chiefly designed to allow scientists working on similar problems to disclose findings quickly and accurately to one another. This disclosure process facilitates refereeing of priority claims to new knowledge and the accumulation of verified hypotheses that constitute scientific knowledge in a field (Committee on the Conduct of Science 1989, pp. 8–16).

The scientific exchange apparatus that originated in Level VI sciences has been modified and used in Level V pre-invention sciences and even to some extent in Level IV activities. At all levels, knowledge must be communicated and exchanged to facilitate the knowledge accumulation process. Scientific papers with their specialized language, usually associated with a discipline and with standards set by scientists themselves, have been a useful vehicle for horizontal exchange at all levels.

Scientific papers have also served as a vehicle for vertical knowledge exchange, but some of the features of scientific papers that facilitate horizontal exchange can hinder vertical exchange. Language and style, for example, facilitate horizontal communication. But across fields and levels of science, they tend to differ significantly. Thus, they can hinder vertical exchange.

Reference citations are a source of evidence about the characteristics of knowledge exchange and the effectiveness of the science system. Table 2.4 reports a list of journals that have been related to science Levels IV, V, and VI shown in Figure 2.3.[15] The *Science Citation Index* and *Social Science Citation Index* (Institute for Scientific Information, or ISI) report inter-journal citations. They report the number of times in a given year that papers from journal A were cited in journal B and vice versa. Using these data one can then construct measures of horizontal (i.e., within-level) and vertical (i.e., between-level) citations, or upstream and downstream citations.

Data for citation patterns for 2 years, 1975 (1978 for journals in ISI, *Social Science Citation Index*) and 1985, are summarized by field of science in Table 2.5. The reader can appreciate that the actual classification of journals into different levels of science contains some arbitrariness. Our objective in classifying was to achieve relative homogeneity of research interest among the categories.

Four categories of general science—chemistry/ biochemistry, biology/molecular biology, microbiology, and economics—are shown. (See Table 2.4 for the journals included in each level of science.) Relatively few of the articles published in these fields and journals are the work of USDA-SAES staff.

The Level V group of journals is important for agriculture. Genetics, entomology, and environmental sciences are fairly distinct fields of science. The crop-related and livestock-related journals in this category were selected in a somewhat more general fashion (e.g., our placing genetics in the Level II group).

Table 2.4 The name and date of establishment of learned American journals in sciences related to agriculture and in agricultural sciences, 1867–1985

I. General Sciences

A. Chemistry, Biochemistry and Biophysics

Year	Journal
1878	Journal of American Chemical Society
1905	Journal of Biological Chemistry
1932	Annual Review of Biochemistry
1941	Biodynamics[ab]
1942	Archives of Biochemistry and Biophysics
1954	Methods of Biochemical Analysis
1958	Journal of Chromatography
1958	Talanta
1959	Biotechnology and Bioengineering
1959	Journal of Lipid Research
1960	Analytical Biochemistry
1960	Biophysical Journal
1962	Biochemistry
1963	Biopolymers
1970	Journal of Cellular Biochemistry
1971	Preparative Biochemistry
1972	Critical Reviews in Biochemistry
1973	Journal of Biological Physics[ab]
1975	Journal of Cyclic Nucleotide & Protein Phosphorylation Research[a]
1976	Applied Biochemistry and Bioengineering
1976	Cell Biophysics
1979	Journal of Applied Biochemistry[a]

B. Biology and Molecular Biology

Year	Journal
1926	Quarterly Review of Biology
1939	Symposium of Society of Developmental Biology[ab]
1949	Survey of Biological Progress[ab] (1962)c
1950	Journal of Biology[ab] (1966)
1962	Life Sciences
1969	Current Topics in Cellular Regeneration[ab]
1972	Adv. in Cyclic Nucleotide Research
1981	Molecular and Cellular Biology[a]

C. Bacteriology and Microbiology

Year	Journal
1916	Journal of Bacteriology
1937	Microbiological Review
1947	Annual Review of Microbiology
1950	Advances in Applied Microbiology
1953	Applied and Environmental Microbiology
1974	Mycotoxin[ab]
1974	Microbial Ecology
1978	Current Microbiology[a]
1977	Experimental Mycology[a]

D. Economics (major journals only)

Year	Journal
1886	Quarterly Journal of Economics
1887	American Economic Review

D. (continued)

Year	Journal
1919	Journal of Political Economy
1933	Rev. of Economics & Statistics
1933	Econometrica
1933	Southern Economic Journal
1933	Rev. of Economic Studies
1934	Journal of Marketing
1952	Economic Development and Cultural Change
1960	International Economic Review
1962	Western Journal of Economics (Economic Inquiry)
1966	Journal of Human Resources
1966	Journal of Finance
1968	Journal of Money Credit and Banking
1972	Journal of Economics and Business
1972	Journal of Econometrics
1974	Journal of Mathematical Economics
1975	Journal of Macroeconomics[a]
1977	Journal of Development Economics[a]
1983	Journal of Labor Economics[a]
1985	Journal of Macroeconomics[a]
1985	Econometric Theory[ab]

(Continues)

Table 2.4 (Continued)

II. Pre-Technology Sciences

A. Genetics

Year	Journal
1910	Journal of Heredity
1916	Genetics
1932	Records of the Genetics Society of America[ab]
1951	Advances in Genetics
1960	Studies in Genetics[ab]
1976	Current Advances in Genetics[ab] (1980)
1979	Developmental Genetics[a]
1981	Journal of Molecular and Applied Genetics[a]

B. Entomology and Environmental Science

Year	Journal
1869	Transactions of the American Entomological Society[ab]
1885	Entomological Americana[ab]
1908	Journal of Econ. Entomology
1908	Annals of Entomological Soc.
1913	Review of Applied Entomology[ab]
1916	Memoirs of American Entomological Society
1920	Ecology (1968)
1933	Pest Control
1947	Coleopyerists Bulletin
1952	Pesticide Review[ab] (1977)
1957	Advances in Pest Control Res.[ab]

B. (Continued)

Year	Journal
1967	Pesticide Monitoring J.[ab] (1981)
1968	International Pest Control[ab]
1971	Insect Biochemistry
1975	Ecological Modeling[a]
1976	Systematic Entomology[a]
1982	Environmental Toxicology and Chemistry[a]

C. Crop-Related Sciences

Year	Journal
1897	Plant World[ab]
1897	Publication of Botanical Society of America
1898	Bryologist
1914	American Journal of Botany
1923	Studies from the Plant Physiology Lab[ab]
1926	Plant Research[ab]
1926	Plant Physiology
1933	Phytologia[ab]
1935	Botanical Review
1947	Economic Botany
1950	Annual Review of Plant Physiology
1961	Phycology[ab]
1965	Journal of Physiology
1976	Systematic Botany[a]

C. (Continued)

Year	Journal
1977	Experimental Mycology[a]
1979	Journal of Plant Nutrition[a]
1981	Plant Molecular Biology[a]
1983	Plant Breeding Review[ab]
1984	Advances in Plant Nutrition[ab]

D. Livestock-Related Sciences

Year	Journal
1898	American Journal of Physiology
1904	Journal of Experimental Zoology
1916	Journal of Immunology
1917	Endocrinology
1940	American Journal of Veterinary Research
1951	Experimental Parasitology
1961	American Zoologist
1961	Advances in Immunology
1963	New Methods of Nutritional Biochemistry[ab]
1967	Methods of Immunology and Immunochemistry
1969	Mammalian Species[ab]
1974	Immunogenetics
1977	Advances in Nutrition Research[ab]
1984	Journal of Leukocyte Biology[a]

III. Applied Sciences

A. Field Crops, Horticultural Crops, and Forestry

1. Production

Year	Journal
1879	Drainage Journal[ab]
1898	American Farm Equipment[ab]
1907	Agronomy Journal

1. (Continued)

Year	Journal
1913	Journal of Agricultural Research (1949)
1916	Soil Science
1924	Wheat Studies[ab] (1944)

. (Continued)

Year	Journal
1926	Fertilizer Review[ab]
1936	Soil Science Society of America Journal
1948	Journal of Range Management

1949   Advances in Agronomy
1951   Irrigation Journal[ab]
1952   Weed Science
1956   J. of Irrig. and Drainage Engineering
1958   Transactions of the Amer. Assoc. of Agricultural Engineers[ab]
1961   Crop Science
1963   Agronomist[ab]
1970   Weeds Today[ab]
1974   Peanut Science[ab]
1976   Journal of Seed Technology[ab]
1978   Timely Soil Topics[ab]
1984   Journal of Fertilizer Issues[a]
1985   Fertilizer Technology[ab]
1899   Rhodora
1903   Journal of American Society of Horticulture Science
1915   Citrography[ab]
1922   American Horticulturalist

1. (Continued)

1923   Bulletin of the American Horticulture Society[ab] (1926)
1935   Journal of Arbiculture[ab] (1974)
1935   Morris Arboretum[ab] Bulletin
1946   Fruit Varieties Journal[ab]
1951   Garden Journal[ab]
1951   Plant Propagator[ab]
1966   Hort Science[ab]
1979   Horticulture Review
1902   Journal of Forestry (For: Quarterly)
1924   Forest Worker[ab] (1933)
1934   Tree-Ring Bulletin[ab]
1934   Timber Statistical Review[ab]
1947   Harvard Forest Papers[ab] (1965)
1951   Forest Product Journal
1951   Journal of Agriculture and Forestry[ab]
1955   Forest Science

1. (Continued)

1957   Journal of Forest History[ab]
1964   International Review of Forestry Research[ab]
1968   Wood Science (Wood and Fiber Science)
1977   Southern Journal of Applied Forestry[ab]

2. Protection-Maintenance

1885   Journal of Mycology[ab]
1909   Mycologia
1911   Phytopathology
1917   Plant Diseases
1928   Studies in Forest Pathology[ab]
1962   Journal of Research on the Lepidoptera[a]
1969   Journal of Nematology
1982   Advances in Plant Pathology[ab]
1983   Journal of Environmental Horticulture[ab]

B. Livestock and Poultry

1. Production

1917   Journal of Dairy Science
1920   International Rev. of Poultry Science[ab] (1939)
1921   Poultry Science
1942   Journal of Animal Science
1949   Proceedings, American Society of Animal Science[ab]
1949   Proceedings, Council on Research of American Meat Institutes[ab]
1965   Dairy Field[ab]
1983   Animal Technology[ab]
1985   Animal Nutrition and Health[ab]

2. Protection-Maintenance

1877   Animal Veterinary (J.Am.Vet.Med. Assoc.)

2. (Continued)

1911   Cornell Veterinarian
1914   Journal of Parasitology
1931   Veterinary Bulletin[ab]
1949   Southeastern Veterinarian[ab]
1954   Journal of Protozoology
1957   Avian Diseases
1958   Veterinary Toxicology[ab]
1959   Journal of Invertebrate Pathology
1959   Journal Am. Veterinary Radiology; Assoc.
1960   Journal of Small Animal Practice
1962   Veterinary Economics[ab]
1963   Veterinarian-International Journal Devoted to Farm Animal Practices[ab]
1964   Veterinary Pathology

2. (Continued)

1965   Bovine Practitioner[ab]
1971   Journal of Veterinary Surgery
1972   Immunology[a] Investigations
1975   Veterinary Parasitology[ab]
1976   Livestock and Veterinary Sciences[ab]
1976   Veterinary Microbiology[a]
1977   Veterinary Clinical Pathology[ab]
1978   Journal of Veterinary Pharmacology and Therapeutics[a]
1979   Veterinary Immunology & Immunopathology[a]
1980   Veterinary Medical Review[ab]
1980   Immunology Today[ab]
1985   Parasitology Today[ab]

(Continues)

Table 2.4 (Continued)

C. Post-Harvest Sciences

1. Food Chemistry, Science and Technology

Year	Journal	Year	Journal
1906	American Food Journal[ab] (1928)		1. (Continued)
1924	Cereal Chemistry	1969	Food Industry Studies[ab]
1937	Journal of Milk and Food Technology[a] (Journal of Food Protection)	1975	Progress in Food and Nutrition Science[a]
1939	Journal of Food Science	1976	Advances in Cereal Science and Technology[a]
1947	Food Technology	1977	Journal of Food Biochemistry[a]
1948	Advances in Food Research[ab]	1977	Journal of Food Distribution Research[ab]
1953	Journal of Agriculture and Food Chemistry	1977	Journal of Food Quality[ab]
1958	Agrichemical Age[ab]	1977	Journal of Food Processing[a] and Preserving
1967	Food Development[ab] (1982)	1981	Dairy and Food Sanitation[ab]

2. Human Nutrition

Year	Journal	Year	Journal
1886	Dietetic and Hygienic Gazette[ab] (1914)		2. (Continued)
		1927	Journal of American Dietetic Association
		1928	Journal of Nutrition
		1942	Nutrition Reviews
		1957	Journal of Applied Nutrition[ab]
		1966	Nutrition Today[ab]
		1969	Journal of Applied Nutrition[a]
		1977	International Journal of Obesity[a]
		1977	Journal of Food Safety[a]
		1978	Nutrition Planning[ab]
		1980	Journal of Nutrition of Elderly[ab]
		1981	Nutrition Research[ab]
		1982	Nutrition and Behavior[ab]

D. Other Applied Sciences

1. Natural Resources and Wildlife

Year	Journal	Year	Journal	Year	Journal
1867	American Naturalist		1. (Continued)		1. (Continued)
1909	American Midland Naturalist	1962	Underwater Naturalists[ab]	1974	Rangelands[ab]
1934	Progressive Fish Culturist	1963	Wildlife Monograph	1974	Agriculture and Environment[a] (1982)
1935	Soil Conservation (1980)	1962	Fisheries Industrial Research[ab] (1970)	1976	Journal of Water Resource Planning and Management[a]
1935	Living Wilderness[ab]	1965	Conservation Research Report[ab]	1979	Prairie Naturalist[ab]
1937	Journal of Wildlife Management	1965	Journal of Wildlife Diseases	1981	Journal of Freshwater Ecology
1939	Great Basin Naturalist	1968	Economics of Clean Water[ab]	1981	North American Journal of Fisheries Management[ab]
1946	Journal of Soil and Water Conservation	1971	International Wildlife		
1950	Game Research Report[ab]	1972	Conservationist		
1956	Southwestern Naturalist	1972	Journal of Environmental Quality		
1958	Water Resource Bulletin	1973	Wildlife Society Bulletin		
1959	Wildlife Disease[ab]	1974	Western Wildllands[ab]		
1961	Natural Resources Journal				

2. Agricultural and Resource Economics

Year	Journal
1919	Journal Farm Economics (Am.J.Agr.Econ.)
1925	Land Economics

2. (Continued)

Year	Journal
1928	*Studies in Land Economics*[b]
1938	*Agricultural Finance Review*[b]
1949	*Agricultural Economics Research*
1956	*Agricultural Marketing* (1971)
1960	*Food Research Institute Studies*[b]
1961	*Agricultural Economic Report*[b]
1966	*Land Economics Monographs*[b]
1969	*Southern Journal of Agricultural Economics*
1972	*Northeastern Journal of Agricultural Economics*
1974	*Journal of Environmental Economics and Management*
1977	*Western Journal of Agricultural Economics*[b]

2. (Continued)

Year	Journal
1979	*North Central Journal of Agricultural Economics*[b]
1979	*Agricultural Law Review*[b]
1983	*Journal of Agribusiness*[b]

3. Rural Sociology

Year	Journal
1936	*Rural Sociology*
1973	*Sociology of Rural Life*[ab]
1980	*Rural Development*[ab]
1981	*Rural Sociologist*[ab]

4. Home Economics (Social Sci. and Textiles)

Year	Journal
1910	*Journal of the Textile Institute*[a]
1912	*Journal of Home Economics*[ab]
1931	*Child Development*
1931	*Textile Research Journal*[a]
1934	*Annals of Child Development*[a]

4. (Continued)

Year	Journal
1939	*Journal of Marriage and the Family*
1952	*Housing*[ab]
1957	*Family Economics Review*[ab]
1960	*Cornell Hotel and Restaurang Quarterly*[ab]
1964	*Advances in Gerentological Research*[ab] (1972)
1967	*Journal of Consumer Affairs*
1969	*Textile Chemistry and Colorist*[ab]
1972	*Home Economics Research Journal*[ab]
1973	*Housing and Society*[ab]
1974	*Advances in Consumer Research*[a]
1974	*Journal of Consumer Research*

[a]The journal is not indexed in the Journal Citation Reports of the 1975 Science Citation Index or the 1978 Social Science Citation Index.

[b]The journal is not indexed in the Journal Citation Reports of the 1985 Science Citation Index or the Social Science Citation Index.

Note: Number in parentheses is date when a journal went out of existence.

Table 2.5 Citation patterns in journals by field of science, 1975 and 1985

Levels of Science	Total Citations	Older than 10 Years	Same Journal	Same Group	Same Level	Upstream Gen. Sci.	Citations Pre-tech.	Downstream Pre-tech.	Citations Applied
I. General Sciences[a] (VI)	35676[b]	.241	.135	.873	.949	—	—	.046	.004
	(171377)[c]	(.224)	(.199)	(.905)	(.975)	—	—	(.023)	(.001)
Chemistry/Biochemistry	319285	.240	.141	.944	.966	—	—	.033	.007
	(143264)	(.226)	(.214)	(.973)	(.993)	—	—	(.006)	(.000)
Biology/Molecular	36859	.180	.054	.361	.740	—	—	.255	.004
	(10365)	(.208)	(.025)	(.140)	(.583)	—	—	(.391)	(.025)
Microbiology	49575	.290	.130	.671	.937	—	—	.035	.027
	(17708)	(.207)	(.177)	(.570)	(.950)	—	—	(.042)	(.007)
Economics	22533	.324	.086	.992	—	—	—	—	.008
	(16148)	(.312)	(.104)	(.993)	(.993)	—	—	—	(.007)
II. Pre-Invention Science(V)	190722	.309	.157	.714	.732	.216	—	—	.053
	(77345)	(.273)	(.157)	(.767)	(.790)	(.126)	—	—	(.082)
Genetics	7623	.361	.130	.712	.786	.098	—	—	.116
	(5479)	(.289)	(.167)	(.691)	(.749)	(.182)	—	—	(.067)
Entomology	10007	.441	.154	.710	.734	.129	—	—	.136
	(4596)	(.367)	(.199)	(.782)	(.861)	(.107)	—	—	(.032)
Environmental Sciences	9914	.450	.101	.624	.676	0	—	—	.323
	(3093)	(.418)	(.141)	(.575)	(.703)	(.007)	—	—	(.288)
Crop-Related Sciences	32597	.336	.112	.524	.554	.350	—	—	.095
	(15662)	(.311)	(.127)	(.736)	(.759)	(.033)	—	—	(.208)
Livestock-Rel. Sciences	139581	.260	.162	.775	.781	.202	—	—	.017
	(48515)	(.241)	(.162)	(.796)	(.801)	(.157)	—	—	(.041)
III. Applied Sciences(IV)									
A. Applied Crop Sciences	49923	.462	.152	.873	.945	.010	.045	—	—
	(20602)	(.384)	(.237)	(.849)	(.916)	(.007)	(.075)	—	—
Crop Protection	13509	.447	.175	.854	.958	.019	.023	—	—
	(6169)	(.336)	(.278)	(.937)	(.983)	(.010)	(.007)	—	—
Field Crop Prod.	22308	.472	.166	.922	.962	.013	.018	—	—
	(11510)	(.373)	(.244)	(.881)	(.926)	(.009)	(.065)	—	—
Hort. Crop Prod.	9623	.464	.103	.648	.807	—	.192	—	—
	(1944)	(.348)	(.087)	(.285)	(.604)	—	(.395)	—	—

Forestry Production	4486 (979)	.448 (.882)	.116 (.177)	.990 (.854)	— (.925)	—	— (.075)
B. Applied Livestock Sciences	78431 (29512)	.358 (.366)	.137 (.230)	.743 (.798)	.816 (.839)	.019 (.026)	.165 (.128)
Livestock Protection	29402 (11455)	.355 (.346)	.090 (.170)	.598 (.734)	.659 (.772)	.042 (.044)	.299 (.183)
Livestock Production	25327 (13202)	.282 (.350)	.271 (.317)	.884 (.884)	.903 (.897)	.007 (.015)	.090 (.088)
Nat. Res.-Wildlife	23702 (4855)	.443 (.451)	.052 (.133)	.614 (.769)	.896 (.821)	.005 (.016)	.098 (.162)
C. Post-Harvest	22634 (10639)	.410 (.319)	.127 (.154)	.726 (.785)	.855 (.952)	.127 (.189)	.018 (.026)
Food Technology	13821 (5810)	.351 (.360)	.069 (.159)	.543 (.548)	.637 (.666)	.293 (.333)	.181
Human Nutrition	—	—	—	—	—	—	—
D. Other Fields	2838 (1620)	.355 (.280)	.139 (.154)	.637 (.552)	.648 (.552)	.351 (.447)	—
Agricultural Economics	1069 (804)	.354 (.407)	.47 (.102)	.999 (.999)	.999 (.999)	—	—
Rural Sociology	—	—	—	—	—	—	—
Home Economics	14364 (6793)	.376 (.299)	.121 (.110)	.993 (.999)	.997 (.999)	.002	—

[a]See Table 2.4 for journals listed in each field group. Economics is excluded in the general science totals.
[b]These data are for 1985.
[c]These data are for 1975 when journals were covered by the ISI, Science Citation Index, Journal Citation Reports and for 1978 when journals were indexed by the ISI, Social Science Citation Index, Journal Citation Reports.

The Level IV group is more applied than the Level V group and frequently leads to the invention of new technology. Virtually all the articles published in Level IV applied science journals are the result of research conducted by USDA-SAES units. The "agricultural sciences" thus encompass both the Level IV sciences and many of the Level V sciences. Effective Level V sciences have been a key part of the USDA-SAES research system.

Citation patterns can be described and summarized in terms of citation indexes. Table 2.5 reports the following indexes:

1. The proportion of citations-to-literature published 10 or more years before the citing publication. This is a measure of the degree of stability on "softness" of a field of science and of its growth potential (Price 1986). Fields of science that cite publications that are 10 years or older at a high rate are undergoing less change and development than fields that primarily cite newer publications.
2. The proportion of self-citations, i.e., to the same journal. This is an index of "inwardness" or intellectual inbreeding of the field. It is affected by the size of the field and the existence of similar journals.
3. The proportion of citations to the same journal or science group (e.g., group 1A). This is also a measure of inwardness or horizontality. This is less affected than (2) by the size and complexity of the journals in the field.
4. The proportion of citations to the same science level. This again is a measure of horizontality.

    Items 2, 3, and 4 above are indices of horizontality or inwardness. Table 2.5 also reports direct measures of verticalness in the form of upstream and downstream citations. (Note that Level VI has no upstream citations, and that Level IV has no downstream citations.)

Certain general consistencies emerge from the cross-journal citation patterns reported in Table 2.5:

1. Level VI fields of science have the lowest level of stability or softness; Level V fields are intermediate; and Level IV fields are most stable, hence softest. Among Levels V and IV sciences, the livestock-related sciences are harder than crop-related sciences.
2. The measures of horizontalness show different patterns according to the indicator. Self-citation does not show strong patterns, although it is low as a proportion of total citations for the molecular biology field. Same-group and same-level citations show stronger patterns. Horizontalness is highest for the general sciences, especially chemistry/biochemistry, and lowest for the Level V sciences. Most applied sciences also have a high degree of horizontalness. This is to some degree governed by the fact that only the pre-invention sciences have both upstream and downstream citing options.
3. The direct evidence for upstream and downstream citations shows patterns more clearly. Upstream citations show science linkage and reflect ties into mother sciences. All pre-invention sciences except the environmental sciences have reasonably strong ties to general science. Upstream linkages are strongest for the crop- and livestock-related groups.

    Among the applied sciences, upstream citation rates are higher for pre-invention sciences than for general science in all fields except food technology, human nutrition, and agricultural and resource economics. (These fields are difficult to classify, and their pre-invention/technology fields may differ from those identified here.) This indicates that the Level V sciences are playing a genuine pre-invention role. Upstream citations in livestock protection are highest for pre-invention sciences. Horticultural sciences also show strong upstream linkages. The forestry field shows weak upstream linkages, while crop protection and crop production show moderate mother science linkages. In the livestock field, linkages to mother sciences have strengthened over time.

4. In the food technology field, science linkages are strong (and, as Chapter 5 will show, downstream linkages to inventors in this field are also strong). Human nutrition sciences are also strongly linked upstream to general science.

5. Agricultural economics shows linkage upstream to economics but not to other applied agricultural science fields. The social science parts of home economics (other than human nutrition) are not science-linked. Home economics and rural sociology have no linkage upstream or downstream.

6. Core science linkages are much stronger to pre-invention sciences than to applied sciences. This evidence is consistent with the role played by pre-invention sciences. The downstream linkages from biology/molecular biology to the pre-invention sciences are especially strong.

The modern system of scientific specialization and knowledge exchange described in Table 2.5 evolved over a long period of time. It represents the outcome of years of conflicts between serving the interests of Levels III and IV of the R&D system (see Figure 2.3). This citation analysis (and the economic studies reported later in Chapter 8) shows that the Level V pre-invention sciences and their linkages to the Level IV sciences have been vital to the effectiveness and productivity of agricultural sciences. Most of the Level V pre-invention sciences would not have been created had the state experiment stations not created them or influenced state universities to erect them. The natural tendencies in science favor Level VI over Level V activities and do not lead to active downstream links. The ultimate viability and effectiveness of Level IV technology invention, however, do depend to a considerable degree on the Level V sciences (see National Research Council, Committee on Opportunities in Agriculture 2003).

Given that knowledge exchange is important for development of a strong science system, it is inevitable that interlevel quality ratings will occur. Quality standards are established within each science field and level that reflect the attributes and relative weights that matter most to the field or level. The weights, and possibly the attributes, differ across the levels of the R&D system. The general and pre-invention science levels place low weight on practicality and policy relevance and place high weight on rigorous methodology and creative thinking. The applied science level places less weight on methodological rigor and creative thinking and greater weight on practicality and policy relevance. It seems inevitable that these types of differences would occur. Sometimes upstream scientists, also, claim that applied research is poor science. This, however, is not necessarily true, but the applied sciences must be linked to good upstream science and regularly expose themselves to possible upstream criticism, if they are to be truly part of a successful R&D system. Agricultural research programs have not always met appropriate standards of scientific rigor and creativity and have justifiably been criticized by upstream scientists.

Evolution of Pre-invention and Applied Agricultural Sciences in the State Land-Grant Colleges

By examining the changes over time in the names of departments and disciplines at the state land-grant colleges the evolution of the pre-invention and applied agricultural sciences can be traced.

Departments and Disciplines

The development process for Level IV and Level V R&D activities can be observed to some extent in data on the organization of university departments. The researchers of the agricultural experiment stations have been listed in the USDA-published handbooks or directories by their disciplinary departmental designation since the 1920s. Information on the academic department appointment of experiment station staff has been collected at 10-year intervals starting in 1930 and ending in 1980. Over this time, the average number of different departments per station increased from 11 to 17. Furthermore, the titles of departments changed, reflecting desegregations of some fields as subspecialties emerged and expanded, aggregation of other departments into one department, and fads in labeling.

The discussion of titles for the broad group of departments covering plants and soils and livestock and poultry disciplines is subdivided into production, maintenance, and general research titles. Additional details are also presented in Huffman 1985.

Science Departments Related to Agriculture

In 1930, the most frequently reported general and pre-invention science departments having agricultural experiment station appointments were agricultural chemistry (40), bacteriology (10), and biology (3). (See Huffman and Evenson 1994, Chapter 5, for more details.) Between 1930 and 1940, the number of stations reporting appointments in physics increased from one to 15 but after 1940 steadily declined. Appointments in agricultural chemistry were replaced by appointments in biochemistry, where the number of stations reporting appointments expanded from three in 1940 to 21 in 1980. Four reported appointments in genetics or agricultural genetics in 1930, and the number did not change much thereafter.

Plants and Soils Departments

In 1930, 51 departments existed in the production-oriented field crops and soils areas. The most frequently reported titles were agronomy, 35; agronomy and soils, five; and soils, four. Other titles included agronomy and genetics, farm crops, field crops and soils, and chemistry and soils. After 1940, the frequency of the two main titles, agronomy and agronomy and soils, decreased. These titles were replaced by the titles of plant science, soils, and plant and soil sciences.

In 1930, there were 37 departments of horticulture, four departments of horticulture and forestry, and seven departments of forestry. Between 1930 and 1940, the number of different forestry departments (schools or colleges) expanded rapidly. After 1960, the number of horticulture and of forestry departments decreased sharply. In the case of horticulture departments, this decrease was not the result of renaming but rather of eliminating the department. Part of the decrease in forestry department titles was, however, due to renaming, and part was due to establishment of schools or colleges of forestry outside colleges of agriculture.

The two main protection-maintenance departmental titles were plant pathology and entomology, although there was a modest number of departments that had the title of botany and plant pathology. Between 1930 and 1980, the number of plant pathology departments increased from 13 to 20, while the number of entomology departments increased by only six.

The number of general science departments decreased from 30 in 1930 to 16 in 1980. The most frequently reported titles in 1930 were botany (20) and geology (2). However, only eight botany departments were represented in 1980. The change between 1930 and 1980 meant that plant scientists in general science departments were now less likely to have station appointments.

Livestock and Poultry Departments

Three main changes have occurred in the organization of departments that focus on livestock and poultry production. First, the number of different departments expanded between 1930 and 1950 and then went into a slow consolidation. Between 1950 and 1980, the number of departments decreased by 30 percent as dairy and poultry husbandry (science) departments were consolidated into animal science departments. In 1980, the most frequently reported departmental titles were animal science (32), dairy science (11), and poultry science (14).

Veterinary medicine or science is the main protection-maintenance department (school or college). The number of stations reporting these appointments decreased slightly between 1930 and 1940. However, between 1940 and 1980, they increased from 23 to 32.

Zoology is the primary general science department associated with livestock research. In 1930, there were 11 combined zoology and entomology departments and seven separate zoology departments. After 1940, the frequency with which station scientists had appointments in these departments declined.

This decline seems to reflect a termination of appointments in this field due to some new specialties moving directly into agricultural departments. Some of the decline may also reflect a change of departmental names, a disaggregation into two separate departments. In 1980, only three stations reported appointments in zoology and entomology and only one in zoology. (Entomology is, however, a department where appointments occur.)

The number of stations reporting appointments in agricultural engineering increased between 1930 and 1950, then decreased slightly between 1970 and 1980.

The relationship between agricultural economics and rural sociology has gone full circle. In 1930, station appointments were made in departments of agricultural economics (37), farm management and rural economics (3), agricultural economics and farm management (3), and rural sociology (2). Between 1930 and 1940, the number of combined agricultural economics and rural sociology departments increased from 1 to 10; but between 1960 and 1970, these joint departments started to split into two. Also, some agricultural economics departments took the new names of agricultural and resource economics and agricultural economics and marketing. The number of stations indicating appointments in sociology, sociology and anthropology, and rural sociology has remained relatively small.

The resource-environmental area is relatively new, and few stations reported appointments in these departments before 1960. Between 1960 and 1970, natural resource departments emerged—five stations reported appointments in 1970. Between 1970 and 1980, the title and emphasis changed to renewable natural resources.

Post-Harvest Departments

The number of stations reporting appointments in nutrition and food technology has been small. In 1930, only two departments reported appointments in nutrition, and none was reported for food technology. There has been a small increase in the number of stations reporting appointments in the nutrition-related areas, and titles have changed to incorporate "science." In 1950, five stations reported appointments in food technology; this number increased to nine in 1960. Between 1960 and 1970, some food technology departments were replaced by food science, food science and nutrition, or food science and technology departments.

Long-term changes in departmental affiliation of scientists who have agricultural experiment station appointments have occurred. A reduction has occurred in the appointments with general science departments, but an increase occurred in the total number of different departments in which general science faculty has appointments in the experiment stations. This has been possible because new and more-specialized departments in the agricultural sciences have been established.

This shift of emphasis away from sciences related to agriculture and toward specialized agricultural sciences may have implications for the long-term success of station research in advancing the science of agriculture. With the shift in private sector R&D activities starting in the 1970s and accelerating in the mid-90s due to strengthening of intellectual property rights (see Chapter 5), the public agricultural research system has withdrawn from some of its traditional applied research areas and slowly shifted research resources to upstream research (National Research Council, Committee on Opportunities in Agriculture 2003).

Conclusion

This chapter has shown that the development of agricultural sciences was facilitated by the university revolution that occurred during the last 25 years of the 19th century. This revolution represented a shift in emphasis away from classical education and no research to utilitarian education with a science focus, advanced or graduate studies, and faculty research. The struggling public land-grant colleges needed the trained scientists from Yale, Harvard, Johns Hopkins, and Chicago to staff teaching and experiment station positions.

The early advances in agricultural sciences in the United States owe much to Samuel Johnson and the Yale Scientific School. They were responsible for the great bulk of the advances during the 1860s to the 1880s. The state universities—i.e., Cornell, Wisconsin—did not emerge as major forces in agricultural research and graduate education until about 1900. Changes in the quantity and quality of scientific activities in the agricultural colleges and experiment stations were hastened by unusual individuals who served as deans of colleges of agriculture and directors of the agricultural experiment stations at Cornell, Wisconsin, and Illinois. Rapid developments at California, Iowa, and Michigan came later.

The agricultural experiment stations have made important contributions to the organization of agricultural research. They established outlying research stations or substations so that their agricultural research could more perfectly meet the diverse geoclimatic conditions that exist in most states.

The formative years in the state agricultural experiment stations and in the USDA were ones of conflict between competing interests, but many of the compromises were partially guided by demands from clientele groups. Clientele groups' interests were most effectively realized at the state level. Numerous applied research units and stations and sub-stations resulted from pressures brought to bear by local clientele groups.

It might be argued (as we do in Chapter 8) that a backlog of scientific potential allowed the very applied research programs in the formative years to make some important advances. However, the years from 1900 to 1920 or so were a period of great stress in the system because few advances were being made. It was during this and later periods that the agricultural sciences came of age. They did so by developing ties with existing upstream sciences but more importantly by investing in the pre-invention (Level V) sciences. They did this by shifting some emphasis within departments in the system from general (Level VI) to pre-invention (Level V) and by expanding and building Level IV departments.[16]

Notes

1. The Division of Microscopy was abolished in 1895 when skills with a microscope became common among biological researchers.
2. In 1927, the Bureau of Soils and the Bureau of Chemistry were combined into a Bureau of Chemistry and Soils. In 1938, all of the soils research was transferred to the Bureau of Plant Industry, except soil erosion work, which was transferred to the Soil Conservation Service.
3. This new institution turned out to be a poor environment for agricultural engineering; consequently, in 1943, the agricultural engineering research was transferred to the Bureau of Plant Industry, and a new Bureau of Plant Industry, Soils, and Agricultural Engineering was established. The Bureau of Chemistry was renamed Bureau of Agricultural and Industrial Chemistry in 1943. This title remained unchanged until 1953, when chemical research was transferred to the Agricultural Research Service (Baker et al. 1963, pp. 470–476).
4. The opposition by private church colleges was strongest in the South. For example, clergymen made strong attacks against the University of Georgia, and they seemed to slow the growth of state funding for the university (Nevins 1962, p. 46).
5. Agricultural interest groups lobbied for the transfer of land-grant benefits from Yale, Brown, and Dartmouth to new agricultural colleges. The Connecticut Grange waged a struggle with Yale Scientific School, complaining that admission standards to Yale were too high. They proposed that the land-grant funds be transferred to Storrs Agricultural School, a private institution that had lower admission standards. In 1893, the legislature changed the name to Storrs Agricultural College and transferred land-grant benefits to it (Brunner 1962, pp. 11–12). In Rhode Island and New Hampshire, farm interest groups played a role in wrestling land-grant funds from Brown and Dartmouth and getting new state agricultural colleges established.
6. At the University of Mississippi, the courses in agriculture were deemed to be so inadequate in 1876 that university officials discontinued even the pretense of providing education for future farmers. Members of the local Grange were outraged, however, and demanded a separate college. In 1878, the state legislature chartered the Mississippi Agricultural and Mechanical College at Starkville (Scott 1970, p. 53). In North Carolina, the University of North Carolina was under attack by the Grange starting in 1876. Later, the North Carolina legislature created North Carolina State College (1887) and transferred the land grant to it. In Kentucky and South Carolina the members of the Grange and Farmers' Alliance lobbied for separate, new state agriculture and mechanic arts colleges (Scott 1970, pp. 53–54).
7. The University of Wisconsin was also attacked by the Grange. In 1868, the university had established a school of agriculture and developed a 3-year course program that built upon the university's Scientific School. It contained primarily courses in botany, zoology, chemistry, and geology. There was one professor of agriculture (Nevins 1962, p. 56). In 1883, the Grange demanded that a new college be established. Attempts to pass a bill in the state legislature failed. University officials then moved quickly to authorize W. A. Henry, a professor in charge of agricultural instruction, to institute a short course program for farmers. Also, the university cooperated with state leaders to organize farmers' institutes (Scott 1970, pp. 54–55).
8. Some states, e.g., New Jersey, maintained two stations for a while—one Hatch-funded and one state-funded.
9. The Hatch Act also aided the establishment of new land-grant colleges in New Hampshire and Rhode Island. In New Hampshire, Dartmouth's loss of land-grant and experiment-station funds was also prompted by the receipt, in 1891, by New Hampshire of a large estate near Durham conditional on the establishment of a state agricultural college. See Kerr 1987 for additional details on the establishment of experiment stations under the Hatch Act.

10. In general, these were the states where the largest diversity, as measured by the number of geoclimatic subregions, existed (see Table 2.2). The number of substations tended to increase in each state until after the mid-1970s.

11. The number of substations or branches of the state agricultural experiment stations expanded very rapidly during 1889–1894, apparently for political reasons. True was convinced that these substations were only peripherally related to research and were a drain on scarce resources. Consequently, in 1894, he ruled that federal funds could no longer be spent on branches or substations. Recent scholars have cast doubt on the wisdom of this policy (see Evenson, Waggoner, and Ruttan 1979). True also ended the policy followed in some states of using station funds to support college teaching.

12. From the perspective of the SAES, the possible benefits are increased by the reduced costs of joint USDA-SAES research, more research on the state's agriculture, and possibly additional teaching and/or dissertation advising services. The primary costs to the SAES are office and possibly laboratory space and the research that the station scientist(s) would have completed in the absence of USDA collaboration. The expected net benefit of collaborators (vs. none) to SAES seems likely to be positive. For the USDA, the net benefit could be positive or negative.

13. The relative importance of the USDA's cotton and tobacco research was undoubtedly closely tied to the high collaboration rates for the Southeast and Southern Plains regions. The Mountain region was first in 1930. In contrast, the Northeast had the lowest collaboration rate in each year.

14. See Deacon and Huffman 1986; and Hefferan, Heltsley, and Davis 1987 for a long-term perspective on important contributions of home economics research to SAES research.

15. The list of journals was compiled from the *Serials Catalogue* in the W. Robert Parks Memorial Library at Iowa State University. The list was restricted to scholarly journals published in the United States, to obtain a better indicator of the pace of science here. Undoubtedly we have missed a few important journals, but we believe that including them would not change our basic conclusion. The date of birth of a journal was the date of the first volume; this was taken from the information given in the *Serials Catalogue* or from the first volume of the journal. If a journal died out, the date of the last volume is listed in parentheses following the journal name.

16. See Figure 2.3 for a specification of fields of science by horizontal layers.

References

Baker, Gladys. 1939. *The County Agent*. Chicago, IL: The University of Chicago Press.

Baker, Gladys, et al. 1963. *Century of Service: The First 100 Years of the United States Department of Agriculture*. Washington, DC: U.S. Government Printing Office.

Bonnen, James T. 1986. "Century of Science in Agriculture: What Have We Learned?" Paper prepared for Fellows Lecture, AAEA Annual Meeting, Reno, NV.

Brunner, Henry S. 1962. *Land-Grant Colleges and Universities, 1862–1962*. U.S. Department of Health, Education, and Welfare, Office of Education Bull. No. 13.

Busch, Lawrence and W.B. Lacy. 1983. *Science, Agriculture, and the Politics of Research*. Boulder, CO: Westview Press.

Committee on the Conduct of Science. 1989. *On Being a Scientist*. Washington, DC: National Academy of Sciences.

Deacon, Ruth E. and W.E. Huffman. 1986. *Human Resources Research, 1887-1987*. Ames, IA: Iowa State University, College of Home Economics.

Eddy, Edward D., Jr. 1957. *Colleges for Our Land and Time: The Land Grant Idea in American Education*. New York: Harper & Brothers.

Evenson, Robert E. 1983. "Intellectual Property Rights and Agribusiness Research and Development: Implications for the Public Agricultural Research System." *American Journal of Agricultural Economics* 65(5): 967–975.

Evenson, Robert E., P.E. Waggoner, and V. Ruttan. 1979. "Economic Benefits from Research: An Example from Agriculture." *Science* 205: 1101–1107.

Hefferan, Colien, M.E. Heltsley, and E.Y. Davis. 1987. "Agricultural Experiment Station Research in Home Economics Celebrates 100th Anniversary." *Journal of Home Economics.* 79:41–44.

Hilgard, Eugene W. 1882. "Progress in Agriculture by Education and Government Aid." *Atlantic Monthly* 49: 531–541, 651–661.

Huffman, Wallace E. 1985. "The Institutional Development of the Public Agricultural Experiment Station System: Scientists and Departments." Unpublished paper, Department of Economics, Iowa State University.

Huffman, Wallace E. 2001. "Finance, Organization, and Impacts of U.S. Agricultural Research: Future Prospects." In *Knowledge Generation and Technical Change: Institutional Innovation in Agriculture,* ed. Stephen Wolf and David Zilberman. Boston, MA: Kluwer Academic Publishers.

Huffman, Wallace E. and R.E. Evenson. 1994. "The Development of U.S. Agricultural Research and Education: An Economic Perspective." Department of Economics, Iowa State University.

Institute for Scientific Information. *Science Citation Index.* Philadelphia, PA. Various annual issues.

Institute for Scientific Information. *Science Citation Index, Guide.* Philadelphia, PA. Various annual issues.

Institute for Scientific Information. *Social Science Citation Index.* Philadelphia, PA. Various annual issues.

Institute for Scientific Information. *Social Sciences Citation Index, Journal Citation Reports.* Philadelphia, PA. Various annual issues.

Joint USDA–NASULGC Committee on the Future of Cooperative Extension. 1983. *Extension in the '80s: A Perspective for the Future of the Cooperative Extension Service.* Cooperative Extension Service, University of Wisconsin, Madison, WI.

Kerr, Norwood Allen. 1987. *The Legacy: A Centennial History of the State Agricultural Experiment Stations, 1887–1987.* Columbia, MO: Missouri Agricultural Experiment Station, University of Missouri–Columbia.

Marcus, Alan I. 1987. "Constituents and Constituencies: An Overview of the History of Public Agricultural Research Institutions in America." In *Public Policy and Agricultural Technology: Adversity Despite Achievement,* ed. Don F. Hadwiger and William P. Browne. London: Macmillian.

Moore, Ernest G. 1967. *The Agricultural Research Service.* New York, NY: Praeger Publishers.

Moores, Richard G. 1970. *Fields of Rich Toil: The Development of the University of Illinois College of Agriculture.* Urbana, IL: University of Illinois Press.

National Research Council, Committee on Opportunities in Agriculture. 2003. *Frontiers in Agricultural Research: Food, Health, Environment, and Communities.* Washington, D.C.: National Academy Press.

Nevins, Allan. 1962. *The State Universities and Democracy.* Urbana, IL: University of Illinois Press.

Office of Technology Assessment. 1981. *An Assessment of the United States Food and Agricultural Research System.* Vol. I. Washington, DC: U.S. Government Printing Office.

Price, Derek. 1968. *The Difference Between Science and Technology.* Detroit, MI: T.A. Edison Foundation.

Price, Derek J. 1986. *Little Science, Big Science . . . And Beyond.* New York, NY: Columbia University Press.

Quisenberry, Karl. 1977. *Agricultural History* 51(1):218–228.

Rossiter, Margaret W. 1975. *The Emergence of Agricultural Sciences: Justus Liebig and the Americans, 1840–1880.* New Haven, CT: Yale University Press.

Ruttan, Vernon W. 1982. *Agricultural Research Policy.* Minneapolis, MN: The University of Minnesota Press.

Scott, Roy V. 1970. *The Reluctant Farmer—The Rise of Agricultural Extension to 1914.* Urbana, IL: University of Illinois Press.

Stadtman, Verne A. 1970. *The University of California, 1868–1968.* New York, NY: McGraw-Hill Book Co.

True, Alfred C. 1928. *A History of Agricultural Extension Work in the United States, 1785–1923.* Washington, DC: U.S. Department of Agriculture, Misc. Publ. No. 15.

True, Alfred C. 1937. *A History of Agricultural Experimentation and Research in the United States, 1607–1925: Including a History of the United States Department of Agriculture.* Washington, DC: U.S. Department of Agriculture, Misc. Publ. No. 251.

True, Charles A. 1929. *A History of Agricultural Education in the United States, 1785–1925.* Washington, DC: U.S. Department of Agriculture, Misc. Publ. No. 36.

U.S. Department of Agriculture. 1933. "The Crop and Livestock Reporting Service of the United States." Washington, DC: U.S. Department of Agriculture, Misc. Publ. No. 171.

U.S. Department of Agriculture. 1980. *Review of the Bankhead-Jones Program: Final Report.* Washington, DC: U.S. Department of Agriculture Science and Education Administration, p. 5.

U.S. Department of Agriculture, Agricultural Research Service. 1983. "The Mission of the Agricultural Research Service." Washington, DC: U.S. Department of Agriculture.

U.S. Department of Agriculture. Various years. *Handbook of Professional Workers in Agricultural Experiment Stations.* Washington, DC: U.S. Department of Agriculture.

Waggoner, Paul E. 1976. "Research and Education in American Agriculture." *Agricultural History* 50(1): 230–247.

3
The Agricultural Scientist

Scientific research, training of new scientists, and higher education are human capital-intensive and productive enterprises, and they seem to be highly complementary activities. This chapter examines the education, nature of the work, and training of U.S. agricultural scientists over the period of institutional development. It also documents the recent contributions of the land-grant universities to the training of scientists in general, pre-invention, and applied sciences for agriculture.

A significant increase in the demand for scientists who could produce discoveries and advances in agricultural technology occurred during the middle and late 19th century because of (1) the increased demand for information to teach in new agricultural college courses, (2) greater local disease and pest problems, and (3) increased inter-regional competition due to reduced transport costs and very different natural fertility of Midwestern and Eastern soils.

Organized training of scientists did not exist in the United States before the 1860s. Early U.S. agricultural scientists were trained in Germany. Completing a Ph.D. degree in the German model during 1800 to 1860 was an indication that an individual had successfully mastered a body of knowledge and skills needed to advance the state of knowledge in the sciences (Chittendon 1928, Vol. I, pp. 86–87). Starting in the 1860s, the training of U.S. agricultural scientists shifted to America. Gradually the training of agricultural scientists evolved into (1) advanced course work in the sciences, (2) training in research methods, and (3) supervised experience in conducting and reporting original research.

Religion, not science, was the training of most college faculty before 1876. At the undergraduate level, Harvard and Yale established carefully segregated "scientific schools" in 1846 and 1847, respectively. These schools developed successful physical, biological, and engineering science research and teaching programs. These two institutions were the main foothold of American science before the establishment of Johns Hopkins University in 1876 and the University of Chicago in 1891. Between 1876 and 1900, science and research became an integral part of all good universities. Faculties were now generally trained in the discipline in which they were teaching. Faculty members were expected to conduct research; and promotion at the best universities required demonstration of significant scholarship (Berelson 1960, pp. 15–19). With the greater emphasis on knowledge creation, clearly requiring a scarce talent, university faculty salaries improved significantly over the pre-1876 levels (see Veysey 1965, p. 6, for a discussion of pre-1876 salaries). University salaries of full-time faculty members were for the first time sufficient to support a family. Salaries of exceptional scientists became quite large. For example, in 1891, the University of Chicago was paying eminent full professors $7,000 per year (Tickton and Ruml 1955, p. 28). Using our research price index to inflate this salary, it is equivalent to $272,300 in 2000 prices.

The first section focuses on the early modern agricultural scientists and the early training of agricultural scientists. In the second section, the number of U.S. agricultural scientists and some of the characteristics that affect their productivity are presented. The third section examines the long-term competition between the public and private sectors for scientists. In the fourth section, a quantitative summary of Ph.D.'s awarded and major characteristics of Ph.D. recipients in agricultural sciences are presented.

The First Agricultural Scientists and Early Training of Agricultural Scientists

Modern agricultural scientists first emerged in the United States in the 1860s, and the training of new agricultural scientists shifted from Europe to the United States by the turn of the century.

Samuel W. Johnson and Yale

Samuel W. Johnson, professor of analytical chemistry at Yale University from 1856 to 1896 and director of the Connecticut Agricultural Experiment Station in New Haven from 1877 to 1900, was the first American agricultural scientist and founder of modern American agricultural science (Browne 1926, p. 179; Rossiter 1975, p. 127). He was part of a small but growing staff at Yale's Scientific School who made it the leading U.S. center of applied physical and biological sciences during the 1860s and 1870s.

Johnson started as a student at Yale in January 1850 and made rapid progress. By the end of the year, John P. Norton, a faculty member renowned for his 33 lectures on scientific agriculture, was advising him to go to Germany for advanced study. Johnson delayed because he lacked funds. During 1852–1853 he was working hard on a new theory of soil science. He was trying to overcome farmers' resistance to agricultural chemistry and to formalize rigorous methods of scientific investigations (Rossiter 1975, pp. 132–135).

In May 1853, Johnson set off for Germany. He spent a year at Leipzig perfecting analytical techniques and publishing papers, and he spent a year at Munich studying with Liebig. The stay with Liebig was not totally satisfactory. Although Liebig was the first outstanding organic chemist, he was slowing down significantly in his work in 1854. Also, he had fallen behind the rapidly advancing frontier in agricultural chemistry. New experimental work was casting serious doubt on some of his generalizations about soil chemistry.

By the end of Johnson's stay in Germany (1855), he was closing in on a new theory of soil science. The crux of the change was that Johnson did not think, as Liebig and Norton had, that scientific agriculture was simple and its problems capable of easy solutions (Rossiter 1975, p. 135). He thought that they could be solved by "patient investigation." The only hope for understanding complex cause-and-effect relationships in agriculture was rigorous and systematic experimentation using scientific controls and accurate measurement. This idea provided the rationale for future U.S. agricultural experiment stations.

John P. Norton died in 1852, and when Samuel Johnson returned to the United States in 1855, he was hired by Yale. In 1856, Yale designated him a professor of analytical chemistry,[1] and agricultural chemistry was added to the professorship in 1857. He was a major force in teaching and research in agricultural and analytical chemistry at Yale for the next 40 years.

Yale waited until the 1860s, when it was well endowed and the demand for Ph.D.'s trained in the physical and biological sciences was growing rapidly, before expanding the staff significantly at the Scientific School and moving into expensive graduate training. An advanced program of scientific study, or a graduate program, was approved by the Yale Corporation for the chemistry section of the

Scientific School in 1860. Furthermore, the degree of Doctor of Philosophy (Ph.D.) was to be awarded to students who successfully completed the program. This degree program was to meet the standards of the German universities at that time. The degree could be conferred on a student who (1) had been enrolled during the year preceding his (or her) examination, (2) had satisfactorily passed an examination in advanced scientific studies, and (3) presented and defended a written thesis containing the results of original chemical or physical investigations (Chittendon 1928, Vol. I, pp. 86–88).

In the 1860s, Johnson reviewed the agricultural chemistry literature and wrote two landmark books: *How Crops Grow* (1868) and *How Crops Feed* (1870). These books bridged the gap between theory and practice and drew upon his carefully prepared class notes on agricultural chemistry and plant physiology. They became classic textbooks. Furthermore, they became available in the United States at an opportune time, because the new land-grant colleges were attempting to establish agriculture programs, and this created a significant new demand for good textbooks and faculty.

Johnson directed the study of three outstanding graduate students: Wilbur O. Atwater (Ph.D. 1869), Henry P. Armsby (Ph.D. 1874), and Edward H. Jenkins (Ph.D. 1874). All became outstanding agricultural chemists (Chittendon 1928, Vol. II, p. 405; Browne 1926). Atwater became known for his human and animal nutrition research and split his career among Wesleyan University, the University of Connecticut-Storrs where he became director of the agricultural experiment station, and the USDA Office of Experiment Stations. Armsby, professor of agricultural chemistry at Pennsylvania State University and later director of the agricultural experiment station, became a leading American authority on the nutrition of farm animals. Jenkins joined the staff of the Connecticut Agricultural Experiment Station, New Haven, and later became director and collaborator on the development of commercial hybrid corn.

Johnson had a major influence on bringing science to agriculture that went beyond the university. During 1856–1859, he led a dramatic campaign in the agricultural press against fertilizer fraud (Rossiter 1975, p. 127 and Chapter 9; Chittendon 1928, Vol. I, pp. 198–204). This was an issue that was of great economic importance to New England farmers who were farming nutrient depleted soils and attempting to compete with the new grain farmers of the fertile Midwestern plains. In 1869, he led a successful lobby in favor of a Connecticut law that required fertilizers to be labeled with a statement of composition. Thus, Johnson became the founder of agricultural regulatory work in America.

Johnson, his colleague William Brewer, and his former student Atwater led the drive which culminated in the establishment of the first U.S. agricultural experiment station in 1875. Johnson offered the state of Connecticut 5 years of free laboratory services at the Yale's Sheffield School as a bribe to move the station from Wesleyan University, where it had initially been located. The state accepted, and Johnson was appointed director.

He fought for funds for expansion of the New Haven station and for research. He also oversaw some of the early pioneering research on vitamins A and B, hybrid corn, etc., that became important to U.S. agriculture in the 20th century.[2]

Harry L. Russell and Wisconsin

Harry L. Russell, professor of dairy bacteriology and later dean of the College of Agriculture at the University of Wisconsin, 1893–1930, made a large impact on agricultural sciences in the land-grant system. He had studied with the best bacteriologists in the world, perhaps even up to the present time—Pasteur in France and Koch in Berlin—and he received a Ph.D. degree from Johns Hopkins.

Louis Pasteur and Robert Koch were exceptional European biological scientists during the last half of the 19th century by any standards. They rapidly elevated the state of knowledge in biology or about life. Pasteur, professor at the University of Strasbourg, 1849–1895, was the first to show that only living things can produce or come from living things. Up to his discovery, a common view was that spontaneous generation of life occurred from dust or dirt. He discovered that bacteria exist almost everywhere and identified them. Some of his discoveries were the sources of modern food processing and preservation methods and biotechnology.

He showed that (1) uncontrolled growth of bacteria causes food spoilage, e.g., causing fruit to spoil, wine to turn to vinegar (1864), milk to sour, (2) heating food in a closed container to a high temperature kills bacteria that cause food to spoil and keeps other bacteria from entering, (3) controlled use of heat and cold kills bacteria, a process called pasteurization, which kills most but not all bacteria in milk, wine, beer, and cheese and extends the product's life without altering taste significantly, and (4) the controlled growth of bacteria could be used successfully in food processing, e.g., fermentation of juices in wine and beer making and cheese making.

Some of Pasteur's discoveries also provided the basis for modern veterinary and human medicine. He showed (1) bacteria frequently cause diseases, i.e., the germ theory of diseases, and (2) an animal's or human's body can be vaccinated with weakened bacteria that cause resistance or immunization. His work during 1865–1885 included vaccines for anthrax in sheep and cholera in chickens. Furthermore, although Pasteur did not get credit for discovering viruses (this was given to Dutch botanist M. Beijerinck in 1898, three years after Pasteur's death), he did develop one of the first successful viral vaccines, one for rabies in humans (1885). Scientific discoveries by Pasteur have been the source of modern water purification and sewage treatment systems that have dramatically reduced the spread of bacterial diseases and helped extend expected length of human life. His discoveries were also the source of the first biotechnical revolution, which occurred more than 100 years ago.

Robert Koch, 21 years younger than Pasteur and professor at the University of Berlin, also worked on the bacterial causes of diseases. He was the first to show that specific bacteria cause certain diseases. He designed a research methodology for scientifically accomplishing this, and it is the foundation of modern pathology. His steps were (1) pathogenic (disease-causing) bacteria are taken from a diseased animal or human, (2) pathogenic bacteria are isolated and grown in the laboratory, (3) the laboratory-grown bacteria are grown in experimental animals, and (4) bacteria are isolated from the diseased experimental animals and shown to be the same kind as the original organisms (*World Book Encyclopedia* 1982). He applied his new methods to anthrax in cattle (1876), cholera, and tuberculosis in humans (1882). Koch received a Nobel Prize in 1905 for his research on tuberculosis.[3]

Russell was hired by the University of Wisconsin as an assistant professor in 1893 to work in dairy bacteriology. Russell had an outstanding research career, but in 1907 he was appointed dean of the College of Agriculture. Under his leadership the College of Agriculture at the University of Wisconsin overcame major criticism by very applied agricultural interests (Nevins 1962; Scott 1970) and became one of the leading American research institutions (Beardsley 1969, pp. 68–70) and a leading Ph.D.-granting institution. Russell was a champion of pure theoretical or basic science research. Although he did not downgrade strictly applied work, he was convinced that the solution to many practical agricultural problems lay in pursuing them through their roots in theory. As dean, he set out to expand agricultural research by getting basic science into the College of Agriculture. Also, see discussion in Rossiter 1986.

The first Ph.D. degree was awarded at the University of Wisconsin in 1890, and in 1905–1906, six graduate students were enrolled in the college. By 1914–1915, the graduate student enrollment had jumped to 114. When Russell stepped down in 1930, the reputation of the College of Agriculture had been firmly established as a first-rate research and graduate student training institution.

Early Training in Land-Grant Colleges

During the early 20th century, graduate education in the agricultural sciences developed rapidly at the land-grant colleges. The leaders were Cornell, Wisconsin, Illinois, Iowa State, and Michigan State. Cornell University and the University of Wisconsin developed rapidly during 1900–1920, but Iowa State College came on very rapidly during the 1920s and 1930s. Some of these programs exhibited real quality. In an American Council on Education survey in 1934, Cornell, Wisconsin, Minnesota, California, Iowa State, and Harvard had one or more graduate programs in the agricultural sciences that ranked "distinguished" (Hughes 1934. See, also, the discussion in Rossiter 1986.)

The successful direction of graduate students is hard work and requires faculty time that might be directed to other activities. By 1910, graduate student enrollment in some colleges of agriculture—e.g., the University of Wisconsin and Cornell—was sizable and directing graduate students was a major teaching activity. The colleges of agriculture were innovative and made supervising and directing graduate students by scientists complementary to experiment station research.

Johns Hopkins University, the University of Chicago, and Harvard University offered fellowships and scholarships to outstanding individuals as an inducement to come to their institutions for graduate studies. The awards were attractive for the students because they were paid to be students and were not asked to teach or conduct research regularly. However, they frequently found it difficult to get actively involved in research because they were not tied directly into the research projects of faculty members. Then as now, getting dissertation research started was frequently a time-consuming and frustrating experience requiring considerable creativity, imagination, and persistence.

The agricultural experiment stations had research funds, and the scientists frequently needed assistance to complete their proposed research objectives. Consequently, about 1910, colleges of agriculture established the institution of awarding research assistantships to outstanding prospective graduate students (Waters 1910). Individuals were expected to enroll as graduate students in one of the departments of the college and to be paid for working under the supervision of a scientist who had an experiment station appointment. In this way, the student could work and learn about scientific research at the same time.

This innovation had the merit of paying the graduate student for work that was related to his (or her) training, of getting him (or her) started learning the art of successful research, and of advancing work on applied research problems. These research assignments generally started as minor activities and built into a major research activity that culminated in a Ph.D. dissertation.

Research assistant manpower has become the hallmark of agricultural experiment station research. The Ph.D. dissertations of research assistants and the new doctorate recipients have become major parts of most experiment stations' research output. The research assistantship has on the whole been a productive experience for everyone concerned.

SAES Scientists

The presidents of land-grant colleges experienced difficulty during 1862–1890 in finding well-trained individuals to staff their agricultural teaching and research programs. During the mid-19th century, only a few fields had established a body of rigorously tested knowledge of a science. These were the natural science fields of geology, physics, and chemistry. Biological sciences, except for botany and entomology, had not yet emerged as established fields. Only a few Ph.D. degrees were being awarded in these fields before 1900 (Table 3.1).[4] Thus, the early land-grant colleges and agricultural experiment stations faced a dilemma between hiring persons trained in the natural science fields or employing practicing agriculturalists who had little formal college training, including knowledge of science.

Data on the number and characteristics of agricultural college staff became available for the first time in 1889 and were published annually thereafter by the USDA. Although these publications present information about the instructional and research staff of land-grant colleges, our presentation focuses only upon the number and selected characteristics of the agricultural experiment station staff. Tabulated information, starting in 1890 and ending in 1980, is presented at 10-year intervals in order to provide a picture of long-term trends in the number of SAES staff and their distribution by science-related characteristics. Busch and Lacy (1983, Chapter 3) also present information on the characteristics of recent scientists.

Numbers

In 1890, there were 250 researchers in the 46 agricultural experiment stations, or an average of about 5 per station. Ninety years later, the number of station researchers was 55 times larger (13,757), or an average of 233 per station[5] (Table 3.2). The number of SAES researchers increased by 40 percent or more per decade from 1890 to 1930. In 1930, there were 2,853 researchers in the SAES system, which is 11.4 times the 1890 number and 21 percent of the 1980 number. The decade of the 1930s was affected by the Great Depression and a slow economic recovery. Consequently, the number of SAES researchers increased only 38 percent then. During the 1940s and 1950s, the rate of increase of SAES staff was about 42 percent per decade. The number of SAES researchers in

Table 3.1   Absolute number of degrees awarded in selected fields, U.S. colleges and universities, 1890–1900

Year	Chemistry[a]		Biological Sciences[b]		Agriculture[c]		
	Master's	Ph.D.	Master's	Ph.D.	B.S.	Master's	Ph.D.
1890	103	28	65	18	425	156	2
1891	77	35	49	22	449	186	3
1892	75	35	48	23	422	179	3
1893	116	40	74	26	534	216	3
1894	126	51	81	33	703	190	4
1895	139	49	89	33	808	177	4
1896	149	48	94	34	793	157	4
1897	141	58	89	38	781	151	5
1898	144	57	91	40	769	138	4
1899	154	63	98	41	834	148	5
1900	155	69	100	46	885	149	6

Source: Adapted from Adkins 1975.
[a]Covers up to seven subfields, including inorganic, organic, physical, and analytical chemistry.
[b]Covers up to 28 fields, including botany, bacteriology, plant pathology, zoology, and entomology.
[c]Covers up to 20 fields, including agronomy, animal husbandry, veterinary medicine.

Table 3.2 Absolute and relative number of SAES research staff, by highest degree attained, 1890–1980

Year	Total, All Degrees		Researchers by Highest Degree Completed			Researchers by Highest Degree Completed (%)		
	Increase over Previous 10 Yrs.	Percentage Change	Ph.D.	M.S.	B.S.	Ph.D.	M.S.	B.S.
1890	250	—	73	78	98	29.4	31.4	39.2
1900	443	57.2	116	160	167	26.2	36.1	37.7
1910	1,032	84.5	203	301	527	19.7	29.2	51.1
1920	1,582	42.7	425	522	635	26.9	33.0	40.1
1930	2,853	59.0	932	1,202	719	32.7	42.1	25.2
1940	4,177	38.1	1,937	1,418	822	46.4	33.9	20.0
1950	6,363	42.1	2,994	2,098	1,271	47.1	33.0	20.1
1960	9,607	41.4	5,788	2,493	1,326	60.2	25.9	13.8
1970	11,771	20.3	8,529	2,261	981	72.4	19.2	8.3
1980	13,757	15.6	10,979	1,756	1,022	79.8	12.8	7.4

Source: U.S. Department of Agriculture 1890, 1901, 1911, 1921, 1930, 1940, 1950, 1960, 1970, 1980.
Note: See Appendix 3.A for details about the classification scheme applied to SAES researchers.

1960 was 70 percent of the 1980 total. The rate of increase of SAES staff was at a much lower rate during 1960–1980 than for earlier decades—20.3 percent for the 1960s and 15.6 percent for the 1970s (Table 3.2). Additional details are also available in Huffman 1985.

Characteristics

Less than one-third of the SAES scientists had completed Ph.D.-level research training before 1930. In contrast, more than 80 percent of current SAES scientists have completed a Ph.D. degree. Researchers in plant sciences and physical sciences each accounted for 30 percent of SAES staff in 1890. The share of scientists in the plant sciences increased and accounted for about 45 percent of the researchers in 1910 and later years (Table 3.3). The share in physical sciences has gradually decreased over time and was 9 percent in 1980. Over time, the geographical distribution of SAES staff has shifted away from the Northeast, which had 30 percent of the total researchers in 1890–1900, to the other production regions.

Education

Although almost all scientists who currently have SAES appointments also have completed Ph.D. degrees, less than one-half of the SAES researchers had a Ph.D. degree before 1950 (Table 3.2). This was significant because the degree Doctorate of Philosophy is a research degree awarded by universities in the sciences to signify excellence in the mastery of a body of scientific knowledge and skills enabling one to perform research through application of the scientific method. Individuals who have not gone beyond a master's degree have limited research skills. Price (1978), the noted historian of science, has cautioned that the rise in the share of researchers who completed Ph.D. degrees overestimates the rise in quality of the staff. The reason is that the distribution of innate mental ability in the population has most likely remained unchanged over time, but a larger share of the more able are obtaining college degrees and more rigorous training in science.

Table 3.3   Absolute and relative distribution of SAES researchers, by field of science, 1890–1980

Year	Plant Sciences	Field or Subject Matter[a] Animal Sciences	Physical Science	Other
1890	74.0	22.5	78.5	75.0
	(29.6)[b]	(9.0)	(31.4)	(30.0)
1900	150.4	61.5	117.5	114.0
	(33.9)	(13.9)	(26.5)	(25.7)
1910	488.5	270.0	181.0	92.5
	(47.3)	(26.2)	(17.5)	(9.0)
1920	783.5	426.5	193.0	179.0
	(49.5)	(27.0)	(12.2)	(11.3)
1930	1,330.0	686.0	334.0	503.0
	(46.6)	(24.0)	(11.7)	(17.6)
1940	1,956.5	952.5	463.0	805.0
	(46.8)	(22.8)	(11.1)	(19.3)
1950	2,831.0	1,414.5	753.5	1,364.0
	(44.5)	(22.2)	(11.8)	(21.4)
1960	4,313.0	2,336.5	929.5	2,028.0
	(44.9)	(24.3)	(9.7)	(21.1)
1970	5,428.0	2,491.0	1,061.0	2,791.0
	(46.1)	(21.2)	(9.0)	(23.7)
1980	6,313.0	2,641.0	1,189.0	3,614.0
	(46.9)	(19.2)	(8.6)	(26.3)

Source: U.S. Department of Agriculture 1890, 1901, 1911, 1921, 1930, 1940, 1950, 1960, 1970, 1980.
[a]See Appendix 3A for a discussion of methods used to group and count SAES researchers.
[b]Numbers in parentheses are shares of the total for a given year.

In 1890, 29 percent of the SAES researchers had completed a Ph.D. degree. But the share holding Ph.D. degrees declined as the SAES system expanded rapidly during the 1890s and first decade of the 20th century. In 1910, only 20 percent of the SAES staff had completed a Ph.D. degree and 51 percent had attained only a bachelor's degree-level knowledge of science, agriculture, or related areas. They undoubtedly had on average acquired substantial useful knowledge since receiving the degrees.

After 1910, the share of SAES researchers holding Ph.D. degrees rose steadily, except for the 1940–1950 period when it remained unchanged. During the decade of the 1950s extensive upgrading of skill levels of SAES staff occurred. The share of researchers who obtained a Ph.D. degree was 13 percent higher in 1960 than in 1950. The Ph.D. share was 60 percent in 1960. Furthermore, the absolute number of individuals who had completed only a B.S. or M.S. degree did not decline until after 1960. Retirements seemed to be a major factor in reducing this number after 1960.

The Nature of Research Work

Given the small body of scientific knowledge and the few fields in which a science had been established, it should not be surprising that a large share of the early SAES scientists were trained in chemistry or botany. In 1889, 51 percent had titles of chemist or botanist (True 1937).[6] An additional 2.2 percent were geologists or physicists. Other individuals on appointment generally had little research training and sometimes their disciplinary information was shallow. Individuals who had the title of agriculturalist or horticulturalist, which comprised 20 percent of the staff in 1889, seemed to have knowledge (and research skills) comparable to good farmers of the time rather than scientists (True 1937, p. 136). As the SAES system expanded rapidly during 1889 to 1900, the share of the staff holding titles such as chemist, botanist, geologist, or physicist decreased 13 percentage points, and the share of agriculturalists, horticulturalists, animal husbandmen, or dairymen increased 16 percentage points. This contrast between SAES researchers trained in general science and agriculturalists with practical farming interests undoubtedly contributed to the tension about the correct mixture of science and applied work in the early years.

The USDA-published *Handbooks* permit a subject-matter classification scheme for the titles or areas of work of the SAES staff starting in 1890. The major categories are plant sciences, animal sciences, physical sciences, social sciences, home economics, food science and technology, and resource-environmental sciences. Although it was impossible to tell which scientists were actually conducting basic or applied research, it was possible to classify them by title or area of research. For example, a researcher who had a title of plant physiologist or an area of work of plant physiology was classified in the plant sciences basic research category. See Appendix 3.A for details of the classification scheme.[7]

In 1890, 60 percent of SAES researchers were classified as working in physical sciences and plant sciences (Table 3.3). Although the total number was approximately the same for these two areas, the physical sciences had a significantly larger share who had completed a Ph.D. degree. Only 9 percent of the SAES researchers were working on animal science research in 1890, and very few researchers were identified as working in social sciences, home economics, food science and technology, or resource and environmental sciences. This latter group was included in the 30 percent of SAES researchers in the "Other" category.

In 1900, the distribution of researchers among major research areas was slightly different than the one in 1890 (Table 3.3). Four percentage points fewer were classified as working in physical sciences and in "Other" areas in 1900 than in 1890. These 8 percent were allocated approximately equally to increases in the share of plant science and animal science researchers.

In 1910 and 1920, the distribution of SAES researchers among major research areas was similar to that of 1900, but the distribution shifted strongly in favor of plant sciences and animal sciences and away from physical sciences and other areas. Almost one-half of all SAES researchers were working in the plant science area, and the animal science area accounted for 27 percent. In all of the research areas, Ph.D.-level researchers were in the minority.

The relative allocation of SAES staff among plant sciences, animal sciences, and physical sciences was approximately the same for 1930, 1940, and 1950—about 45 percent, 23 percent, and 11 percent, respectively.

The number of researchers in the "Other" area increased by about 9 percentage points over 1920. This increase occurred because of the Great Depression and expansion of social science research, which was aided by passage of the Purnell Act in 1926 and by advances in the social sciences (Schultz 1941). In 1930, 7.9 percent (226) of the SAES staff was classified as being in agricultural economics research, 1.4 percent in other social science research (primarily rural sociology), and 4.1 percent (116) in home economics research. By 1950, the agricultural economics and home economics shares had edged up slightly to 8.3 percent and 5.1 percent.

The distribution of SAES staff among areas of work shifted slightly in the later years. The share working in plant sciences increased slightly and in animal sciences decreased slightly. The share in physical sciences decreased gradually and was 3 percentage points lower in 1980 than in 1950. This decrease was reflected in larger shares for food science and technology and resource-environmental science areas.

Table 3.4 Regional distribution of SAES researchers and share holding a Ph.D. degree, United States, 1900–1980

Production Regions	Distribution by Region					Percent Holding Ph.D. Degree				
	1900	1920	1940	1960	1980	1900	1920	1940	1960	1980
Lake States	—	12.8	11.1	8.4	8.1	29.6	26.7	48.4	75.2	93.0
Corn Belt	11.0	15.7	16.3	13.9	13.2	18.4	27.8	49.6	61.8	80.5
Delta States	6.5	3.7	4.6	6.3	7.7	13.8	23.7	44.3	39.7	59.9
Southern Plains	3.4	3.5	4.8	7.2	6.1	13.3	26.8	42.6	53.1	74.9
Northern Plains	9.7	8.6	6.8	7.0	7.7	16.3	26.5	38.7	52.2	73.8
Southeast	5.4	3.2	6.4	8.6	9.7	45.8	28.0	41.4	55.1	82.3
Appalachia	8.6	8.3	9.0	9.8	10.4	26.3	19.1	35.0	58.7	74.2
Pacific	5.7	12.1	12.7	14.0	13.5	39.2	20.3	51.4	68.8	85.7
Mountain	13.5	11.9	9.3	9.4	9.3	15.0	25.4	37.9	52.6	73.7
Northeast	30.0	20.1	19.1	14.0	14.2	34.6	34.9	54.4	66.7	85.4
Total	100.0	100.0	100.0	100.0	100.0	26.2	26.9	46.3	60.2	79.8

Source: U.S. Department of Agriculture 1901, 1921, 1940, 1960, 1980.
Note: See Appendix 3A for discussion of the method used to count SAES researchers.

Regional Distribution

The Northeast region was a major center of agricultural research during the late 19th century. This was reflected in the fact that the region, comprised of 10 states, had 30 percent of the total SAES staff in 1900. Over time, that share steadily declined and equaled 14 percent by 1980 (Table 3.4).

The regional distribution of SAES scientists shifted away from the Northeast, Mountain states, and Northern Plains in favor of the Southern Plains, Southeast, and Appalachia over time. In the Lake states and Corn Belt, the regional share increased until 1920 and 1940, respectively, and then declined. For the Pacific region, the share of scientists jumped from 5.7 to 12.1 percent between 1900 and 1920 and has remained between 12 and 14 percent since then.

The research training as reflected in the share of SAES staff holding Ph.D. degrees also varies across the regions. In the Delta and Mountain states, every year the share holding Ph.D. degrees was several percentage points below the U.S. average. In contrast, the Pacific region and Lake states were above the national average every year except 1920 (Table 3.4). The regional quality differences of agricultural scientists may have affected agricultural productivity during the 20th century.

Competition for Scientists among Sectors

The public and private sectors of our economy compete for science and technology personnel. Unfortunately our data for the private sector about characteristics of scientists are very poor. On the whole, the private sector is much more heavily involved in technology invention (Level IV) science (Figure 2.3), new technology development, and marketing (Level III activity) than the public sector. These are activities to which individuals with master's- and bachelor's-level training can contribute. Over time the share of private sector scientists that have completed a Ph.D. degree is much lower than for the public sector. One indication is that for 1982, 36 percent of the scientists involved in plant breeding activities in the private sector held Ph.D. degrees (Table 3.5). Among SAES scientists working in plant science, more than 80 percent had completed Ph.D. degrees.

Table 3.5   Plant breeding activity in U.S. private sector by major crop categories, 1982

Major Crop Categories	Companies with Breeding Program	Full-time Equivalent Scientists	
		Ph.D.	M.S. and B.S.
Corn	66	155	302
Soybeans	26	36	58
Wheat	21	23	46
Grain sorghum	21	22	44
Barley, oats, rye, triticales, millet	11	7	8
Rice	5	7	6
Forage legumes, mainly alfalfa	14	23	35
Forage and turf grasses	13	11	14
Cotton and other fiber crops	13	17	30
Sugar beets	5	14	8
Sunflowers	16	15	26
Safflower	3	2	2
Tobacco	3	1	2
Vegetables and fruits	46	96	158
Flowers and ornamentals	9	5	17
Total	—	434	756

Source: Adapted from Kalton and Richardson 1983.

Data on the sector of employment for agricultural scientists are available only after 1967. These data show that more than 60 percent of the Ph.D.'s employed in applied agricultural sciences were working in educational institutions (Table 3.6).[8] This share, however, decreased 12.3 percentage points between 1973 and 1985. The private R&D sector absorbed almost all of this distributional shift.

A significant share—20 to 30 percent—of Ph.D.'s employed in applied agricultural sciences completed Ph.D. degrees in majors outside the applied agricultural science fields. Also, there has been significant mobility of Ph.D.'s among sectors of employment. Since 1973 the net sectoral migration has been from educational institutions and government to business and industry, including self-employment. Salaries of Ph.D.'s employed in business-industry and government were significantly higher than for the educational sector, even when adjustments were made to a 12-month equivalent (National Research Council 1988). Although information was not available on the amount of consulting income earned by Ph.D.'s, important differences are expected to exist across the three sectors of employment.

Sector of Employment

In 1973, there were approximately 11,000 Ph.D.'s employed in applied agricultural sciences. In 1985, this number was twice as large (Table 3.6).[9] The number of Ph.D.'s employed in agriculture-related basic sciences is two to four times larger than the number employed in applied agricultural sciences. This is important for two reasons: (1) The amount of new knowledge being produced by agriculture-related general and pre-invention sciences was relatively large and was an input into applied agricultural research (Chapter 2). (2) Ph.D.'s employed in these sciences could switch sectors

Table 3.6 Absolute number of employed Ph.D.'s by science field and employment sector, 1973 and 1985

Science Field of Employment[a]	Year	Educational Institution[b]		Business Industry[c]		Government		Total
Applied Agricultural Sciences	1973	6,902	(63.0)	1,953	(17.8)	2,109	(19.2)	10,964
	1985	11,801	(50.7)[d]	7,265	(31.2)	4,214	(18.1)	23,280
Plant & Soil Sciences	1973	2,585	(64.8)	568	(14.2)	759	(19.0)	3,989
	1985	3,211	(60.2)	1,282	(24.0)	831	(15.6)	5,332
Animal Sciences	1973	1,509	(65.5)	495	(21.5)	266	(11.6)	2,303
	1985	2,459	(63.8)	1,087	(28.2)	308	(8.0)	3,854
Food Sciences	1973	484	(42.9)	495	(43.9)	129	(11.4)	1,128
	1985	677	(25.4)	1,755	(65.9)	232	(8.7)	2,664
Natural Resource & Environmental	1973	870	(54.6)	168	(10.5)	511	(32.1)	1,593
	1985	2,027	(33.3)	1,954	(32.1)	2,114	(34.7)	6,095
Agricultural Economics	1973	872	(68.3)	120	(9.4)	269	(21.1)	1,277
	1985	1,903	(71.6)	307	(11.5)	449	(16.9)	2,659
Other Applied Agricultural Sciences	1973	582	(65.3)	107	(12.0)	175	(19.6)	891
	1985	1,524	(56.8)	880	(32.8)	280	(10.4)	2,684
Agri-Related Gen. & Pre-tech. Sciences	1973	24,830	(63.1)	6,966	(17.7)	4,934	(12.5)	39,370
	1985	31,328	(68.2)	9,597	(20.9)	4,980	(10.8)	45,912
Biological Sciences[e]	1973	24,641	(64.4)	6,589	(17.2)	4,414	(11.5)	38,267
	1985	34,570	(68.3)	10,739	(21.2)	5,322	(10.5)	50,631
All Natural Sciences[f]	1973	71,148	(53.4)	38,452	(28.9)	16,795	(12.6)	133,139
	1985	104,989	(53.2)	67,350	(34.1)	24,987	(12.7)	197,413

Source: National Research Council, Survey of Doctorate Recipients, special tabulations. Also see National Research Council 1988, p.24.
[a]Definitions of Fields of Science and Fields of Employment of Ph.D.'s are as follows. Plant and Soil Sciences: Agronomy, Soils, Soil Sciences, Horticulture, Hydrobiology, Plant Breeding and Genetics, and Plant Sciences — Other. Animal Sciences: Animal Husbandry, Animal Science, Animal Nutrition, Veterinary Medicine, Animal Breeding, and Animal Genetics. Food Sciences: Food Science and Technology, Agriculture and Food Chemistry. Natural Resource and Environmental: Hydrology and Water, Environmental Sciences — General and other, Fish and Wildlife, and Forestry. Other Applied Agricultural Sciences: Agricultural Engineering. Agriculture — General and Other. Agricultural-related General and Pre-tech. Sciences: Biochemistry, Biophysics, Biometrics, Biomathematics and Statistics, Cytology, Embryology, Ecology, Molecular Biology, Genetics — Unspecified and Human, Biological Sciences — General and Other, Immunology. Animal and Plant Physiology, Zoology, Entomology. Natural Sciences: Specialties include applied agricultural sciences, but covers specialties listed by numbers 540 through 599 in National Research Council 1988. Natural Sciences: Specialties include applied agricultural and biological sciences, as well as specialties listed by numbers 000 through 399 and 500 through 999 in the National Research Council 1985. See Huffman and Evenson 1987. Appendix 4.B for added definitions.
[b]Does not include post-doctoral students.
[c]Includes self-employed Ph.D.s.
[d]The numbers in parentheses are shares of the total employed in a field.
[e]Does not include biological applied agricultural sciences.
[f]Does include applied agricultural sciences.

84

of employment and bring their skills to bear on problems in agriculture, e.g., microbiologists shifted to developing new technology for profit during the 1980s.

In 1973, Ph.D.'s employed in applied agricultural sciences made up 8.1 percent of all Ph.D.'s in the natural sciences. In 1985, their share had risen to 11.4 percent. A majority of applied agricultural scientists were employed in educational institutions, but the share was 12.3 percentage points lower in 1985 than in 1973 (Table 3.6). In 1973, about 18 percent of all Ph.D.'s employed in applied agricultural sciences were employed in each of the government and business-industry sectors. In 1985, the share employed in the government sector had decreased 1 percentage point, but the share employed in business-industry increased 13.4 percentage points. Seventy-two percent more applied agricultural science Ph.D.'s were employed in private R&D than in the government sector in 1985. The number of applied agricultural science Ph.D.'s employed in private R&D was 3.7 times larger in 1985 than in 1973. Furthermore, the absolute increase between 1973 and 1985 was larger than the increase for either of the other two sectors.

The change between 1973 and 1985 in the importance of private R&D as an employer of Ph.D.'s in the applied agricultural sciences was truly remarkable. Care must be taken, however, about projecting these past trends into the future. During 1973–1980, total real net farm income was high compared with historical levels, and agricultural business firms were taking advantage of this prosperity by expanding the services they supplied. Starting in 1981, the economic picture for agriculture and agricultural business firms became temporarily gloomy. The data, however, show that the economic downturn during 1981–1985 did not reduce the share of applied agricultural scientists employed in the private R&D.

The Ph.D.'s employed in the applied agricultural sciences were desegregated into five broad areas for further comparisons: plant and soil sciences, animal sciences, food sciences, agricultural economics, and other. In the plant and soil science field approximately 4,000 Ph.D.'s were employed in 1973. This number was 33 percent larger in 1985. Between 1973 and 1985, the share of these Ph.D.'s employed in educational institutions decreased from 64 percent to 60 percent, but the share employed in the business-industry sector increased 10 percentage points (Table 3.6).

In the animal science area, the total number of Ph.D.'s employed was approximately 2,300 in 1973 and in 1985 was 3,850. The share employed in educational institutions was 65.5 percent in 1973 and 2 percentage points lower in 1985. Private R&D accounted for 21.5 percentage points of the employment in this field in 1973 and 28.2 percent in 1985. Historically, the number of Ph.D.'s employed in the animal science area has been small relative to the number employed in the plant science area.

The number of Ph.D.'s employed in food sciences increased by 136 percent between 1973 and 1985. The food science area was unusual because of the relatively large share of Ph.D.'s who were employed in the business-industry sector—43.9 percent in 1973 and 65.9 percent in 1985. The share employed in educational institutions was equal to the share employed in private R&D in 1973, but dropped to 25.4 percent in 1985.

The 1970s was a decade when natural resource and environmental issues became important nationally. Although approximately 1,590 Ph.D.'s were employed in the natural resource(environmental area in 1973, the number was 3.8 times larger in 1985. In 1973, 54.6 percent of these Ph.D.'s were employed in educational institutions, 10.5 percent in business-industry, and 32.1 percent in the government sector. The distribution had changed dramatically by 1985 when employment was split approximately equally among the three sectors. The 22 percentage point increase of the business-industry share between 1973 and 1985 indicates that resource and environmental issues were taken seriously by the private sector during the 1970s.

In the agricultural economics field, the number of Ph.D.'s employed in 1973 was 1,277 and two times larger in 1985. The distribution of agricultural economists among sectors was approximately the same for 1973 and 1985. In 1973, 68.3 percent were employed in educational institutions, 9 percent in business-industry, and 21 percent in government.[10]

The distribution, by sector of employment, for Ph.D.'s in agriculture-related basic sciences and biological sciences was similar to that of Ph.D.'s employed in the animal science area (Table 3.6).

Field Mobility

Individuals holding Ph.D. degrees generally are employed in the same field in which they obtained their doctorate. However, a significant share switch fields. Also, see National Research Council 1988. This is due primarily to economic opportunities and interests, given that most biological sciences are interrelated. The new area of biotechnology that developed during the 1980s was unusual in that Ph.D. scientists in microbiology could and did successfully develop new technologies that were commercially successful. Skills obtained in one Ph.D. field can be employed productively in other related fields. Outside of biological sciences, some fields are also closely related, e.g., economics and agricultural economics.

In 1973, 78 percent of the Ph.D.'s employed in the applied agricultural sciences, excluding agricultural economics, received a Ph.D. degree in one of the applied agricultural sciences (see Table 3.7). Twelve percent received a Ph.D. degree in agriculture-related general and pre-invention sciences and 5 percent in other natural sciences. In 1985, the share where the employment and degree fields matched dropped to 61 percent. The share of Ph.D.'s who were employed in applied agricultural sciences and who received doctorates in agriculture-related sciences increased to 20 percent, and in other natural sciences to 13 percent.

Thus, between 1973 and 1985, a significant shift in the training of Ph.D.'s employed in the applied agricultural sciences (excluding agricultural economics) occurred. The employers were drawing more heavily from the general and pre-invention sciences in 1985 than in 1973. This is one method for the applied agricultural sciences to establish upstream linkages to pre-technology and general sciences (also see Chapter 2).

Among Ph.D.'s employed in the field of agricultural economics, 66 percent had received a Ph.D. degree in agricultural economics in 1973 and 72 percent in 1985. Although 6 percent had obtained a Ph.D. in other applied agricultural science fields, a quarter were from other sciences, i.e., economics (Table 3.7).

Among Ph.D.'s employed in the fields of agriculture-related general and pre-technology sciences and other natural sciences, only a small percentage had received a Ph.D. degree in applied agricultural science fields. Thus, the upstream sciences do not have "downstream" linkages for scientific staff.

The Production of New Agricultural Scientists

The doctoral degree is the highest recognized level of educational training. The skills associated with doctorates in the sciences are different from other areas. The Doctor of Philosophy, a research degree, is awarded in the sciences to signify excellence in the mastery of an appropriate body of scientific knowledge and in skills to perform research through application of the scientific method.

Table 3.7   Field of doctorate distribution of all employed U.S. Ph.D.'s, 1973 and 1985

Field of Employment[a]	Year	Science Field of Doctorate					
		Applied Agr. Excl. Ag. Econ.	Agricultural Economics	Agr. Rel. Gen. & Pre-tech. Sci.	Other Natural Sciences	Other	Total
Appl. Agr. Sci. excl.	1973	78[b]	0	12	5	5	100
Agr. Econ.	1985	61	1	20	13	6	100
Agricultural	1973	6	66	0	1	27	100
Economics	1985	3	72	0	0	25	100
Agr. Rel. Gen.	1973	7	0	74	14	5	100
& Pretech. Sci.	1985	3	0	87	9	1	100
Other Natural	1973	2	0	4	82	12	100
Sciences	1985	1	0	10	78	12	100
Other	1973	1	0	1	6	92	100
	1985	·1	0	2	8	90	100

Source: National Research Council, Survey of Doctorate Recipients, special tabulations.
[a]See Table 3.6, footnote a for definition of the fields.
[b]For example, among Ph.D.'s employed in the applied agricultural science fields (excluding agricultural economics), 78% received a Ph.D. in one of the these applied agricultural science fields and 12% received a Ph.D. in an agriculture-related basic science field.

Doctorates awarded in the humanities signify mastery of a body of knowledge having primarily a cultural character. Doctorates awarded in education and the professions (medicine, law, dentistry, etc.) signify mastery of a body of knowledge and skills appropriate for practicing a particular profession. These latter two doctorates are non-research degrees.

The Early Leaders in Doctorate Production

The first doctoral degree was conferred by Yale University in 1861. Currently more than 300 U.S. universities grant doctoral degrees. In 130 years, over 740,000 doctorates have been awarded by U.S. institutions in all fields (sciences, engineering, humanities, education, and professions). About 65 percent of these doctorates have been in science and engineering fields with most awarded after 1960. The universities leading in total number of doctoral degrees awarded have changed over time from the Ivy League private universities to the large state universities, i.e., University of California—Berkeley, University of Wisconsin, University of Illinois—Urbana, and University of Michigan (Harmon 1978).

The number of Ph.D. degrees awarded in the applied agriculture and pre-invention science field group grew rapidly in the 10-year period 1920–1929. But the Great Depression of 1929–1933 slowed this growth rate, and World War II caused a sharp reversal starting in 1942. The growth rate of new Ph.D.'s awarded in applied agricultural production sciences was somewhat less affected by both the Great Depression and entry into World War II compared with the growth rate for new Ph.D. degrees awarded in pre-invention and general sciences for agriculture. The number of Ph.D. degrees awarded in 1946 was only 44 percent of the prewar, 1940 number (Huffman and Evenson 1993).

The end of World War II, however, brought large enrollments of war veterans in colleges because of favorable G.I. Bill educational benefits.[11] The number of Ph.D. degrees awarded in 1950–1954 was 2.7 times larger than the previous high for any preceding 5-year period. This growth slowed temporarily in 1955, turned sharply negative in 1958, and continued at a slower average annual rate of over 1950 to 1971. A temporary peak was reached over 1970–1980 before further growth occurred.

Number of Doctorates in Agricultural and Pre-Invention Sciences

This section presents a quantitative summary of the Ph.D. output and major characteristics of Ph.D. recipients in 14 applied agricultural science fields, including home economics, and nine pre-invention science fields associated with applied agricultural sciences. The numbers reported are for all U.S. universities, starting in 1920 and continuing to 2000.

Totals and by Field

In these 23 fields, a total of only 10,361 doctoral degrees were awarded by U.S. institutions during 1920–1959, or an average of 259 per year (see Table 3.8). A large share of these degrees were awarded in the immediate post–World War II era, supported by the GI benefits for education (Harmon 1978). For the period 1960–1979, 35,262 doctorates, or an average of 1,763 per annum, were awarded in these fields. This number increased to 49,112, or 2,338 per year, over 1980–2000. Although data on the characteristics of Ph.D. recipients do not begin until about 1960, they show some significant changes in the gender and national origin of recipients during 1960–1980.

Turning to the applied agricultural science fields, a total of 4,778 doctorates were awarded over 1920–1959, or an average annual rate of only 119 with most of them occurring in the last 15 years. This number rose to 16,949 over 1960–1979, or an average of 847 per annum. In the latter period, 1980–2000, the number of applied agricultural science doctorates reached 26,286, or 1,252 per annum.

Although doctorates in the fields of agronomy, horticulture, and animal (including dairy and poultry) husbandry dominated the total number of doctorates awarded by U.S. universities over 1920–1959 and accounted for 75 percent of the total when the science of agriculture was being developed, major growth occurred in the number of doctorates awarded in other fields over 1960–1979—in soils and soil science, forestry, agricultural engineering, agricultural economics, food science and technology, and fish and wildlife management (see Table 3.8).[12] In this second period, doctorates in agronomy, horticulture, and animal husbandry (or science) accounted for only 45 percent of the total, and in the last period, 1980–2000, the share of the three references doctorate fields in the total number of doctorates was only 32 percent.

Women

Although the gender composition of U.S. doctorates was not tracked before 1960, relatively few women were pursuing doctorates in this era. Over 1960–1979, Table 3.9 shows that a total of 2,928 doctorates were awarded women in the 23 fields of applied agricultural and pre-invention sciences, or an average of 146 per annum. Only 756, or 38 per annum, were awarded in applied agricultural science fields. Over 1980–2000, the total doctorates to women for the 23 fields increased dramatically to 11,562, or 550 per annum. In applied agricultural sciences, the total degrees for this period

Table 3.8 Number of Ph.D. degrees awarded by U.S. universities in applied agricultural and pre-invention science fields, 1920–2000

Science Fields	1920–1959	1960–1979	1980–2000
Applied Agricultural Sciences			
Agronomy (and plant breeding)	1,752	3,060	4,030
Soils, soil science, soil chemistry and microbiology	—	821	1,989
Horticulture (science)	828	1,234	1,463
Forestry (science, biology, engineering, mgt, wood science)	338	1,258	2,126
Animal (dairy, poultry) husbandry and (or) animal science, breeding, nutrition	986	3,348	2,947
Veterinary medicine	91	486[e]	433[f]
Agricultural engineering	101	829	1,562
Agricultural economics and business	—	1,680	3,250
Food science, engineering and technology	—	960	2,981
Agricultural and food chemistry	382	451	—
Fisheries and wildlife, conservation/renewable natural resources	122	796	1,947
General and other agricultural sciences	178[a]	1,393	963
Nutrition, nutritional science or dietetics	—	282[c]	2,595
(Other) home economics	—	351	—
(Subtotal)	(4,778)	(16,949)	(26,286)
[Average per year]	[119]	[847]	[1,252]
Pre-Invention Sciences			
Genetics (plant, animal and human)	933	2,484	3,835
Plant physiology	2,147	1,358	1,237
Animal and human physiology	—	5,179	5,296
Plant pathology (phytopathology)	789	1,814	1,248
Entomology	1,456	2,948	2,940
Meteorology and atmospheric science[d]	171	992	2,255
Hydrology and water resources	—	280	530
Environmental sciences	—	420	1,271
Probability, statistics and econometrics	90[b]	2,838	4,194
(Subtotal)	(5,583)	(18,313)	(22,826)
Total	(10,361)	(35,262)	(49,112)

Source: Survey of Earned Doctorates, various years.
[a]Includes agricultural economics, food sciences, soil sciences and animal sciences.
[b]Statistics not included.
[c]Field started in 1976.
[d]This field includes meteorology up to 1976 and atmospheric physics and chemistry, atmospheric dynamics, and atmospheric science-other for 1976 on.
[e]1960–74 only.
[f]1980–88 only.

was 6,138, or 292 per annum, which is 8 times larger than for the earlier period. Clearly, women have invaded the sciences of agriculture over the past two decades.

Over 1960–1979 and in the pre-invention sciences, women accounted for 12 percent of all doctorates but for 22 percent of the doctorates in genetics. In applied agricultural sciences, women were awarded only 4.5 percent of the doctorates. However, in fields most closely associated with food—food science and technology, agricultural and food chemistry, and nutrition or nutritional sciences—they accounted for more than 13 percent, and for 78 percent of the doctorates in nutrition or nutritional sciences. Over the second period, women were awarded 24 percent of the doctorates in

Table 3.9   Number of Ph.D. degrees awarded to women by U.S. universities in applied agricultural and pre-invention science fields, 1960–2000

Science Fields	1960–1979		1980–2000	
	No.	%	No.	%
Applied Agricultural Sciences[a]				
Agronomy(and plant breeding)	46	1.5	571	14.2
Soils, soil science, soil chemistry and microbiology	20	2.4	294	14.8
Horticulture(science)	45	3.6	344	23.5
Forestry(science, biology, engineering, mgt, wood science)	15	1.2	330	15.5
Animal (dairy, poultry) husbandry and (or) animal science, breeding, nutrition	81	2.4	554	18.8
Veterinary medicine	NA	NA	NA	NA
Agricultural engineering	2	0.2	120	7.7
Agricultural economics and business	45	2.7	584	18.0
Food science, engineering and technology	163	17.0	1,074	36.0
Agricultural and food chemistry	60	13.3	—	—
Fisheries and wildlife, conservation/renewable natural resources	17	2.1	331	17.0
General and other agricultural sciences	42	3.0	143	14.8
Nutrition, nutritional science or dietetics	220	78.0	1,793	69.1
(Other) home economics	NA	NA	NA	NA
(Subtotal)	(756)	(4.5)	(6,138)	(23.4)
[Average per year]	[38]		[292]	
Pre-Invention Sciences				
Genetics (plant, animal and human)	550	22.1	721	18.8
Plant physiology	175	12.9	410	32.9
Animal and human physiology	839	16.2	1,964	37.1
Plant pathology (phytopathology)	101	5.6	139	7.7
Entomology	173	5.9	610	20.7
Meteorology and atmospheric science	37	3.7	107	4.7
Hydrology and water resources	2	0.7	94	17.7
Environmental sciences	51	12.1	381	30.0
Probability, statistics and econometrics	244	8.6	998	24.0
(Subtotal)	(2,172)	(26.1)	(5,424)	(23.8)
Total	(2,928)	(8.3)	(11,562)	(23.5)

Source: Survey of Earned Doctorates, various years.
[a]See table 3.8 for total doctorates by field.

the pre-invention sciences, accounting for more than 30 percent of the degrees in plant physiology, animal and human physiology, and environmental sciences. In applied agricultural sciences, the share of doctorates going to women in this second period was 23.4 percent. Women accounted for more than 30 percent of the doctorates in food science (engineering) and technology and (human) nutrition or nutritional sciences, but also for more than 20 percent of the doctorates in horticulture.

Non-U.S. Citizens

The share of all academic Ph.D. degrees awarded by U.S. Ph.D.-granting institutions to individuals who are not U.S. citizens has increased steadily, at least since 1960. The National Research Council started collecting data on citizenship of doctoral degree recipients in that year. Earlier data can be derived from an analysis of the baccalaureate source of Ph.D. recipients. These data show a long-term upward trend in the number of Ph.D. recipients from U.S. institutions who earned their bachelor's degrees abroad. The rate is 7–9 percent until the 1960s. Then the trend is sharply upward (Harmon 1978, p. 47). In agriculture sciences and engineering, the foreign doctorate share has been high relative to other areas of science.

Over 1960–1979, 8,934 doctorates were awarded by U.S. universities to non-U.S. citizen in the 24 fields in the applied agricultural sciences and pre-invention sciences, or an average of 447 doctorates per annum (see Table 3.10) or 25 percent of the total. The share of non-U.S. citizens obtaining doctorates in the 15 applied agricultural science fields was higher, at 31 percent for the pre-invention science fields (21.5 percent). Given the needs of most low-income and developing countries to increase food production, it is not surprising that a smaller share of Ph.D. degrees are awarded in pre-invention sciences to foreigners than occurs in the applied agricultural science field group. In the pre-invention sciences, 29 percent or more of the doctorates were awarded to non-U.S. citizens in plant pathology, in hydrology and water resources, and in probability, statistics, and econometrics. Even in this early period, the importance of non-U.S. citizen doctorates to total doctorates awarded in particular fields was quite high in soils and soil science, agricultural economics and business, food science (engineering) and technology, agricultural and food chemistry, and nutrition or nutritional sciences, exceeding 39 percent and being more than half of the total in soils and soil science and agricultural and food chemistry (Table 3.10).

Over 1980–1991, a somewhat shorter period, 8,622 doctorates, or 410 per annum, were awarded to non-U.S. citizens in these 23 fields in the applied agriculture and pre-invention sciences (Table 3.10). The number of doctorates awarded foreigners in the pre-invention sciences fell over this shorter second period relative to the first period but rose for the applied agricultural science fields. For the latter fields, the average rate of foreign doctorate awards increased by 64 percent compared to the earlier period. In the pre-invention sciences, the share of foreign doctorates was lower in the second than for the first period in the fields of genetics (plant, animal, and/or human), plant physiology, plant pathology, and entomology. In contrast, in the fields of meteorology or atmospheric sciences and statistics and econometrics the foreign share increased sharply for the second period relative to the first. In the applied agricultural science fields, non-U.S. citizens accounted for 41 percent of all doctorates. The share was over half for the fields of animal science, breeding or nutrition; agricultural engineering; and food science (engineering) and technology and the rise in the foreign share was striking for the first two fields. Only in forestry and nutrition or nutrition sciences did the foreign share of doctorates fall significantly from the first to second period (Table 3.10).

The rising share of foreign doctorates in the applied agricultural science fields has helped maintain the size of many U.S. graduate programs in the face of a declining domestic demand for doctorates for the academic market in the U.S. in the early 1980s and 1990s. Non-U.S. citizens generally bring their own financial aid when they initially enroll in U.S. Ph.D. programs. But given that university tuition rates do not cover the marginal cost of training large numbers of students, the states seem to be providing a sizeable amount of aid to foreign countries in the form of Ph.D. training of non-U.S. citizens. Furthermore, one means by which science and technology is transferred internationally is through Ph.D.-level scientists trained outside their home country. However, in some fields, immigrant Ph.D. recipients have once again become an important source of U.S. science manpower and contributed to U.S. R&D activities.

Table 3.10   Number of Ph.D. degrees awarded by U.S. universities to non-U.S. citizens in applied agricultural and pre-invention science fields, 1960–1991

Science Fields	1960–79[a]		1980–91	
	No.	%	No.	%
Applied Agricultural Sciences				
Agronomy(and plant breeding)	1,142	37.3	1,034	39.2
Soils, soil science, soil chemistry, and microbiology	415	50.5	515	44.9
Horticulture(science)	452	36.6	318	35.1
Forestry(science, biology, engineering, mgt, wood science)	297	23.6	153	13.1
Animal (dairy, poultry) husbandry and (or) animal science, breeding, nutrition	303	9.1	716	51.9
Veterinary medicine	NA	NA	NA	NA
Agricultural engineering	295	35.6	471	55.1
Agricultural economics and business	669	39.8	820	43.2
Food science, engineering and technology	447	46.6	817	52.4
Agricultural and food chemistry	261	57.9	NA	NA
Fisheries and wildlife, conservation/renewable natural resources	88	11.1	207	23.2
General and other agricultural sciences	510	36.6	285	39.4
Nutrition, nutritional science or dietetics	115	40.8	324	23.4
(Other) home economics	NA	NA	NA	NA
(Subtotal)	(4,994)	(31.0)	(5,660)	(40.8)
[Average per year]	[250]		[472]	
Pre-Invention Sciences				
Genetics (plant, animal and human)	543	21.9	283	16.3
Plant physiology	350	25.8	168	23.3
Animal and human physiology	592	11.4	436	13.5
Plant pathology (phytopathology)	640	35.3	185	18.6
Entomology	676	22.9	312	17.4
Meteorology and atmospheric science	171	17.2	330	31.9
Hydrology and water resources	82	29.3	80	33.6
Environmental sciences	54	12.9	83	14.8
Probability, statistics and econometrics	832	29.3	1,085	48.9
(Subtotal)	(3,940)	(21.5)	(2,962)	(23.6)
Total	(8934)	(25.3)	(8622)	(32.6)

Source: Survey of Earned Doctorates, various years.
[a]See Table 3.8 for total doctorates by field.

Land-Grant Universities

As indicated in Chapter 2, land-grant universities became a major source of doctorates in the agricultural science fields during the 20th century. Up to 1920, when the National Research Council records on earned doctorates began (Harmon 1978), Cornell University had a sizeable lead on the total number of doctorates awarded in the agricultural sciences (Rossiter 1986, p. 41).

In 1920,[13] eight of the land-grant universities founded under the 1862 Land-Grant Act (University of California—Berkeley, University of Illinois—Urbana, Iowa State University, University of Missouri, University of Minnesota, Rutgers University, Cornell University, and the University of Wisconsin) awarded Ph.D. degrees in applied agricultural and pre-invention science fields associated with agriculture and home economics (Table 3.11). Among doctorates awarded in

Table 3.11   Top 15 LGUs, by number of Ph.D. degrees awarded in applied agricultural and pre-technology science fields associated with agriculture and home economics, 1920–1979

University	Year Founded	Year First Ph.D. Awarded	Ph.D. Degrees Awarded[b] 1920–79		1965–79	
University of Wisconsin-Madison	1836	1920[a]	3,624	(9.3)	1,622	(6.7)
Cornell University	1865	1920[a]	3,154	(8.1)	1,350	(5.6)
University of Minnesota	1851	1920[a]	2,392	(6.1)	1,052	(4.4)
University of Illinois-Urbana	1867	1920[a]	2,339	(6.0)	1,314	(5.4)
Michigan State University	1855	1927	2,148	(5.5)	1,465	(6.1)
Iowa State University	1885	1920[a]	2,104	(5.4)	983	(4.1)
Purdue University	1865	1930	1,835	(4.7)	1,204	(5.0)
University of California-Berkeley	1868	1920[a]	1,806	(4.6)	812	(3.4)
Ohio State University	1870	1921	1,760	(4.5)	907	(3.8)
University of California-Davis	1908	1949	1,458	(3.7)	1,112	(4.6)
Texas A&M University	1876	1940	1,157	(3.0)	876	(3.6)
Rutgers University	1766	1920[a]	1,042	(2.7)	497	(2.1)
University of Missouri	1839	1920[a]	1,022	(2.6)	660	(2.7)
Oregon State University	1868	1935	1,010	(2.6)	695	(2.9)
North Carolina State University	1887	1947	986	(2.5)	743	(3.1)
Total All LGUs			39,095	(100.0)	24,137	(100.0)

Source: National Research Council, Survey of Earned Doctorates, various years. Also, see Huffman 1986.
[a]The first year for which data are available is 1920.
[b]Degrees awarded in genetics were excluded at the time that tabulations were completed. The numbers in parentheses are the percentage of all degrees awarded by LGUs in fields associated with agriculture and home economics.

agricultural sciences up to 1939, the land-grant colleges accounted for not more than 89 percent (Rossiter 1986, p. 49). Up to 1939, leading non-land-grant institutions were Johns Hopkins University and the University of Michigan. Fifty years later, all 51 of the land-grant universities were awarding Ph.D. degrees in these fields. The long-term compound average growth rate in number of Ph.D. degrees awarded by land-grant universities in these fields was 6.9 percent (1920–1979). Over the 60-year period 1920–1979, these universities awarded more than 39,095 Ph.D. degrees in these fields. Sixty-two percent were awarded during the 15-year period, 1965–1979.

The Institutional Distribution

A large share of the Ph.D. degrees awarded by U.S. land-grant universities in applied agricultural and pre-invention science fields is concentrated in the top 20 percent of the doctorate-granting land-grant universities. For the 60-year period 1920–1979, about 6 in 10 of all Ph.D.'s awarded in the applied agricultural and pre-invention science fields by land-grant universities were awarded by the top 10 Ph.D.-producing universities. For the 15-year period, 1965–1979, their share is only slightly lower at about one-half (see Table 3.11). For the 60-year period 1920–1979 and 1965–1979, the University of Wisconsin is the leading producer of Ph.D. degrees in the applied agricultural and pre-invention science fields. For the 60-year period, Cornell University, University of Minnesota, University of Illinois, and Michigan State University are the second-through-fifth leading producers of doctorates in these fields, respectively. Over 1965–1979, Michigan State University moves up to second, Cornell University drops to third, University of Illinois remains fourth, and Purdue

University rises to fifth.[14] These same universities rank high in total Ph.D.'s awarded in all academic fields. Also, see discussion in Huffman 1984 and Rossiter 1986.

The relative number of Ph.D. degrees awarded in applied agricultural and pre-invention science fields to the total number of Ph.D. degrees awarded in all academic fields varies widely across land-grant universities. For the University of California-Berkeley, Ohio State University, and the University of Illinois, Ph.D. degrees awarded in these fields are a small share (6–13 percent) of total academic Ph.D.'s awarded in the 55-year period 1920–1974. The University of California-Davis is an exception, with one-half of total Ph.D. degrees awarded during this period being in these fields. For Cornell University, University of Minnesota, Michigan State University, and Purdue University, the share is 17–24 percent.

The Geographical Distribution

Because much of agriculture research, especially crops research, is location specific (Ruttan 1982, Chapter 7; Evenson 1989) and Ph.D. research frequently focuses on issues of local importance, it is useful to consider the geographic distribution of Ph.D. degrees awarded. For this purpose we grouped states by the 10 USDA-ERS farm production regions.

Since 1920, the production of Ph.D. degrees in applied agricultural science and pre-invention science fields is concentrated geographically in the Midwestern states. The Corn Belt region is the leading producer for the 60-year period 1920–1979, with over 9,000 Ph.D. degrees awarded; and for the latest 15-year period 1965–1979, with more than 5,000 Ph.D.'s awarded. The Lake states are second, Northeast (due largely to Cornell's Ph.D. output) is third, and the Pacific region is fourth (see Table 3.12). The Appalachian region, Southern Plains, Mountain, Northern Plains, Southeast, and the Delta states are regions that produce relatively few Ph.D. degrees.

Over 1980–2000, the Northeast region has moved up to be the largest producer of doctorates in the applied agricultural and pre-invention science fields (see Table 3.13). The Corn Belt is second, and the Pacific states region is third. The Northern Plains and Delta states regions produce relatively few doctorates in these fields. The Northeast region also ranks first in the number of doctorates awarded in the general (or core) science fields and the Corn Belt states region ranks second. The Lake region is a close third.

Post-Ph.D. Employment

Historically, more than 50 percent of U.S. citizen new Ph.D. recipients have taken immediate post-Ph.D. employment in colleges and universities where they can combine research and teaching.[15] However, during the late 1960s and 1970s, an increasing share of new Ph.D. recipients continued their research training through postdoctoral study. In the sciences, those choosing postdoctoral study increased from 14 percent in 1960–1964 (Harmon 1978, p. 78) to 42 percent in 1978–1982 (Doctorate Recipients from United States Universities Summary Report 1983).

Postdoctoral study historically was restricted to a few outstanding scholars or scientists. As a rule, the objective was to obtain research experience under the guidance of a professor recognized for his (or her) research achievements and ability to communicate knowledge, technique, or approach to scholars and scientists. However, when the academic labor market became depressed during the late 1960s, Ph.D. recipients in some fields were in excess supply for available assistant professor positions. This reduced the opportunity cost of additional training. An increasing number

Table 3.12 Number of Ph.D. degrees awarded in applied agricultural and pre-invention science associated with agriculture and home economics, by USDA farm production regions, 1920–1979, and 1965–1979

Regions	Ph.D. Degrees Awarded 1920–1979	Ph.D. Degrees Awarded 1965–1979	Regions	Ph.D. Degrees Awarded 1920–1979	Ph.D. Degrees Awarded 1965–7979
Corn Belt	9,060	5,068	Southern Plains	1,769	1,358
University of Illinois	2,339	1,314	Texas A&M University	1,157	876
Iowa State University	2,104	983	Oklahoma State University	612	482
Purdue University	1,835	1,204	Mountain	1,679	1,493
Ohio State University	1,760	907	Colorado State University	488	468
University of Missouri	1,022	660	University of Arizona	420	372
Lake States	8,164	4,139	Utah State University	325	248
University of Wisconsin	3,624	1,622	University of Wyoming	163	139
University of Minnesota	2,392	1,052	University of Idaho	135	133
Michigan State University	2,148	1,465	Montana State University	114	99
Northeast	6,930	3,602	New Mexico State University	28	28
Cornell University	3,154	1,350	University of Nevada	6	6
Rutgers University	1,042	497	Northern Plains	1,534	1,184
Pennsylvania State University	971	579	Kansas State University	707	512
University of Maryland	745	404	University of Nebraska	540	410
University of Massachusetts	493	309	North Dakota State University	182	176
University of Connecticut	178	149	South Dakota State University	105	86
University of Vermont	80	74	Southeast	1,449	1,263
University of New Hampshire	72	60	University of Florida	631	505
University of Rhode Island	69	64	University of Georgia	415	391
University of Delaware	68	58	Auburn University	267	238
University of Maine	58	58	Clemson University	136	129
Pacific	4,894	3,034	Delta States	907	716
Univ. of California-Berkeley	1,806	812	Louisiana State University	557	396
Univ. of California-Davis	1,458	1,112	Mississippi State University	274	251
Oregon State University	1,010	695	University of Arkansas	76	69
Washington State University	620	415	Others	307	284
Appalachian	2,402	1,996	University of Alaska	18	18
North Carolina State Univ.	986	743	University of Hawaii	289	266
Virginia Polytechnic Institute	446	403			
University of Tennessee	435	384			
University of Kentucky	315	291	Overall Total	39,095	24,137
West Virginia University	220	175			

Source: National Research Council, Survey of Earned Doctorates, various years. Also, see Huffman 1986.
Note: Degrees in the field of genetics were excluded at the time tabulations were completed.

Table 3.13 Number of Ph.D. degrees awarded in general science and applied agricultural and pre-invention science, by USDA farm production regions, 1980–2000

Regions	Ph.D. Degrees Awarded		Regions	Ph.D. Degrees Awarded	
	General Science[a]	Applied Science[b]		General Science[a]	Applied Science[b]
Corn Belt	8,166	13,000	Southern Plains	1,999	4,632
University of Illinois	2,774	5,164	Texas A&M University	1,413	3,618
Iowa State University	1,136	2,428	Oklahoma State University	586	1,014
Purdue University	1,606	3,241	Mountain	4,220	6,721
Ohio State University	1,760	907	Colorado State University	881	1,847
University of Missouri	890	1,260	University of Arizona	1,874	2,177
Lake States	7,282	11,547	Utah State University	413	753
University of Wisconsin	2,942	4,666	University of Wyoming	389	370
University of Minnesota	2,455	3,958	University of Idaho	124	513
Michigan State University	1,885	2,923	Montana State University	109	234
Northeast	12,117	16,826	New Mexico State University	184	575
Cornell University	2,199	4,180	University of Nevada	246	252
Rutgers University	1,652	2,408	Northern Plains	1,330	2,660
Pennsylvania State University	1,663	2,872	Kansas State University	459	1,147
University of Maryland	2,182	2,292	University of Nebraska	788	1,061
University of Massachusetts	1,387	1,664	North Dakota State University	31	340
University of Connecticut	1,050	1,256	South Dakota State University	52	112
University of Vermont	253	367	Southeast	3,300	5,593
University of New Hampshire	299	295	University of Florida	1,409	2,660
University of Rhode Island	617	454	University of Georgia	1,196	1,271
University of Delaware	650	760	Auburn University	479	822
University of Maine	174	278	Clemson University	216	840
Pacific	6,200	12,194	Delta States	1,333	2,448
Univ. of California-Berkeley	4,121	5,607	Louisiana State University	895	1,117
Univ. of California-Davis	619	3,753	Mississippi State University	171	705
Oregon State University	523	1,686	University of Arkansas	267	626
Washington State University	937	1,148	Others	1,536	828
Appalachian	3,953	8,205	University of Alaska	169	61
North Carolina State Univ.	820	2,818	University of Hawaii	1,367	767
Virginia Polytechnic Institute	920	2,480			
University of Tennessee	1,005	1,320			
University of Kentucky	815	998	Overall Total	52,547	87,602
West Virginia University	393	589			

Source: National Research Council, Survey of Earned Doctorates, various years.
[a] General science includes physics, geosciences, astronomy, mathematics, statistics, psychology, social sciences.
[b] Applied agricultural and pre-technology science includes engineering, computer, biological & agricultural sciences.

of Ph.D. recipients chose the next best available university position, postdoctoral study, as a way to raise their probability of obtaining an assistant professorship in the future. Also, see National Research Council 1988.

For fields associated with agriculture and home economics, data on four different immediate post-Ph.D. activities of U.S. citizens are reported: employment in an educational institution, government, or other (business, nonprofit organization, and other) and postdoctoral study. During 1960–1964, 47 percent of the new Ph.D.'s in the fields associated with agriculture and home economics (52 percent, excluding recipients in biochemistry and genetics) chose employment in educational institutions.

During 1960–1964, government-sector employment was the choice of 14 percent of all Ph.D. recipients, and other employment was the choice of 12 percent. Two Ph.D. fields, forestry and fish and wildlife, had 36 percent of their Ph.D. recipients accepting employment in government. Fields that had a relatively high frequency of other employment were agricultural and food chemistry (39 percent), agriculture (general and other, 25 percent), and biochemistry (23 percent).

In the fields associated with agriculture and home economics, new Ph.D.'s during 1975–1979 had significantly lower employment rates in the educational sector and higher rates of postdoctoral study than new doctorates of 1960–1964. For all fields combined, the proportion of recipients taking employment in educational institutions decreased to 40 percent (47 percent, excluding biochemistry and genetics). The share taking postdoctoral study increased to 32 percent (21 percent, excluding biochemistry and genetics).

For Ph.D. recipients in 1975–1979, postdoctoral study was the immediate post-Ph.D. activity for more than half of the recipients in biochemistry (81 percent), animal physiology (67 percent), genetics (62 percent), and plant physiology (52 percent). Educational institutions continued to employ at least a majority of new Ph.D. recipients in home economics (87 percent); horticulture (67 percent); agricultural engineering (65 percent); forestry (54 percent); animal husbandry, animal science, and nutrition (53 percent); agricultural economics (54 percent); fisheries and wildlife (52 percent); and nutrition or dietetics (50 percent).

In almost all fields, the share of new Ph.D. recipients employed by the government declined from 1960–1965 to 1975–1979. Agricultural economics, which had 29 percent of Ph.D. recipients during 1975–1979 being employed by the government, seems to be an exception. Although data are not available for 1960–1965, the share of new doctorates in agricultural economics during 1975–1979 taking employment in the government sector was 23 percent. Among Ph.D. recipients receiving degrees during 1975–1979, the proportion taking "other" employment from the fields of agronomy (including soils and soil sciences), horticulture, plant physiology, and agricultural engineering more than doubled over the 1960–1965 level; but the proportion from biochemistry and genetics choosing the other category dropped by more than 50 percent.

The data on post-Ph.D. plans of non-U.S. citizens first began to be collected in 1970, so less information is available on their activities. Initial employment in an educational institution is considerably lower, and in the government sector it is considerably higher for non-U.S. citizens than for U.S. citizens. For doctorates during 1975–1979, about 39 percent obtained employment in educational institutions in the United States or abroad, 25 percent chose postdoctoral study, and 22 percent chose government-sector employment. Postdoctoral study, which is one method for non-U.S. citizens to prolong their stay in the United States, does not seem to be at an unusually high rate.

Conclusion

The early training of agricultural scientists could advance only after a new science system was created and in place. To establish this system, methods were borrowed from general sciences, e.g., chemistry, physics, botany, but others were developed to meet the special circumstances associated with agricultural and home economics research in land-grant colleges and universities, e.g., research assistants and assistantships. The creation of this new system occurred largely between 1862 and 1920 (also see Chapter 2). By 1920, a half-dozen state colleges had established good graduate programs in the agricultural sciences.

Conflicts among SAES researchers over the science versus applied nature of their work have a long history. It is a natural—and probably healthy—outcome when SAES researchers come from different general, pre-invention, or applied science backgrounds and work on a wide range of projects, many of which are associated with local clientele problems. These conflicts continue today and are expected to continue.

Colleges and universities have competed with other organizations for individuals trained in science for more than 140 years. Before 1920, most of the competition was with the government, primarily the USDA. Both college and government scientists were largely engaged in similar activities. During the 1930s, private industry started hiring better-trained individuals for research and development of new agricultural technologies (see Chapters 2 and 5). As private property rights were strengthened (see Chapters 1, 5, and 9), the private sector has expanded its own research and development activities and hired larger numbers of Ph.D.-level scientists. This trend seems likely to continue as the potential for applications of biotechnology for profit grows.

U.S. universities, including land-grant universities and the agricultural experiment stations, have been training most of the new agricultural scientists since 1920. Over the 40 years, 1920–1959, U.S. universities produced about 10,000 Ph.D. degrees in the applied agricultural science fields and in pre-invention science fields associated with agriculture and home economics. Slightly less than one-half of these degrees were awarded in the applied agricultural science fields. Over 1960–1980, the total number of doctorates in these fields increased to 35,000 with almost one-half being in the applied agricultural science fields. In the latest period, 1980–2000, the total number of doctorates increased to about 49,000 with more than one-half being in applied agricultural sciences.

Data on gender and citizenship of U.S. doctorates start in 1960, and they show a rising share of women and non-U.S. citizens receiving doctorates in the applied agricultural science fields up to at least 2000. The share of women receiving doctorates in the pre-invention science fields has also been rising, but the share of non-U.S. citizens receiving doctorates in these fields after 1979 was lower than over 1960–1979.

In the 20th century, the land-grant universities matured into a major graduate training system and this will continue in the 21st century. They became an important source of the scientists needed for staffing applied agricultural (and pre-invention) science fields. Although a large share of the total degrees awarded are concentrated in the top 10 universities over 1900–1979, the concentration in the top 10 has decreased since then. Over the long term, the University of Wisconsin-Madison has awarded the largest number of degrees in these fields and Cornell University ranks second.

Hiring new scientists largely from the land-grant agricultural experiment station system has the advantage of obtaining trained individuals who have learned to work on problems that have been important to agriculture. However, the extent of specialization may have progressed too far in the early 1980s, because the advance in the scientific methods of biotechnology were from general science and represented a significant break with the immediate past research traditions of most agricultural experiment stations.

Appendix 3A. A Guide to the Classification of Scientists in State Agricultural Experiment Stations

The primary objective of the classification scheme was to obtain a rough count of the number of persons, by major subject matter areas of work, who were on permanent appointment primarily as researchers in the state agricultural experiment stations. Staff listed in the *Handbook* but not counted as researchers were the director(s) of the station and nonresearch personnel (foremen, inspectors, librarians, secretaries, treasurers, soil surveyors, editors, illustrators, extension personnel, and staff engaged in feed and fertilizer [regulatory or control]). The superintendents of SAES substations and substation farms are counted as researchers, and their field of research is designated as "other." The *Handbooks* do not indicate the share of time a staff member allocates to research versus teaching and/or extension. Staff who have an administrative title and a field of research designated are arbitrarily assumed to be 50 percent in administration and 50 percent in station research. They are coded as 0.5 of a station researcher. All other persons are coded as one station staff researcher. Researchers listed as being in two of our major fields of research are split 50:50 between them and counted as 0.5 of a researcher in each field.

The researchers are classified in the plants and soils research area and livestock research area based upon their type of research specialty or title. The following guideline was followed.

Plants and Soils

Production areas: Agronomy (general plants and soils), soils (general), horticulture, forestry, silviculture, viticulture, olericulture, pomology, agricultural engineering—irrigation, drainage and mechanization, soil fertility (general), seed analyst, range ecology, taxonomy.

Protection-maintenance: Entomology, plant pathology, mycology, biological control.

Pre-technology and general science: Plant breeding, plant physiology, botany (general), plant genetics, plant nutrition, soil chemistry, fertilizer chemistry, geology, soil physics, genetics, biology, plant chemistry.

Livestock and Poultry

Production areas: Animal husbandry (science or industry), dairy husbandry (science or industry), poultry husbandry (science or industry), agricultural engineering(farm buildings).

Protection-maintenance: Veterinary medicine (general); animal pathology, parasitology, histology; (dairy) bacteriology.

Pre-technology and general science: Animal physiology, animal nutrition, animal breeding, feed chemistry, animal genetics, dairy chemistry, zoology (except entomology), genetics, and biology.

A few arbitrary decisions were made in classifying researchers into other fields. Food technology is assigned 0.5 to home economics-nutrition and 0.5 to chemistry. Physics is assigned to chemistry except for soil physics.

The highest degree attained by researchers was translated into one of three types: Ph.D., M.S., B.S., or their equivalent. We have coded M.D.'s and D.V.M.'s at the Ph.D. level; and Vet. Sci. is coded at M.S. level. In chemistry, C.A. and A.C. are bachelor's-level degrees. A Ph.C. is coded as Ph.D. in chemistry. For engineering, M.E. and C.E. are coded as bachelor's degrees. D.Eng. is coded as Ph.D. degree. In law and divinity, LL.D. and D.D. are coded at M.S. level. The degree D.Sc. is coded as Ph.D. degree. In forestry, B.F. and F.E. are coded as bachelor's degrees.

Notes

1. In 1856 the Scientific School had six professors.
2. During his career, Johnson published seven books and 172 articles on agriculture and agricultural chemistry (Browne 1926, p. 179). He was the third president of the American Chemical Society and was named to the National Academy of Sciences in 1866 (Chittenden 1928, Vol. I, p. 202).
3. He also introduced the effective technique of sterilization by dry heat.
4. Another indication of the small size of the land-grant college system at this time is the small number of faculty and students. In 1889–1890, the 36 functioning land-grant colleges had 611 instructors who were in collegiate departments. There were 158 graduate students and 6,191 undergraduates enrolled in all departments (U.S. Education Bureau 1893, p. 1027).
5. The number is for individuals who are on research appointments in the agricultural experiment station. Administrators or directors of the experiment stations or nonprofessional staff were not counted as researchers. See details in Appendix 3A.
6. A significant amount of early work in chemistry was public service-oriented. Chemical analysis of soils was frequently conducted as a service to farmers. Other services were of a regulatory or inspection nature. Chemists checked fertilizer and feed materials for accuracy of labeling, and food was checked for adulteration.
7. This classification scheme is admittedly rough, but we believe that it contains useful information about the focus of agricultural research and how it changed over the long run.
8. The field of science groups are defined in Table 3.6.
9. The science field of employment and of the doctorate are not always the same for a Ph.D. This issue is discussed later.
10. Among Ph.D.'s employed as agricultural economists, the share employed in the business-industry sector is slightly (but probably not significantly) larger than in 1981 and 1983. See Huffman and Orazem 1985 for an econometric model of the market for new Ph.D.'s in agricultural economics.
11. A total of 14.4 percent of World War II veterans obtained some college-level training that was paid for by the G.I. Bill; and 706,821 graduate degrees have been awarded to G.I. Bill beneficiaries (1945–1982).
12. Exactly where Ph.D. degrees awarded in agricultural economics are recorded before 1969 is unclear. Some were undoubtedly reported in economics. Others may have been counted in the field called "other agriculture." Also, Ph.D. recipients at Iowa State University and North Carolina State University can obtain their degree in either economics or agricultural economics.
13. The 1890 land-grant colleges and universities are not included in this study. They did not award Ph.D. degrees.
14. Note that the University of California-Davis awarded its first Ph.D. degree in 1949. This is much later than for any of the other major Ph.D.-granting universities.
15. Data collection on employment plans of Ph.D. recipients started in 1960.

References

Adkins, Douglas L. 1975. *The Great American Degree Machine: An Economic Analysis of the Human Resource Output of Higher Education.* Berkeley, CA: Carnegie Commission on Higher Education.

Beardsley, Edward H. 1969. *Harry L. Russell and Agricultural Science in Wisconsin.* Madison, WI: The University of Wisconsin Press.

Berelson, Bernard. 1960. *Graduate Education in the United States.* New York, NY: McGraw-Hill Book Co.

Browne, Charles A. (ed.). 1926. "A Half-Century of Chemistry in America, 1876–1926." *Journal of American Chemical Society,* Vol. 48,(8a), Part 2.

Busch, Lawrence and W.B. Lacy. 1983. *Science, Agriculture, and the Politics of Research.* Boulder, CO: Westview Press.

Chittendon, Russell H. 1928. *History of the Sheffield Scientific School of Yale University, 1846–1922.* Vols. I and II. New Haven, CT: Yale University Press.

Doctorate Recipients from United States Universities Summary Report. 1983. "Postgraduation Study and Employment Plans of Doctorate Recipients." Office of Scientific and Engineering Personnel National Research Council. Washington, D.C.: National Academy Press.

Evenson, Robert E. 1989. "Spillover Benefits of Agricultural Research: Evidence from U.S. Experience." *American Journal of Agricultural Economics* 71:447–452.

Harmon, L.R. 1978. *A Century of Doctorates.* Washington, D.C.: National Academy of Sciences.

Huffman, Wallace E. 1985. "The Institutional Development of the Public Agricultural Experiment Station System: Scientists and Departments." Unpublished paper, Department of Economics, Iowa State University.

Huffman, Wallace E. 1986. "The Supply of New Agricultural Scientists by U.S. Land-Grant Universities." In *Agricultural Scientific Enterprise: A System in Transition,* ed. L. Busch and W.B. Lacy. Boulder, CO: Westview Press.

Huffman, Wallace E. and R. E. Evenson. 1987. "The Development of U.S. Agricultural Research and Education: An Economic Perspective, Part II." Ames, IA: Iowa State University, Department of Economics Staff Paper No. 169.

Huffman, Wallace E. and R.E. Evenson. 1994. *The Development of U.S. Agricultural Research and Education: An Economic Perspective.*' Ames, IA: Department of Economics, Iowa State University. (Available through the Parks Library at Iowa State University.)

Huffman, Wallace E. and R.E. Evenson. 1993. *Science for Agriculture: A Long-Term Perspective.* Ames, IA: Iowa State University Press. 1st Ed.

Huffman, Wallace E. and Peter Orazem. 1985. "An Econometric Model of the Market for New Ph.D.'s in Agricultural Economics in the United States." *American Journal of Agricultural Economics* 67: 1207–1214.

Hughes, R.M. 1934. "Report of the Committee on Graduate Instruction." *Education Record* 15: 192–234.

Kalton, R.R. and P. Richardson. 1983. "Private Sector Plant Breeding Programs: A Major Thrust in U.S. Agriculture." *Diversity* 5: 16–18.

National Research Council. 1981. "Special Tabulations from Surveys of Earned Doctorates, 1920–1979." Washington, D.C.: National Academy Press.

National Research Council. 1985. *Summary Report: 1984 Doctorate Recipients from United States Universities.* Washington, D.C.: National Academy Press.

National Research Council. 1988. *Educating the Next Generation of Agricultural Scientists.* Report of NRC Committee on Evaluation of Trends in Agricultural Research at the Doctoral and Postdoctoral Level. Washington, D.C.: National Academy Press.

National Science Foundation. 1983. *Science and Engineering Doctorate Awards: 1960–82.* Washington, D.C.: NSF83-328.

National Science Foundation. 1993. *Science and Engineering Doctorate Awards: 1960–1991.*Washington, D.C.: NSF93-301.

National Science Foundation. 2001. *Science and Engineering Doctorate Awards: 1991–2000.*Washington, D.C.: NSF02-305.

Nevins, Allan. 1962. *The State Universities and Democracy.* Urbana, IL: University of Illinois Press.

Price, Derek. 1978. "Ups and Downs in the Pulse of Science and Technology." *Sociological Inquiry* 48: 162–171.

Rossiter, Margaret W. 1975. *The Emergence of Agricultural Sciences: Justus Liebig and the Americans, 1840–1880.* New Haven, CT: Yale University Press.

Rossiter, Margaret W. 1986. "Graduate Work in the Agricultural Sciences, 1900–1970." *Agricultural History* 60: 37–57.

Ruttan, Vernon W. 1982. *Agricultural Research Policy.* Minneapolis, MN: The University of Minnesota Press.

Schultz, Theodore W. 1941. *Training and Recruitment of Personnel in the Rural Social Studies.* Washington, D.C.: American Council on Education.

Scott, Roy V. 1970. *The Reluctant Farmer—The Rise of Agricultural Extension to 1914.* Urbana, IL: University of Illinois Press.

Tickton, Sidney G. and Beardsley Ruml. 1955. *Teaching Salaries Then and Now: A 50 Year Comparison with Other Occupations and Industries.* (Report for the Fund for the Advancement of Education.) New York, NY: The Seventh Company, Inc.

True, Alfred C. 1937. *A History of Agricultural Experimentation and Research in the United States, 1607–1925: Including a History of the United States Department of Agriculture.* Washington, D.C.: U.S. Department of Agriculture, Misc. Publ. No. 251.

U.S. Department of Agriculture. 1890. *Organizational Lists of the Agricultural Experiment Stations, and Agricultural Schools and Colleges in the United States, 1890.* U.S. Department of Agriculture, Office of Experiment Stations, Bull. 5.

U.S. Department of Agriculture. 1901. *Organizational Lists of the Agricultural Colleges and Experiment Stations in the United States, with a List of Agricultural Experiment Stations in Foreign Countries, 1900.* U.S. Department of Agriculture, Office of Experiment Stations, Bull. 74.

U.S. Department of Agriculture. 1911. *Organizational Lists of the Agricultural Colleges and Experiment Stations in the United States, 1910.* U.S. Department of Agriculture, Office of Experiment Stations, Bull. 233.

U.S. Department of Agriculture. 1921. *Lists of Workers in Subjects Pertaining to Agriculture and Home Economics, 1920–1921.* Part 2. Washington, D.C.: U.S. Government Printing Office.

U.S. Department of Agriculture. 1930. *Workers in Subjects Pertaining to Agriculture in Land Grant Colleges and Experiment Stations.* Washington, D.C.: U.S. Department of Agriculture, Misc. Publ. 67.

U.S. Department of Agriculture. 1940. *Workers in Subjects Pertaining to Agriculture in Land Grant Colleges and Experiment Stations.* Washington, DC: U.S. Department of Agriculture, Misc. Publ. 378.

U.S. Department of Agriculture. 1950. *Professional Workers in State Agricultural Experiment Stations and Other Cooperating State Institutions.* Washington, D.C.: U.S. Department of Agriculture, Agricultural Handbook No. 3.

U.S. Department of Agriculture. 1960. *Professional Workers in State Agricultural Experiment Stations and Other Cooperating State Institutions.* Washington, D.C.: U.S. Department of Agriculture, Agricultural Handbook No. 116.

U.S. Department of Agriculture. 1969–1980. *Dictionary of Professional Workers in State Agricultural Experiment Stations and Other Cooperative State Institutions.* Washington, D.C.: U.S. Department of Agriculture, Agricultural Handbook No. 305.

U.S. Education Bureau. 1893. *Report of the Commissioner of Education, 1889–1890.* Washington, D.C.: U.S. Government Printing Office.

Veysey, Lawrence R. 1965. *The Emergence of the American University.* Chicago, IL: The University of Chicago Press.

Waters, H.J. 1910. "The Function of Land Grant Colleges in Promoting Collegiate and Graduate Instruction in Agriculture." (Proceedings of 23rd Annual Conference of the Association of American Agricultural Colleges and Experiment Stations.) U.S. Department of Agriculture, Office of Experiment Stations, Bull. 228.

The World Book Encyclopedia. 1982. Chicago, IL: World Book Inc.

4

Resources and Their Allocation

Inventions or advances in scientific knowledge require inputs of labor or time of skilled individuals and other resources. These resources are costly because they have alternative public and private uses. However, exactly how much has been spent during some past periods on inventions and advances in knowledge is difficult to determine.

Although we showed in Table 1.2 that patenting of agricultural technologies was significant in the United States during the first decade of the 19th century and averaged more than 150 per annum during the early 1830s, the U.S. data on private sector R&D expenditures do not start until more than a century later, in 1953 (Griliches 1990). We, however, will use the average real cost of patents in 1956, the patenting rate by decade for earlier periods, and other information to derive estimates of U.S. private agricultural research expenditures for the period 1890–1955.

Although agricultural research at the federal level started in the Patent Office during the 1840s and was transferred to the newly established USDA in 1862 (see Chapters 1 and 2), we will focus on resources for the USDA's own research activities starting in 1888. Public agricultural research in the states started with the establishment of the Connecticut Agricultural Experiment Station in 1875 and grew as agricultural experiment stations were established in other states. We, however, start our data series in 1888 when the Hatch Act provided a large increase in available funds for state agricultural experiment stations. Likewise, resources were allocated to agricultural extension before the Smith-Lever Act of 1914, but our data series on Cooperative Extension expenditures starts in 1915.

This chapter first focuses on the amount and growth rates of public and private agricultural research starting in 1888 and public agricultural extension expenditures starting in 1915 and continuing to 2000. Most of the examination focuses on major subperiods: before 1915, 1915–1945, 1945–1980, and 1980–2000. These subperiods represent spans of time that have been important in the development and growth of agricultural science and education in the United States. The second section presents compositional characteristics of agricultural research and extension expenditures for selected dates, and the third section shows some of the variation in public research and extension expenditures by state for recent years.

Long-Term Trends

The average long-term rate of growth of U.S. total public agricultural research expenditures (in constant prices) over the 55 years 1900–1955 was a 5.5 percent compound rate per annum, and over 1955–2000 it grew at 6.2 percent per annum (see Figure 4.1 and 4.2 or Table 4.1). However, over 1980–2000, the annual average rate of growth of public agricultural research expenditures was only 0.2 percent per annum. Over these same time intervals, U.S. total private agricultural research grew at 4.5 percent, 4.0 percent, and 2.5 percent per annum.

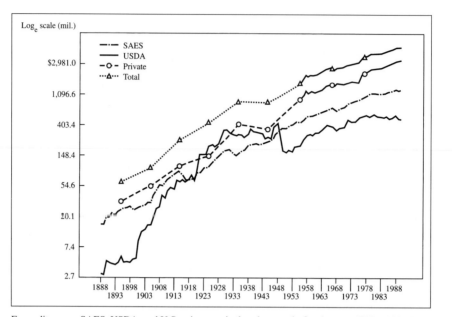

Fig. 4.1. Expenditures on SAES, USDA, and U.S. private agricultural research, fiscal years, 1888–1990 (thousands of constant 1984 dollars). (From Table 4.1.)

We estimate that a significant amount of real resources were being invested in private agricultural research in the 1880s when the Hatch Act was passed. Up to about 1906, U.S. private agricultural research expenditures exceeded public sector expenditures (USDA and SAES system combined). The public expenditures grew rapidly with the public research institution building that occurred through 1915, and total public agricultural research expenditures moved ahead of total private expenditures about 1906 and continued to be larger until 1950. In the late 1940s, the USDA's expenditures on its own research in constant prices dropped dramatically, and at the same time private

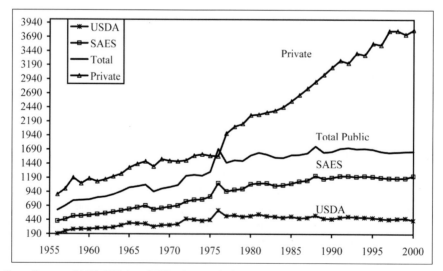

Fig. 4.2. Expenditures on SAES, USDA, and U.S. private agricultural research, fiscal years, 1956–2000 (thousands of constant 1984 dollars). (From Table 4.1.)

Table 4.1 Total U.S. public (USDA and SAES) and private agricultural research funds by performing organization, 1888–2000 (millions of 1984 dollars)

Year	Price Index for Agr. Research (1984 = 1.0)	Public Agricultural Research			Private Agricultural Research[a]
		USDA	SAES	Total	
1888	0.0472	3.093	15.254	18.347	
1889	0.0472	3.030	15.254	18.284	
1890	0.0472	4.767	19.513	24.280	
1891	0.0469	4.435	19.446	23.881	
1892	0.0458	4.214	22.620	26.834	
1893	0.0471	4.119	20.658	24.777	
1894	0.0444	4.392	23.086	27.477	
1895	0.0454	5.374	24.361	29.736	32.400
1896	0.0452	4.469	26.394	30.863	
1897	0.0452	4.558	26.372	30.929	
1898	0.0468	4.423	27.201	31.624	
1899	0.0493	5.051	24.665	29.716	
1900	0.0521	5.067	24.434	29.501	
1901	0.0522	8.927	26.284	35.211	
1902	0.0547	11.865	27.148	39.013	
1903	0.0571	12.907	28.161	41.068	
1904	0.0557	14.794	30.844	45.637	
1905	0.0567	14.797	30.459	45.256	53.100
1906	0.0583	18.971	41.252	60.223	
1907	0.0605	25.256	45.934	71.190	
1908	0.0596	26.510	55.084	81.594	
1909	0.0617	39.287	54.587	93.874	
1910	0.0632	35.570	61.487	97.057	
1911	0.0612	45.033	65.523	110.556	
1912	0.0640	50.016	70.188	120.203	
1913	0.0664	47.666	76.596	124.262	
1914	0.0655	63.771	86.107	149.878	
1915	0.0668	60.419	86.482	146.901	101.600
1916	0.0753	65.511	75.219	140.730	
1917	0.0933	59.893	64.652	124.544	
1918	0.1020	64.490	65.147	129.637	
1919	0.1091	74.601	65.060	139.661	
1920	0.1225	63.184	66.947	130.131	
1921	0.1016	90.039	81.093	171.132	
1922	0.1062	149.369	81.742	231.111	
1923	0.1087	148.289	92.355	240.644	
1924	0.1089	151.598	96.814	248.411	
1925	0.1130	195.858	96.664	292.522	140.000
1926	0.1133	206.346	110.477	316.823	
1927	0.1126	190.107	119.547	309.654	
1928	0.1153	199.922	133.591	333.513	
1929	0.1158	250.924	144.940	395.864	
1930	0.1116	326.093	164.095	490.188	
1931	0.1064	328.769	173.571	502.340	
1932	0.1018	304.715	173.320	478.035	

(Continues)

Table 4.1 *(Continued)*

Year	Price Index for Agr. Research (1984 = 1.0)	Public Agricultural Research			Private Agricultural Research[a]
		USDA	SAES	Total	
1933	0.0993	289.940	159.980	449.919	
1934	0.1003	276.670	143.918	420.588	
1935	0.0992	285.121	153.972	439.093	405.400
1936	0.1012	281.611	164.526	446.136	
1937	0.1067	257.460	167.994	425.455	
1938	0.1028	277.986	195.039	473.025	
1939	0.1029	340.214	202.634	542.847	
1940	0.1035	318.406	207.362	525.768	
1941	0.1077	307.697	210.594	518.292	
1942	0.1134	296.984	202.019	499.004	
1943	0.1176	293.980	207.993	501.973	
1944	0.1252	250.072	217.236	467.308	
1945	0.1247	260.634	227.466	488.099	346.300
1946	0.1371	251.014	243.027	494.041	
1947	0.1584	358.794	261.521	620.316	
1948	0.1731	415.881	295.904	711.785	
1949	0.1711	282.548	331.473	614.021	
1950	0.1821	164.618	357.062	521.680	
1951	0.1985	157.149	352.932	510.081	
1952	0.2038	168.391	375.020	543.410	
1953	0.2106	160.095	385.408	545.503	
1954	0.2183	186.702	409.372	596.074	
1955	0.2263	188.785	435.024	623.809	
1956	0.2395	194.418	419.415	613.833	890.600
1957	0.2517	239.050	449.670	688.721	994.100
1958	0.2490	271.940	508.446	780.386	1189.900
1959	0.2640	279.091	511.504	790.595	1086.700
1960	0.2717	274.056	523.905	797.961	1175.300
1961	0.2788	294.756	540.219	834.975	1120.800
1962	0.2896	295.328	555.435	850.763	1159.400
1963	0.2970	309.165	581.313	890.478	1210.500
1964	0.3092	343.797	605.049	948.845	1258.600
1965	0.3236	384.778	631.100	1015.878	1367.800
1966	0.3416	376.259	661.212	1037.471	1431.800
1967	0.3570	372.756	691.476	1064.232	1476.200
1968	0.4049	319.331	624.194	943.524	1386.200
1969	0.3983	345.205	649.608	994.813	1517.100
1970	0.4183	350.210	673.653	1023.863	1486.000
1971	0.4328	365.016	692.740	1057.756	1479.500
1972	0.4519	459.533	765.751	1225.284	1494.400
1973	0.4837	445.979	795.127	1241.106	1579.700
1974	0.5286	424.410	801.901	1226.311	1602.700
1975	0.5654	440.021	852.821	1292.842	1577.800
1976	0.5921	607.782	1091.464	1699.247	1569.600
1977	0.6253	507.660	948.698	1456.359	1973.500
1978	0.6656	524.340	974.890	1499.231	2092.700
1979	0.7240	495.119	991.779	1486.898	2146.000

1980	0.7484	510.750	1075.402	1586.152	2300.100
1981	0.8183	541.317	1091.845	1633.163	2311.300
1982	0.8716	508.608	1092.585	1601.193	2348.800
1983	0.9518	500.718	1046.763	1547.481	2380.800
1984	1.0000	482.492	1059.343	1541.835	2444.700
1985	1.0531	502.702	1088.075	1590.877	2550.100
1986	1.0944	471.241	1125.848	1597.089	2660.100
1987	1.1383	482.894	1141.861	1624.754	2774.800
1988	1.1210	521.967	1225.893	1747.860	2894.500
1989	1.2729	468.344	1170.289	1638.633	3019.300
1990	1.3379	458.988	1193.254	1652.242	3149.500
1991	1.376	486.608	1220.447	1707.055	3270.066
1992	1.411	502.429	1219.532	1721.960	3223.450
1993	1.444	490.339	1209.504	1699.843	3401.144
1994	1.497	487.055	1219.696	1706.752	3369.189
1995	1.541	480.642	1211.531	1692.174	3575.113
1996	1.583	462.337	1192.869	1655.206	3548.246
1997	1.636	452.895	1183.894	1636.789	3797.736
1998	1.701	462.898	1181.240	1644.138	3801.257
1999	1.767	470.616	1182.104	1652.719	3734.496
2000	1.825	435.572	1221.765	1657.337	3814.650

Source: Tables 4.2, 4.3, and Appendix Table A4.1.
[a]Estimates of private agricultural research expenditures were derived as a decade average for the period 1890–1950, and for 1985 and later, the numbers for private agricultural research are also an estimate.

sector research expenditures grew more rapidly than before. Consequently, private agricultural research expenditures pushed ahead of total public expenditures about 1950 and maintained this lead through 2000. In 2000, U.S. private agricultural research expenditures exceeded U.S. public agricultural research expenditures by a large margin, approximately 2.3 times as much (Table 4.1).

Real public expenditures on Cooperative Extension grew very rapidly during 1915–1921 with the institution building that was occurring primarily in the states, and real expenditures on extension increased at 18.7 percent per annum compound. During 1921–1980, the long-term trend rate of growth of real expenditures on extension was a much lower 2.9 percent per annum compound (see Figure 4.3). Starting in the 1980s very little growth in real expenditures on Cooperative Extension occurred in the United States, and since the mid-1990s our judgment is that there has been a decline in real resources. However, with the dwindling federal support for Cooperative Extension, the USDA has largely lost control of the documentation of the aggregate funding of Cooperative Extension in the United States, which makes an assessment difficult.

Agricultural Research Expenditures

Our assessment is that for the 100 years 1900–2000 U.S. real private agricultural research and development expenditures have grown on average more rapidly than U.S. public agricultural research expenditures, and both have grown at significant positive average rates. For some subperiods, however, public agricultural research expenditures have grown more rapidly than private expenditures. Before going any further, however, we need to describe the process used to derive our estimates of U.S. private agricultural research and development expenditures before 1955, when official data on private R&D expenditures were not collected.

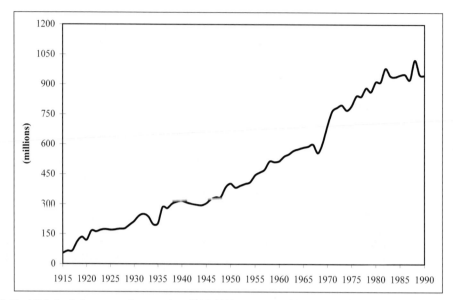

Fig. 4.3. Total U.S. funds for cooperative extension, 1915–1990 (constant 1984 dollars). (From Table 4.1 and Appendix Table A4.3.)

Estimating Private R&D Expenditures

No data series on private R&D expenditures exists before 1953, and one major task undertaken for this book was to devise a method for constructing estimates based upon data that are available for a much longer time period. The primary resource for this process is the annual data on patents granted in agricultural-technology patent classes by the U.S. Patent Office (also see Chapter 1). Data are available, starting in 1839, on the total number of patents and assigned patents to businesses or industrial firms.[1]

The first step was to estimate the average cost of a private sector agricultural patent for the decade of the 1950s. Total private agricultural research expenditures in 1956 were estimated by adjusting the total applied research expenditures on food and kindred products, textile mill products, agricultural chemicals, drugs and medicine, and farm machinery obtained from *Science Indicators* (National Science Foundation). The adjustments were based on information about private agricultural research given for detailed subject areas in 1961 (U.S. Department of Agriculture 1962; Huffman and Evenson 1994, Appendix Table 3), and converted to 1984 dollars using our research price index (see Table 4.1). The adjusted real private agricultural R&D expenditure for 1956 (in 1984 prices) was divided by the average annual number of assigned patents in agricultural technology fields for 1950–1959. This number is then an estimate of the average real private agricultural R&D expenditure per assigned patent for the decade of the 1950s.

One could use this average cost in 1956 (at the 1984 price level) and annual data on assigned patents in agricultural technology classes to directly estimate the earlier real private R&D expenditures. Griliches (1990), however, has shown that the efficiency of the private sector R&D process and of the U.S. Patent Office has changed over time in a way that has raised the real cost of patents granted.[2] Thus, our latest estimates, which are reported in Table 4.1, are estimates of U.S. private R&D adjusted for these two trends. The net result is to reduce the estimate of real private R&D expenditures for the pre-1955 period from what they would be if we ignored these trends.[3]

Data on Public Agricultural Research

The expenditures by the USDA and SAES on agricultural research are directly available in, or can be derived from, public documents. For the USDA research agencies, the data on their federal appropriations have been reported in *U.S. Statutes at Large* for 1888–1922, *Budget of the U.S. Government* for 1922–1968, and the *Inventory of Agricultural Research* for 1968–2000.

The Office of Experiment Stations was established in the USDA in 1888 and initiated an accounting and supervision system as a fiscal check on the individual SAES. This scheme did not include any information about the type of research activities undertaken, except capital versus current operating expenditures. This system continued up to the early 1970s.

Basic data on expenditures, by state, of the SAES system are reported in "Report of the Director of the Office of Experiment Stations" for 1888–1893; "Statistics of Agricultural Colleges and Experiment Stations," in OES *Circular* Nos. 27, 29, and 35 for 1894–1896; "Statistics of Land-Grant Colleges and Agricultural Experiment Stations in the United States," in OES *Bulletin* Nos. 51, 64, 78, 97, and 114 for 1897–1901; *Annual Report, Office of the Experiment Stations* for 1902–1941; *Report on the Agricultural Experiment Stations* for 1943–1953; *Report on State Agricultural Experiment Stations* for 1954–1960; and *Funds for Research at State Agricultural Experiment Stations* for 1961–1968.

In 1948, the Office of Experiment Stations implemented a supplemental research classification scheme for the SAES that contained some detail about research expenditures by commodity and field of science. These data, which exist for 1948–1966, were used only for internal evaluation purposes and were not published. The Current Research Information System (CRIS), which came into existence in 1967 as a computerized record system, was the outgrowth of a 1966–1967 national evaluation and long-range planning effort (U.S. Department of Agriculture and the Association of State University and Land-Grant Colleges 1966). The initial proposal for CRIS had a three-way classification system: (1) activities, (2) commodities or resources, and (3) fields of science. The modified system that was adopted and is in current use has these three categories plus a fourth: research problem areas (RPAs). The research problem areas were built around the first nine goals of the National Programs (U.S. Department of Agriculture and the Association of State University and Land-Grant Colleges 1966). These data are published in the *Inventory of Agricultural Research*. The CRIS research classification system underwent a major revision, which was implemented in 1998 (see U.S. Department of Agriculture 1993 and 2003 for comparison).

Trends during 1888–1915

In 1888, we estimate that expenditures on U.S. private agricultural R&D expenditures were significantly larger than total public (USDA plus SAES) agricultural research expenditures, but after about 1906 their relative size reverses (Figure 4.1 or Table 4.1). In 1895, U.S. private R&D expenditures were $32 billion (1984 price level). This is 1.3 percent of their level in 1984, and it is about 9 percent larger than total public agricultural research expenditures in that year. During this subperiod, which was a time of rapid institution building in the public sector, real public agricultural research expenditures were growing more rapidly than private expenditures—7.7 percent versus approximately 5.7 percent per annum compound.

When the Hatch Act was passed in 1887, there was a large jump in funding for SAES research, but the Hatch Act did not provide any future increases in funding (not even to compensate for inflation), and many of the state governments were poor, so growth in real SAES expenditures during this period was somewhat constrained for a system trying to develop rapidly (also see Chapter 2). Furthermore, during most of this subperiod, James Wilson, as Secretary of Agriculture, was developing a large and high-quality agricultural research capacity (see Chapter 2). Thus, it is not too surprising that during this subperiod agricultural expenditures grew more rapidly in the USDA than in the SAES system, 7.7 percent versus approximately 5.7 percent per annum compound.

Trends during 1915–1945

This subperiod is one of multidimensional transitions. First, there were the post–World War I adjustments which led into the agricultural recession of the 1920s, followed by the Great Depression of 1929–1933. The resources and other adjustments, including the high unemployment rates, extended to the start of World War II (see Huffman, 1996). Between 1941 and 1945, U.S. resources in all sectors were redirected to winning the war.

In agricultural sciences, however, the transition is from institution building to one of establishing the framework of modern agricultural sciences (also see Chapter 3). Furthermore, the first grand technology success of U.S. public agricultural research occurred. This was the development of the first commercially successful (double cross) hybrid seed corn variety at the Connecticut Agricultural Experiment Station and sale. During the 1930s and 1940s, hybrid seed corn replaced open-pollinated seed corn in the U.S. Corn Belt and fringes (see Griliches 1960; and Chapter 6). Hybrid seed corn, however, was not a one-time technological change but the successful application of a new process of technology development. In the Corn Belt and other regions, new hybrid varieties have continually been developed since about the 1930s. Some have been successful but for a relatively short period of time—generally 5–10 years, and then they have been replaced by another new and relatively superior variety.

For all of this subperiod, total public agricultural research expenditures are larger than private R&D expenditures, and from about 1924 to 1932 the USDA's research expenditures exceeded U.S. total private R&D expenditures (Figure 4.1 or Table 4.1). Also, after 1920, the USDA's expenditures on agricultural research are larger than for the SAES system. This short period from about 1925–1932 seems to represent the peak of relative importance of the USDA's own research in the whole U.S. agricultural research system (also see Chapter 2). Marcus (1987, pp. 24–25) points out important changes in the organization of agricultural research in the USDA research agencies that helped it assume a major role in U.S. agricultural research and development during this period.

During this subperiod, all growth rates for real agricultural research expenditures are lower than for the preceding subperiod. The growth rate for private R&D and public agricultural research expenditures are approximately equal at 4.0 percent per annum. The research expenditures of the USDA continue to grow more rapidly during this subperiod than for the SAES system, 4.9 percent versus 3.2 percent per annum. At the end of this subperiod, the USDA's research expenditures accounted for about 53 percent of total public agricultural research expenditures.

Funding during 1945–1980

The subperiod 1945–1980 might be described as one when the U.S. agricultural research system first became a mature, successful, and relatively well-integrated science system. First, the relative importance of the USDA's (national government's) own research activity declined dramatically. In 1956 the USDA's research expenditures accounted for only 12.9 percent of total U.S. public and private agricultural research and development expenditures, which were down from 31.2 percent in 1945. Second, the state agricultural experiment stations (the state system) attained the majority role in U.S. public agricultural research, accounting for 68.3 percent of total public agricultural research expenditures in 1956 compared to only 46.6 percent in 1945. Third and perhaps most important, private R&D attained the majority position in modern U.S. agricultural research and development activity. We estimate private R&D's share of total U.S. agricultural research and development expenditures to be 66.8 percent in 1956, compared to only 41.5 percent in 1945.

During 1945–1980, real private agricultural R&D expenditures grew at a rate of about 3.4 percent per annum, which is lower than for the preceding subperiod. The rate of growth of real public agricultural research expenditures is less than for the private rate by an average of about 4 percentage points per year. The rate of growth of the USDA's real research expenditures for this period drops to 2.7 percent per annum, which is little more than one-half the rate for the preceding subperiod and about one-sixth the average rate for the first subperiod. In contrast, the rate of growth of SAES expenditures on average is 6.2 percent per annum, which is higher than the rate for the preceding subperiod. Furthermore, after the realignment of relative importance of private R&D, USDA research, and SAES research expenditures about 1950, the new relative positions were maintained through 2000 (see Figure 4.2).

Thus, we conclude that the dramatic privatization of U.S. agricultural research and development started during and accelerated after 1975. Furthermore, many developing countries do not seem to ever attain the stage of modern agricultural research and technology development the United States attained during 1945–1974 where private sector R&D expenditures became large relative to public research and development efforts.

Trends during 1980–2000

During the early 1970s, public concerns were registered about methods for setting the research agenda and lack of breadth in new technology development. These issues were precipitated by the abrupt rise in fossil fuel energy prices, a greater awareness of the potential for resource and environmental degradation from production technologies, a falling population density in rural areas, and other changes in our society. New research issues surfaced and gradually shifted research priorities away from agricultural productivity toward rural development, energy and natural resource conservation, environmental pollution and quality, and food safety (e.g., see Committee on Research Advisory to the U.S. Department of Agriculture 1972; Hightower 1973; Rockefeller Foundation 1982; Office of Technology Assessment 1981; Heimlich 2003, Ch. 5.2, p. 3). We will argue later (in Chapters 8 and 9) that these issues have affected the availability of funds for agricultural research. Also, during the 1980s and 1990s, the technology potential offered by advances in biotechnology and personal computers and information technologies started to affect the whole U.S. agricultural research and development process.

During this subperiod, real private agricultural research expenditures grew more than twice as fast as real public agricultural research expenditures, 2.5 versus 0.9 percent per annum (Figure 4.2). This meant that the private sector had obtained a larger share of U.S. total agricultural research activity in 2000 than it had coming into this subperiod. The private sector's share of the U.S. total was approximately 70 percent in 2000, which was 11 percentage points higher than in 1980. (Also, see Heimlich 2003, Ch. 5.2; and Klotz, Fuglie, and Pray 1995 for private agricultural R&D data over 1990–2000.)

The growth rates for USDA and SAES real research expenditures during this subperiod were small compared to the earlier subperiods. Very little growth of real USDA agricultural research expenditures occurred during this subperiod (see Figure 4.2). The average growth rate was only 0.2 percent per annum, and the average growth rate during 1981–1990 was negative (1.8 percent per annum). In contrast, the average rate of growth of real SAES expenditures was 0.6 percent. In 2000, the SAES system accounted for 74 percent of total U.S. public agricultural research expenditures, and this large share shows the dominant role the SAES system has attained in the U.S. public agricultural research system during the last half of the twentieth century.

Federal Funding of SAES Research

After 1887, the state agricultural experiment stations received funds from four important source types: (1) the USDA's Office of Experiment Stations (OES), which later became the Cooperative State Research Service (CSRS), (2) other federal government funds, primarily contracts, grants, and cooperative agreements with research agencies in the USDA and contracts and grants from other federal agencies, e.g., the National Science Foundation, the National Institute of Health, the Department of Defense,[4] (3) state government appropriations, and (4) other nonfederal government sources, e.g., foundations, private corporations, farm and trade associations, commodity groups, and fees for services performed and sales of research products or by-products.

In historical documents of the OES, the funds in the first category are labeled "federal funds," and the remaining three categories are aggregated together and labeled "nonfederal funds." Thus, the amount of funds in these historical documents that are labeled "federal funds" is generally less than total funds received from the USDA or from the federal government. Furthermore, SAES "nonfederal funds" generally contain a significant amount of funds that are from agencies of the federal government, including the USDA research agencies.

Furthermore, we are certain that not all funds received by SAES researchers from non-OES- or non-CSRS-administered sources are generally counted in the official category of "nonfederal funds." The reasons are that accounting methods and philosophies differ across agricultural experiment stations and across time periods. However, when the matching requirement for OES- or CSRS-administered funds came into existence in 1935, all non-OES-administered funds could be used to match "federal funds." Thus, during time periods when a state agricultural experiment station is barely meeting its matching requirement, the comprehensiveness of the accounting for all research funds of SAES researchers by OES or CSRS records seems likely to be more complete than at other times. During the post–World War II period, almost all state agricultural experiment stations have had non-OES or CSRS administered funds that exceeded by a large margin the matching requirement (also see Chapter 8).

During the early years of the SAES system but after 1887, OES-administered funds provided a large share of SAES support, 82.6 percent in 1888. Although new legislation was occasionally passed that increased federal support, non-OES funds or primarily state appropriations grew more rapidly. By 1913, non-OES funds accounted for 67.1 percent of the total SAES funds (see Figure 4.4).

During these early years, the Hatch Act provided all of the OES-administered funds (see Table 4.2). However, over the first 67 years of SAES history, 1887–1954, the Hatch Act provided generally a *very small share* of total OES-administered funds for SAES; for example in 1955, its share was only 4 percent (Table 4.2). This was due to a proliferation over time of separate pieces of federal legislation that provided funding for the SAES system. This legislation was occasionally consolidated, e.g., Amended Hatch Act of 1955. However, OES-administered funds never provided more than 20 percent of total SAES funds after 1915.

Fig. 4.4. Share of total SAES system funds from OES- or CSRS-administered sources, 1888–1990. (From Tables 4.2 and 4.3.)

The Hatch Act treated all states (and territories) equally; they were each entitled to $15,000 per year. Thirty-nine states and territories obtained funding in 1888, 48 in 1894. In each of 1930, 1931, and 1935, a new U.S. territory was included in the set of eligible states, giving a total of 51 states and territories in 1935. The Hatch Act legislation fixed the payment to states in nominal terms, so the purchasing power eroded as inflation in the cost of resources for research increased (see Table 4.1 for our research price index). Federal funds for activities of the Office of Experiment Stations did start in 1894 and grew (see Table 4.2).[5] Much of this expansion occurred as A. C. True expanded the role of his office.

New legislation was sought to increase federal funding of agricultural research. In 1903, A. C. True and Congressman H. C. Adams (Wis.) cooperated to propose new legislation for SAES funding. True proposed an (a maximum) additional $15,000 per year per state of SAES support following the original Hatch Act guidelines, except the funds were to be spent solely on original research.

The bill, known as the Adams Act, was passed by Congress and signed by President Roosevelt in 1906. It doubled the OES-administered funds available to each state, although the increase was phased in over 6 years. However, the OES-administered share of total SAES support dropped to a new all-time low of 16 percent in 1925 (Figure 4.4).

Federal research funding for the SAES system was subsequently raised by the Purnell Act of 1925. It enlarged the scope of federally supported station research to include efficiency of the agricultural industry and social economic factors to improve the rural home and rural life, i.e., economics, sociology, and home economics research. The funding increases under the Purnell Act were phased in over 5 years and provided a maximum payment to each state of $30,000 per year.

In 1935, the Bankhead-Jones Act provided additional federal funding of SAES. Funds were for the first time allocated on a formula basis—each state's share of the U.S. rural population—rather than as an equal payment to each state (U.S. Department Agriculture 1986, p. 111). States were also required for the first time to match Bankhead-Jones funds with nonfederal or non-OES-administered

Table 4.2 Total funds available in current dollars of U.S. state agricultural experiment stations, including the Office of Experiment Stations, by major sources of funding, fiscal 1888–1955 ($1,000)

| Year | Hatch Act | Administered by Office of Experiment Stations (OES)[a] | | Not Adm. by OES[d] | Total |
		Other[b] Programs	Office of Experiment Stations[c]		
1888	$585	—	$10	$125	$720
1889	—	—	10	125	720
1890	585	—	15	321	921
1891	660	—	15	227	912
1892	705	—	23	308	1,036
1893	705	—	23	245	973
1894	720	6	25	274	1,025
1895	720	16	25	345	1,106[d]
1896	720	30	30	413	1,193
1897	720	27	30	410	1,192
1898	720	33	30	490	1,273
1899	720	43	30	423	1,216
1900	720	70	33	450	1,273
1901	720	108	33	511	1,372
1902	720	126	33	606	1,485
1903	720	154	37	707	1,608
1904	720	170	40	788	1,718
1905	720	172	40	795	1,727
1906	720	577	51	1,057	2,405
1907	720	726	55	1,278	2,779
1908	720	854	61	1,648	3,283
1909	720	778	65	1,805	3,368
1910	720	899	74	2,193	3,886
1911	720	988	80	2,222	4,010
1912	720	1,050	94	2,628	4,492
1913	720	1,055	97	3,214	5,086
1914	720	1,074	107	3,739	5,640
1915	720	1,092	119	3,846	5,777
1916	720	887	163	3,894	5,664
1917	720	908	202	4,202	6,032
1918	720	931	219	4,775	6,645
1919	720	961	265	5,152	7,098
1920	720	1,002	288	6,191	8,201
1921	720	990	309	6,220	8,239
1922	720	996	280	6,685	8,681
1923	720	980	286	8,053	10,039
1924	720	925	99	8,799	10,543
1925	720	930	132	9,141	10,923
1926	720	1,901	105	9,791	12,517
1927	720	2,394	126	10,221	13,461
1928	720	2,878	125	11,680	15,403
1929	720	3,366	130	12,568	16,784
1930	735	3,847	155	13,576	18,313
1931	735	3,854	163	13,716	18,468

1932	750	3,837	169	12,888	17,644
1933	750	3,758	161	11,217	15,886
1934	765	3,694	149	9,827	14,435
1935	765	3,688	137	10,684	15,274
1936	765	4,299	156	11,430	16,650
1937	765	4,924	162	12,074	17,925
1938	765	5,537	162	13,586	20,050
1939	765	5,843	162	14,081	20,851
1940	765	6,167	162	14,368	21,462
1941	765	6,181	162	15,573	22,681
1942	765	6,244	162	15,738	22,909
1943	765	6,252	166	17,277	24,460
1944	765	6,336	156	19,941	27,198
1945	765	6,343	176	21,081	28,365
1946	765	6,541	154	25,859	33,319
1947	765	6,611	173	33,876	41,425
1948	765	9,156	211	41,089	51,221
1949	765	10,059	197	45,694	56,715
1950	765	12,313	232	51,711	65,021
1951	765	11,956	236	57,100	70,057
1952	765	11,664	—	64,000	76,429
1953	765	11,676	—	68,726	81,167
1954	765	12,689	—	75,912	89,366
1955	765	18,189	—	79,492	98,446

Source: Roland Robinson, U.S. Department of Agriculture, Cooperative States Research Service, from special archive work on CSRS records 1887–1955 for funds administered by Office of Experiment Stations. For funds not administered by Office of Experiment Stations, 1888–1905, Circulars and Bulletins, U.S. Government Printing Office; 1905–1939, Report, U.S. Experiment Station Office, Office of Experiment Stations; 1940–1955, Report on the Agricultural Experiment Stations, U.S. Department of Agriculture, Agricultural Research Service.

[a]The Office of Experiment Stations administered all SAES federal funds, supervised the agricultural experiment station located in the U.S. territories (1896–1940), and published research results (1888–1906).

[b]Includes expenditures on agricultural-experiment stations in the U.S. territories and expenditures on research investigations by the Office of Experiment Stations. Data exclude expenditures on extension work that was administered by OES during 1916–1923.

[c]There are administration costs of OES.

[d]These data are frequently labeled as being "nonfederal", but this means that they were not administered by OES.

federal funds.[6] Before 1935, $60,000 per year was the maximum federal payment that any one state could receive; and for the 51 states and territories this amounted to $3.06 million. Under Bankhead-Jones, a maximum of $3 million of additional funds were authorized for the SAES system, phased in over 5 years.

Substantial additional changes were introduced in new SAES funding provided by the Research and Marketing Act of 1946. It emphasized marketing and utilization research and made provisions for support of regional research by the states. For nonregional research, part was split equally among the states and territories, and part was allocated based upon a new formula (see Table 1.6).

Federal appropriations to the SAES were consolidated in 1955 under the Amended Hatch Act. The total funds that could be allocated was left open-ended. Proportions were, however, established for the share of the appropriation that was to be divided equally among states, allocated by formula, and allocated to marketing research and to regional research (see Table 1.6). Amended Hatch Act funds accounted for 98 percent of the OES-administered funds for SAES in 1956 (see Table 4.3).

Table 4.3 Total funds in current dollars for U.S. state agricultural experiment stations by major funding sources, fiscal 1956–2000 (thousands of dollars)

Year	CSRS-Administered Funds					Funds Not Administered by CSRS			Total Funds (9)
	Regular (1)	Regional (2)	Special Research Grants (3)	Competitive Research Grants (4)	Other USDA Adm. Funds (5)	USDA Contracts, Grants, Coop. Agreements (6)	Other Federal Funds (7)	Nonfederal Funds (8)	
1956	19,528	4,176	—	—	513	—	76,233	—	100,450
1957	22,970	5,381	—	—	463	—	84,368	—	113,182
1958	23,582	5,560	—	—	503	—	96,958	—	126,603
1959	24,446	5,900	—	—	515	—	104,176	—	135,037
1960	24,446	5,900	—	—	428	—	111,571	—	142,345
1961	25,166	6,149	—	—	487	—	118,811	—	150,613
1962	27,326	6,899	—	—	475	—	126,154	—	160,854
1963	28,809	7,410	—	—	449	—	135,982	—	172,650
1964	30,429	7,977	—	—	724	—	147,950	—	187,081
1965	34,569	9,414	—	—	1,942	—	158,299	—	204,224
1966	36,729	10,164	—	—	3,690	—	175,286	—	225,870
1967[b]	39,639	10,164	—	—	5,376	—	191,678	—	246,857
1968[a]	38,580	10,132	1,109	—	2,467	—	200,448	—	252,736
1969	39,249	10,282	1,316	—	2,842	7,749	27,576	169,725	258,739
1970	40,961	10,926	1,325	—	2,988	6,825	26,440	222,514	281,789
1971	45,273	12,381	1,333	—	3,548	6,320	25,852	205,112	299,818
1972	50,150	13,598	2,170	—	3,805	6,850	27,299	242,171	346,043
1973	51,983	14,527	2,837	—	3,879	7,476	28,730	265,168	384,603
1974	52,838	14,754	3,253	—	4,637	8,614	31,170	308,619	423,885
1975	57,586	16,428	2,665	—	5,866	10,761	34,178	354,701	482,185
1976[c]	79,417	22,662	4,837	—	8,715	12,342	47,112	471,171	646,256
1977	72,436	21,642	4,582	—	7,284	11,739	51,433	425,105	593,221
1978	80,650	24,401	5,831	—	8,225	14,772	53,028	461,980	648,887
1979	80,643	24,355	7,610	—	11,987	19,141	60,904	513,409	718,048
1980	87,147	26,636	9,627	—	13,445	24,361	67,420	576,195	804,831
1981	94569	29,150	12,160	—	13,721	29,808	77,740	636,309	893,457

Year									
1982	134,635	—	12,887	5,529	13,777	30,979	77,763	676,727	952,297
1983	141,355	—	12,858	6,040	13,955	33,149	73,350	715,601	996,309
1984	144,819	—	15,758	6,093	14,280	33,327	81,719	763,346	1,059,343
1985	148,231	—	19,152	7,729	14,595	31,038	90,318	834,894	1,145,957
1986	141,122	—	19,885	11,872	13,357	30,826	110,798	904,268	1,232,128
1987	140,885	—	21,810	16,824	12,889	33,018	114,907	959,447	1,299,780
1988	147,384	—	23,189	19,272	16,406	37,329	114,956	1,015,690	1,374,226
1989	147,594	—	29,122	21,923	17,300	43,834	130,395	1,099,493	1,489,661
1990	146,893	—	39,654	20,039	17,004	49,462	143,860	1,179,543	1,596,455
1991	155,391	—	44,662	19,759	18,971	53,498	154,108	1,232,946	1,679,355
1992	160,626	—	52,034	25,820	19,757	53,849	172,187	1,236,484	1,720,759
1993	160,969	—	50,538	33,220	20,627	58,883	189,695	1,232,592	1,746,524
1994	162,975	—	53,883	41,826	21,970	68,346	210,250	1,266,635	1,825,885
1995	162,368	—	53,237	44,143	22,049	68,916	213,791	1,302,466	1,866,970
1996	159,316	—	49,090	44,831	20,615	69,810	216,069	1,328,579	1,888,311
1997	158,309	—	46,963	44,193	18,202	75,550	215,095	1,378,537	1,936,850
2000[d]	200,900	—	47,000	44,700	—	75,000	285,400	1,270,700	2,229,700

Source: Roland Robinson, U.S. Department of Agriculture, Cooperative State Research Service, revised reports of expenditures of agricultural experiment stations, OD-1044, 1956–68. U.S. Department of Agriculture, Inventory of Agricultural Research 1969–97. CRIS summary reports 2000.

[a]Expenditures for research by Schools of Forestry, Schools of Veterinary Medicine, and the 1890 Land-Grant Colleges are excluded from the members reported in this table. See Appendix Tables A4.2 and A4.3 for expenditures on research by Schools of Forestry and Schools of Veterinary Medicine, respectively.

[b]Total expenditures reported by state agricultural experiment stations on form OD-1044 do not exactly match the amounts reported for similar items in the Inventory of Agricultural Research. This is especially true during the transition period (1966–68) to the Current Research Information System (CRIS). The transition year in our table is 1968. The amount for CRS-administered funds is from form OD-1044. The amount reported in column (9), total funds, is an interpolation between totals for 1967 and 1969. Total funds not administered by CSRS are computed residually.

[c]Funding is unusually large for 1976 because there are five quarters in the year due to a change in the date for beginning new fiscal years.

[d]Data for 2000 are rounded numbers.

With the amendment, the Adams Act, Purnell Act, and Title I of the 1946 Agricultural Marketing Act were repealed. In 1977, the requirement that 20 percent of Hatch funds be allocated to marketing research was dropped. The new formula for allocating USDA's federal formula funds to the states was set as follows: 20 percent of each year's appropriation was to be allocated equally among states; 26 percent allocated according to a state's share of the U.S. rural population as established in the most recent Decile Census of Population; 26 percent according to a state's share of the U.S. farm population as determined by the most recent Decile Census of Population; and 25 percent for cooperative regional (later called multistate) research. The remaining 3 percent is to go for federal administration of the program. This formula remains in effect in 2004.

In 1965, the Research Facilities Act and the Special Research Grants Act provided additional funding for agricultural research. The Facilities Act funding was to be used to improve facilities—new construction, remodeling, and new equipment. The Special Research Grants Act authorized grants to the state stations, other public institutions, and individuals to perform research on problems of concern to the USDA. Title V of the 1972 Rural Development Act authorized funding of research on rural development and small farms, but relatively little money has been appropriated.

Title XIV of the 1977 Farm Bill continued most previous Amended Hatch programs. The requirement that at least 20 percent of the funds be spent on marketing research was dropped. In the 1981 Farm Bill, the requirements for allocation of Hatch Regional Research funds were modified, and these funds were basically combined with regular Hatch funds. Also, a Competitive Research Grants program was added. This grant program represented a fairly dramatic change from earlier legislation dealing with SAES research in that researchers in all colleges, universities, federal agencies, and private institutions were eligible to participate. Recipient institutions were not required to match these funds with other funds.

Although the relative importance of OES-administered funds for the SAES system increased from 1925 to 1941 and 1954 to 1957, the long-term trend in this share has been negative (see Figure 4.4). Furthermore, the OES-, CSRS-, or CSREES-administered share of total SAES funds has fallen steadily from 25.4 percent in 1957 to 13 percent in 2000. The National Research Initiative (NRI) Competitive Grant Program was initiated in 1990, and it accounted for 2 percent of SAES funds in 2000. Thus, in 2000, the CSREES-administered share of SAES funds is small, and as a share of total public and private agricultural research expenditures, it is very small, about 3 percent.

Agricultural Extension

The passage of the Smith-Lever Act in 1914 and subsequent development of the Cooperative Extension Service relieved the research agencies of the USDA and the SAES system from the obligation of distributing information on applied problems to local clientele groups. The trends in the growth of funding for Cooperative Extension can be seen in Figure 4.5. The expenditures are expressed in real terms before graphing. In converting these expenditures to real terms we chose to use our research price index (Table 4.1), which expresses expenditures in 1984 prices.[7]

Starting in 1914, real total U.S. expenditures on public extension increased rapidly up to 1921, an average of 18.7 percent per annum. This was due in part to a 24.4 percent per annum growth rate in real federal funds. From 1921 to 1980 the growth rate was surprisingly steady at an average rate of about 2.9 percent per annum, and then the growth rate turned negative.

Cooperative Extension obtains its funds from (1) the federal government, (2) state government appropriations, (3) county government appropriations, and (4) non-tax or other sources (Appendix

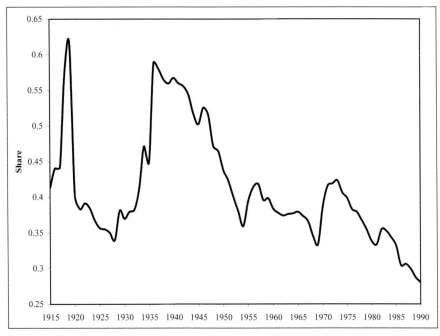

Fig. 4.5. Share of total cooperative extension funds from federal sources, 1915–1990. (From Appendix Table A4.3.)

Table A4.3). In 1915, the federal share was 41.3 percent, and the state and county government share was 50.7 percent. Since 1915, the long-term trend of the federal share has been negative (Figure 4.5), and the trend in the state and county government share has been positive. The federal share did shoot up dramatically during 1928–1936, but except for short-term reversals during 1954–1957 and 1968–1973, the trend in the federal share has been strongly negative. Since 1973, the federal share has fallen from 42.4 percent to 28.1 percent in 1990, and the state and county government share accounts for 67.7 percent. Thus, in 1990 the financial resources for Cooperative Extension are primarily from state and county governments and not from the federal government.

Composition of Agricultural Research and Extension

This section presents comparisons of the composition of research and extension. First, USDA, SAES, and U.S. private agricultural research scientist-man-year (SMY) allocations are presented for 1961 and compared. These data were taken from an excellent U.S. Department of Agriculture study that was never distributed because of political pressures. The data are unique in the sense that no other estimate exists on private agricultural research activity by subject matter or commodity. Other comparisons are presented for 1920 and 1935, 1951 and 1964, and 1969, 1984, and 1997. Data also are presented on the distribution of public extension activities for selected dates.

The Commodity Focus

A large share of agricultural research is applied and has a particular agricultural commodity as its focus. Because agricultural commodities have unique biological, physical, and/or behavioral science characteristics and the production of plants and animals to a lesser extent are highly geoclimate-specific (see the discussion in Chapter 7), it seems useful to disaggregate agricultural research expenditures by commodity classification.

Only one snapshot exists of the whole U.S. agricultural research system—USDA, SAES, and private business or industry. It was taken in 1961. For 1969 and 1984, information is presented for the public agricultural research system as summarized in USDA-CRIS records. The classification scheme for the 1961 data contains 38 different subject-matter areas. The classification scheme for 1969 and 1984 (and 1997) is a modified version of the CRIS commodity classification scheme (see U.S. Department of Agriculture 1993). The CRIS commodities cover total research expenditures. We have tried to make the two classification schemes comparable. (See Huffman and Evenson 1994, Appendix Table 14, for the translation of CRIS commodities into our research commodities.)[8]

In comparing the magnitudes of public and private agricultural research expenditures, we do not claim that public and private sectors are doing the same types of research. The research of the private sector is primarily oriented to the development of marketable products, which are frequently protected by patent rights (see Chapters 2 and 5). The research of the public sector spans basic advances in science, solutions to applied problems, and some technology development. Very little of public-sector research outcome is patentable, and a large share of it is published in professional journals.

Sector Allocations: 1961

In 1961, we conclude there were approximately 12,056 scientist-man-years of research in private-sector agricultural research and 9,014 in the public sector. In these data, we define four mutually exclusive and comprehensive categories of agricultural research: crop-production-related, livestock-production-related, post-harvest, and economics and statistics (see Table 4.4).

Using this categorization scheme, a majority (51.1 percent) of the SMYs in SAES research was allocated to crop-production-related research in 1961, and the majority (52.5 percent) of private sector SMYs was allocated to post-harvest research. For the USDA, none of the four categories was dominant; 41 percent was allocated to crop-production-related research and 33 percent to post-harvest research. Across the institutional types, the SAES system allocated the largest share of SMYs (23 percent) to livestock-production-related research in 1961 (see Table 4.4).

Looking at the individual subject-matter categories, the following eight differences across institutional types are apparent:

1. Although the private sector, SAES, and USDA allocated a similar number of SMYs to research on cereals, the private sector allocated a much smaller share of its science resources to cereals than did the public sector.
2. In 1961, all of the soybean and other oilseed research was in the public sector, split about equally between the USDA and SAES systems.
3. The private sector allocated 2.3 times as many SMYs to plant nutrition research as did the public sector. Most of the public sector research was in the SAES system.
4. Virtually all of the soil and water research was in the public sector.
5. SAES SMYs for beef and dairy cattle research were large relative to those of the USDA and private sector. The private sector led in SMYs allocated to swine and poultry research.
6. The private sector allocated more SMYs to research on livestock insects and disease control than did the public sector, but only 2.6 percent more. Thus, this area was not dominated by the private sector.

Table 4.4 Absolute and relative agricultural research investment by USDA, SAES, and business-industry by subject matter: United States, 1961

Subject Matter	Research Sector					
	USDA SMYs (%)		SAES SMYs (%)		Industry[a] SMYs (%)	
Crop Breeding-Management						
Corn-Sorghum	41.3	(1.0)	107.2	(2.3)	107	(0.9)
Wheat-Rye	58.1	(1.4)	57.8	(1.2)	20	(0.2)
Other Cereals	37.7	(0.9)	64.4	(1.3)	17	(0.1)
Cotton & Other Fibers	86.1	(2.1)	66.9	(1.4)	21	(1.7)
Forage, Pasture and Range	126.9	(3.1)	146.3	(3.0)	46	(0.4)
Tobacco	29.5	(0.7)	42.7	(0.9)	44	(0.4)
Sugar	51.8	(1.3)	31.6	(0.6)	96	(0.3)
Soybeans	17.1	(0.4)	20.2	(0.4)	1	(—)
Other Oilseeds	34.6	(0.8)	16.6	(0.4)	3	(—)
Potatoes	13.6	(0.3)	47.9	(1.0)	55	(0.5)
Vegetables	35.1	(0.9)	184.2	(3.7)	163	(1.4)
Citrus Fruits	17.0	(0.4)	69.1	(1.4)	35	(0.3)
Other Fruits	35.4	(0.9)	184.5	(3.7)	31	(0.3)
Tree Nuts	11.0	(0.3)	7.2	(0.1)	4	(—)
Ornamentals	18.0	(0.4)	76.8	(1.6)	117	(1.0)
Crop-Related						
Insect Control	257.1	(6.3)	287.4	(5.8)	357	(3.0)
Disease Control	24.0	(0.6)	57.6	(1.2)	58	(0.5)
Weed Control	22.7	(0.6)	102.0	(2.1)	370	(3.1)
Plant Nutrition	88.0	(2.2)	175.0	(3.5)	600	(5.0)
Soils-Water	546.0	(13.4)	416.0	(8.4)	70	(0.6)
Plant Physiology-Botany	41.7	(1.0)	212.5	(4.3)	91	(0.8)
Livestock Breeding-Mgt.						
Beef Cattle	30.8	(0.7)	183.1	(3.7)	84	(0.7)
Dairy Cattle	54.3	(1.3)	173.8	(3.5)	86	(0.7)
Swine	16.6	(4.1)	75.8	(1.5)	102	(0.9)
Sheep and Other	15.5	(0.4)	54.4	(1.1)	32	(0.3)
Poultry	40.9	(1.0)	182.8	(3.7)	215	(0.2)
Livestock-Related						
Insect Control	101.5	(2.5)	66.1	(1.3)	157	(1.3)
Disease Control	217.0	(5.3)	211.7	(4.3)	455	(3.8)
Other	24.2	(0.6)	110.6	(2.2)	4	(—)
Agricultural Engineering						
Machine	75.6	(0.2)	109.2	(2.2)	592	(4.9)
Buildings-Structures	41.5	(1.0)	56.7	(1.1)	157	(1.3)
Land	27.3	(0.7)	32.0	(0.6)	51	(0.4)
Post-Harvest Utilization of Agr. Commodities	914.0	(22.4)	298.0	(6.0)	3900	(32.3)
Human Nutrition-Clothing-Nutri.	61.0	(1.5)	138.0	(2.8)	200	(1.7)
Household Econ.	119.0	(2.9)	173.0	(3.5)	625	(5.2)
Transportation and Mkt. Quality	266.0	(6.5)	125.0	(2.5)	1600	(13.3)
Economics & Statistics						
Farm, Market, Stat. Economic Analysis	469.0	(11.5)	566.0	(11.5)	840	(7.0)
Statistical Reporting	17.0	(0.4)	1.0	(—)	680	(5.6)

(*Continues*)

Table 4.4   *(Continued)*

Subject Matter	USDA SMYs (%)		SAES SMYs (%)		Industry[a] SMYs (%)	
Summary						
Crop Prod.-Related	1689	(41.4)	2521	(51.1)	2949	(24.5)
Livestock Prod.-Rel.	542	(13.3)	1115	(22.6)	1262	(10.5)
Post-Harvest	1360	(33.4)	734	(14.9)	6325	(52.5)
Economics and Statistics	486	(11.9)	567	(11.5)	1520	(12.6)
Grand Total	4077	(100.0)	4937	(100.0)	12056	(100.0)

Source: Computations prepared by Robert E. Evenson using information in U.S. Department of Agriculture, 1962.
[a]In the private sector an SMY is equivalent to $25,000 in current dollars or $89,670 in 1984 prices.

7. SMYs allocated to machinery research were heavily concentrated in the private sector.
8. The private sector allocated 3,900 SMYs to research on utilization of agricultural commodities and 1,600 SMYs to research on transportation and market quality. The SMYs of the public sector in these subjects look small by comparison, although the USDA allocated 22 percent of its SMYs to utilization research in 1961.

Public Sector Allocations: 1969, 1984, and 1997

The data for total public (all USDA and state institutions) agricultural research expenditures, by research commodity, for 1969, 1984, and 1997, are presented in Table 4.5. All expenditures for 1969 are expressed in constant 1984 prices to facilitate comparisons across the 3 years. Insects and general science were not included in the CRIS classification scheme in 1969, so these commodities had zero expenditures for that year. In 1969, expenditures on insects and general science are included primarily in the "other commodity" category.

In 1969, $817 million (1984 prices) was spent for research on 16 crop and livestock commodities; 64.2 percent of this total was allocated to research on crops. In 1984, the total amount spent on crop and livestock research was 21.2 percent larger, or $1.01 billion. In 1997, the total was 53.7 percent. Although the total expenditures on crop research commodities increased between 1969 and 1997, the crop-livestock share allocated to crop commodities decreased to 61 percent.

Fruit and vegetable research accounted for 30 percent of the research expenditures on crop and livestock commodities in 1969, but dropped to about 21 percent in 1984 and 1997. Beef cattle and dairy were the second and third most important research commodities in both years. Research on beef cattle accounted for 9.5 percent of the 1969 total and 12.8 percent in 1984 (but only 10.4 percent in 1997). Dairy accounted for 9.0 percent in 1969, 9.2 percent in 1984, and 8.6 percent in 1997.

In 1969, $457 million was spent on the other 13 research commodities. This amount increased to $788 million in 1984 (and $1,477 million in 1997). Forestry research—the largest item—accounted for 25 percent of this latter total in 1969, but for only 20 percent in 1984 and 1997.

Disaggregating Public Institutions

Agricultural research in the public sector has been primarily an activity of the USDA's Agricultural Research Service and Economic Research Service after 1953 and the SAES and vet-med schools in the state sector, but other public institutions do engage in a small amount of agricultural research.[9]

Table 4.5 Absolute and relative support for agricultural research, by commodity for federal and state institutions, 1969, 1984 and 1997 (thousands of 1984 dollars)

Research commodities	1969 Total	1969 Percent	1984 Total	1984 Percent	1997 Total	1997 Percent
Crops:						
Corn	32,483	2.6	47,667	2.7	101,391	3.2
Wheat	27,399	2.2	40,872	2.3	82,452	2.6
Other cereal	25,616	2.0	40,238	2.2	68,454	2.1
Soybeans	19,193	1.5	53,433	3.0	85,078	2.7
Other oil seed	17,316	1.4	18,365	1.0	35,257	1.1
Cotton	57,546	4.5	47,100	2.6	76,031	2.4
Forage	54,822	4.3	83,280	4.7	117,561	3.7
Tobacco-sugar	33,545	2.6	30,955	1.7	32,432	1.0
Potato	13,153	1.0	17,757	1.0	37,906	1.2
Fruit-vegetable	244,770	19.2	210,375	11.8	376,946	11.8
New & misc. crops	10,188	0.8	10,340	0.6	30,835	1.0
Subtotal, (crops)	(536,031)	(42.1)	(600,383)	(33.6)	1,044,343	(32.8)
Livestock:						
Beef cattle	77,389	6.1	129,371	7.2	178,463	5.6
Swine	34,545	2.7	62,801	3.5	104,493	3.3
Sheep and wool	40,300	3.2	59,363	3.3	136,142	4.3
Dairy	74,182	5.8	92,478	5.2	147,473	4.6
Poultry	54,651	4.3	56,826	3.2	101,324	3.2
Subtotal, Livestock	(281,067)	(22.1)	400,839	(22.4)	667,894	(20.9)
Subtotal (Crops & Live.)	($817,098)	(64.2)	(1,001,222)	(56.0)	(1,712,237)	(53.7)
Other:						
Soils	46,503	3.7	82,108	4.6	154,604	4.8
Water	68,913	5.4	81,037	4.5	172,699	5.4
Insects[a]	—	—	33,105	1.9	47,891	1.5
Structure and inputs	14,333	1.1	9,688	0.5	8,306	0.3
Farm management	6,709	0.5	8,854	0.5	12,643	0.4
Forestry	112,404	8.8	160,851	9.0	297,390	9.3
Recreation-wildlife	21,220	1.7	54,476	3.0	146,531	4.6
Economics	21,857	1.7	34,589	1.9	31,517	1.0
Family life	40,782	3.2	80,686	4.5	175,030	5.5
Food and textiles	20,084	1.6	45,818	2.6	84,342	2.6
General science[a]	—	—	180,121	10.1	307,086	9.6
Research	—	—	16,073	0.9	35,597	1.1
Unclassified:	104,070	8.2	138	0.0	2,960	0.1
Subtotal, Other	(456,879)	(35.9)	(787,545)	(44.0)	(1,476,596)	(46.3)
TOTAL	$1,273,977	100.0	$1,788,767	100.0	$3,188,833	100.0

Source: U.S. Department of Agriculture, Current Research Information System.
[a]Not a designed CRIS research commodity in 1969.

Our data on the institutional location of public agricultural research were derived from CRIS records (see Table 4.6) and so do not cover in-house research by federal agencies other than the USDA.

In 1969, the USDA conducted about 41 percent of total public agricultural research (measured by expenditures), and the Agricultural Research Service and Economic Research Service accounted for 79 percent of the USDA total (see Table 4.6). In 1984, the USDA's share of U.S. public agricultural research had decreased to 34.8 percent, although ARS and ERS continued to comprise 80 percent

Table 4.6 Distribution of agricultural research expenditures by performing institutions and commodity, 1969, 1984 and 1997 (percentage of total expenditures on a commodity)

Commodities	1969[a] ARS-ERS[b]	Other USDA[c]	SAES & Vet Med	Other State	1984[a] ARS-ERS[b]	Other USDA[c]	SAES & Vet Med	Other State	1997[a] ARS-ERS	Other USDA	SAES & Vet Med	Other State
Crops:												
Corn	46.0	0.1	53.9	0.1	29.4	0.0	70.2	0.4	38.4	0.0	58.9	2.7
Wheat	57.5	0.0	42.5	0.0	36.0	0.1	63.4	0.5	35.4	0.0	64.1	0.5
Other Cereals	42.0	0.0	58.0	0.0	26.4	0.1	73.0	0.5	24.9	0.0	73.5	1.6
Soybeans	43.4	0.0	51.5	5.1	31.2	0.0	65.1	3.7	29.8	0.0	67.0	3.3
Other Oil Seeds	66.2	0.0	33.7	0.1	36.7	0.0	60.7	2.5	36.5	0.0	60.4	3.0
Cotton	67.9	0.1	32.0	0.1	60.8	0.0	38.7	0.5	51.8	0.0	48.1	0.1
Forage	27.6	7.3	64.9	0.3	29.0	4.1	65.5	1.4	31.7	3.7	62.4	2.2
Tobacco-Sugar	58.0	0.0	42.0	0.0	46.2	0.0	52.7	1.1	33.0	0.0	66.2	0.8
Potatoes	41.7	0.0	58.3	0.0	27.6	0.0	72.0	0.4	36.3	0.0	63.3	0.3
Fruit-Vegetables	21.0	0.0	79.0	0.0	24.7	0.0	73.7	1.5	25.7	0.0	72.7	1.6
New & Misc. Crops	40.2	0.0	59.8	0.0	41.6	0.0	58.4	0.0	28.7	0.0	68.4	2.9
Livestock:												
Beef Cattle	34.6	0.1	65.2	0.1	24.8	0.0	74.3	0.9	22.0	0.0	77.1	1.0
Swine	37.7	0.0	62.1	0.1	31.2	0.0	67.2	1.6	26.5	0.0	71.0	2.5
Sheep & Wool	42.4	0.0	57.6	0.0	23.2	0.0	73.3	3.5	12.9	0.0	82.8	4.2
Dairy	33.9	0.0	66.1	0.0	25.1	0.0	74.3	0.5	19.5	0.0	79.9	0.6
Poultry	30.7	0.0	69.3	0.0	24.8	0.0	73.4	1.9	25.1	0.0	71.7	3.3
Other:												
Soils	37.7	1.1	60.6	0.5	33.1	2.4	62.0	2.5	25.0	7.2	63.2	4.6
Water	50.3	17.9	30.5	1.2	38.6	12.6	46.1	2.7	27.5	9.7	54.6	8.1
Insects	—	—	—	—	50.4	0.0	48.9	0.7	9.4	3.3	82.4	4.9
Structures & Inputs	22.6	33.9	43.6	0.0	18.9	10.6	67.8	2.8	9.0	2.5	82.2	6.3

Farm Management	46.5	0.0	53.5	0.0	11.6	0.0	84.3	4.1	34.1	0.0	63.5	2.5
Forestry	2.2	71.0	21.0	5.9	2.3	54.3	30.3	13.1	2.0	42.2	29.0	26.8
Recreation-Wildlife	1.0	23.9	70.9	4.2	2.1	21.2	69.9	6.8	7.1	13.1	68.8	10.9
Economics	60.1	8.4	31.4	0.1	61.9	6.1	31.0	1.0	16.6	3.3	77.2	2.9
Family Life	29.4	0.2	69.6	0.8	44.5	3.9	45.7	5.9	40.3	0.9	52.7	6.2
Food & Textiles	27.1	0.0	72.8	0.0	27.6	11.5	57.5	3.3	39.5	0.0	57.8	2.6
General Science	—	—	—	—	25.3	0.0	72.0	2.7	11.0	0.5	84.6	3.9
Research	—	—	—	—	15.5	1.3	81.8	1.5	8.7	3.7	78.3	9.2
Other	26.8	0.3	71.4	1.5	100.0	0.0	0.0	0.0	100.0	0.0	0.0	0.0
TOTAL	32.5	8.6	57.9	0.9	27.7	7.1	62.2	3.1	22.9	5.8	65.6	5.7

Source: U.S. Department of Agriculture, Current Research Information System.

[a]For each year, the row total is 100 percent.

[b]Includes research completed by Agricultural Research Service, Economic Research Service, Statistical Reporting Service, and Soil Conservation Service.

[c]Includes research completed by Forest Service, Farm Coop. Service.

of the USDA total. In 1969, state institutions performed 58.8 percent of total public agricultural research (measured by expenditures), and the SAES and veterinary medicine schools comprised 98.5 percent of the state total. In 1997, the USDA's share had fallen to 28.7 percent of the total public agricultural research expenditures. In 1984, state institutions performed 65.3 percent of the public agricultural research, with the SAES and vet-med schools' share being 92 percent in both years. In 1997, state institutions accounted for a larger 71.3 percent of the total public agricultural research expenditures, but the SAES and vet-med schools retained a constant 92 percent of state part.

Commodity Focus

Although the SAES system and vet-med schools performed more than 50 percent of the public agricultural research in 1969, 1984 and 1997 (based on expenditures), the USDA performed more than 50 percent for some crop and natural resource commodities. For example, in 1969, ARS and ERS accounted for 67.9 percent of cotton and cotton seed research, 66.2 percent of the research on other oilseeds (peanuts, other oilseeds, and oil crops excluding cottonseed and soybeans), 57.5 percent of wheat research, and 50.3 percent of water research. They also performed 60.1 percent of economics (excluding farm management) research. Other USDA agencies (i.e., the Forest Service) performed 71 percent of total public forestry research in 1969.

In 1984, the number of commodities where the USDA performed more than 50 percent of total public agricultural research expenditures dropped to four, and one was a new research commodity (insects) added after 1969. They were cotton (60.8 percent), forestry (54.3 percent), insects (50.4 percent), and economics (61.9 percent). In 1997, cotton is the only one of 30 research commodities where the USDA performed more than 50 percent of the public research. For the cotton and economics research commodities there was a dramatic fall in the share of public agricultural research performed by the USDA's research agencies between 1984 and 1997 (see Table 4.6).

General or Basic versus Applied Research

Advances in basic biological sciences and their applications have historically proved to be a major source of new technologies for agriculture. The importance of this research has been re-emphasized with recent biotechnology advances. Furthermore, a 1977 General Accounting Office (GAO) report expressed a concern that relatively few resources were being allocated to basic agricultural science research. Three high-priority research areas were identified—photosynthesis, biological nitrogen fixation, and cell culture studies.

The CRIS reporting forms ask scientists to designate the percentage of each research project that is basic research, applied research, and development. It is our assessment that this basic research category matches our concept of pre-invention research more closely than general science research. CRIS-designated basic biological science research (see Table 4.7 for the fields of science included) composed 21.2 percent of the total public agricultural research expenditures in 1969, 28.2 percent in 1984, and 31.3 percent in 1997 (Table 4.8). Hence, this aspect of public agricultural research has received an increasing portion of the funding over the past three decades. Furthermore, the USDA's own agricultural research and non-SAES state research had less of a tie to basic biological sciences than the SAES system. In 1969, only 13.2 percent of public agricultural research in the USDA system, vet-med schools, and other non-SAES state institutions was classified as basic biological science research. This magnitude, however, increased to 22 percent in 1984 and to 28.3 percent in

Table 4.7   Absolute (1984 dol) and relative support for agricultural research by field of science, all SAES, 1969, 1984, and 1997

Field of Science	1969 $ thousands	1969 %	1984 $ thousands	1984 %	1997 $ thousands	1997 %
Basic Biological Sciences	198,777	30.6	323,024	30.5	389,548	33.0
Biochemistry and Biophysics	52,063	8.0	70,623	6.7	72,025	6.1
Molecular Biology	3,346	0.5	15,896	1.5	94,390	8.0
Genetics	69,761	10.7	122,099	11.5	106,343	9.0
Microbiology	18,518	2.9	37,679	3.6	45,248	3.8
Physiology	55,090	8.5	76,727	7.3	71,542	6.1
Applied Biology-						
Production Systems	227,878	35.1	383,768	36.3	382,185	32.4
Bio-Environmental Systems	111,015	17.2	203,705	19.3	209,341	17.8
Entomology	40,587	6.2	69,230	6.5	74,639	6.3
Nutrition and Metabolism	76,276	11.7	110,833	10.5	98,205	8.3
Applied Biology-Diseases	71,583	11.0	115,732	10.9	114,399	9.7
Immunology and Virology	14,461	2.2	28,841	2.7	32,167	2.7
Nematology	3,863	0.6	8,165	0.8	10,331	0.9
Parasitology	4,136	0.6	4,986	0.5	6,660	0.6
Pathology-Pharmacology	49,124	7.6	73,740	7.0	64,594	5.5
Physical Sciences	88,891	13.7	144,594	13.7	169,193	14.4
Chemistry	37,573	5.8	54,577	5.2	61,080	5.2
Engineering	29,510	4.5	48,589	4.6	56,263	4.8
Geophysical Sciences	11,887	1.8	19,532	1.8	26,384	2.2
Other Physical Sciences	9,921	1.6	21,896	2.1	25,458	2.2
Social Sciences	57,926	8.9	90,423	8.6	123,502	10.5
Economics	40,172	6.2	61,636	5.8	73,900	6.3
Sociology	12,799	2.0	12,665	1.2	23,015	2.0
Other Social Sciences	4,955	0.7	16,123	1.5	26,587	2.3
Other Fields of Science[a]	4,552	0.7	—	—	—	—
Total	649,608	100.0	1,057,542	100.0	1,178,827	100.0

Source: Computed from data tapes of U.S. Department of Agriculture, Current Research Information System.
[a]Also includes unclassified research.

1997. Hence, by 1997 there was little difference in the relative importance of basic biological science research between the SAES and ARS-ERS USDA public agricultural research systems.

The basic biological science share of research was larger for crop and livestock (including poultry) research than for other commodities (Table 4.8). We still argue, however, in Chapters 8 and 9 that there may not be a clear distinction among the levels of science for livestock research. In 1969, 29.8 percent of the total expended on the 16 crop and livestock research commodities was tied to basic biological sciences. This increased to 40.0 percent in 1984 and 41.7 percent in 1997. In 1969, a larger share of the research on livestock than on crops seems to have been basic biological science research. The ordering reversed in 1984 relative to 1969 but then returned to the original ordering in 1997. Thus, over 1969–2000 the emphasis on CRIS-designated basic biological science research

Table 4.8   Absolute and relative expenditures for basic biological science research, by commodity for federal and state institutions, 1969, 1984 and 1997(thousands of 1984 dollars)

Commodity	1969 Total	1969 % of Comm. Res. Exp.	1984 Total	1984 % of Comm. Res. Exp.	1997 Total	1997 % of Comm. Res. Exp.
Crops:						
Corn	12,051	37.1	17,797	37.3	39,068	38.5
Wheat	11,059	40.4	17,736	43.4	37,169	45.1
Other cereals	10,636	41.5	16,774	41.7	34,610	50.6
Soybeans	5,494	28.6	20,596	38.5	40,273	47.3
Other oil seeds	4,340	25.1	6,758	36.8	13,540	38.4
Cotton	11,073	19.2	12,551	26.6	17,386	22.9
Forage	13,894	25.3	20,024	24.0	30,867	26.3
Tobacco-sugar	11,277	33.6	11,646	37.6	12,943	39.9
Potatoes	4,131	31.4	6,836	38.5	14,150	37.3
Fruit-vegetable	45,986	18.8	65,122	31.0	128,420	34.1
New and misc. crops	2,835	27.8	4,006	38.7	11,382	36.9
Livestock:						
Beef cattle	28,880	37.3	48,738	37.7	77,426	43.4
Swine	15,577	45.1	24,434	38.9	52,564	50.3
Sheep and wool	14,015	34.8	19,031	32.1	60,471	44.4
Dairy	30,843	41.6	37,765	40.8	66,367	45.0
Poultry	21,468	39.3	19,958	35.1	47,714	47.1
subtotal	($243,559)	(29.8)	($280,643)	(40.0)	($544,548)	(41.7)
Other						
Soils	4,954	10.7	5,673	6.9	13,468	8.7
Water	3,769	5.5	3,902	4.8	8,543	4.9
Insects[a]	—	—	6,647	20.1	11,543	24.1
Structure and inputs	582	4.1	350	3.6	145	1.8
Farm management	0	0.0	0	0.0	55	0.4
Forestry	15,917	14.2	21,722	13.5	48,428	16.3
Recre-wildlife	1,210	5.7	6,521	12.0	26,236	17.9
Economics	0	0.0	0	0.0	0	0.0
Family life	0	0.0	10,223	12.7	4,770	2.7
Food and textiles	0	0.0	7,666	16.7	11,609	13.8
General science[a]	—	—	89,651	49.8	182,875	59.6
Research	0	0	1,777	11.1	7,549	21.2
Other	0	0	0	0.0	0	0.0
Total	$269,990	21.2	$503,903	28.2	$999,571	31.3

Source: U.S. Department of Agriculture, Current Research Information System.
[a]Not a designated research commodity in 1969.

increased for crops but fell for livestock over 1969 to 1984, only to rise over 1984 to 1997 and to be higher for each livestock research category in 1997 than in 1969.

Research Foci

Public agricultural research can be classified into several major foci, as reflected in the CRIS Research Problem Areas (RPAs). We have defined five major research foci by aggregating RPAs into relatively homogeneous groups and constructing research expenditures for each of them. They are biological efficiency, mechanization, protection-maintenance, management, and post-harvest (see Huffman and Evenson 1994, Appendix Table 16, for details). There is a sixth residual category for RPAs that did not seem to fit into our classification scheme.

This research-foci classification scheme was conceived as a way to separately identify research on biological efficiency of plants and animals, protection and maintenance of past biological efficiency gains, mechanization of production, management, and post-harvest research. For the 16 crop and livestock commodities, roughly one-third of the research was focused on biological efficiency, one-third on protection and maintenance, and one-fourth on post-harvest, management, or mechanization. The residual RPAs accounted for the remainder (see Table 4.9; also Heimlich 2003, Ch. 5.2, p. 3).

Although public research on agricultural mechanization has occasionally received attention from the popular press—e.g., the tomato harvester developed by the California Agricultural Experiment Station (Schmitz and Seckler 1970)—only a tiny share of public agricultural research funding has been spent on agricultural mechanization research. In 1969, less than 2.5 percent of the 16 crop and livestock commodities focused on mechanization. The share has declined over time. Another surprise was the reduced expenditures on post-harvest research between 1969 and 1984, but the post-harvest emphasis was higher in 1997 than in 1984.

Forages topped the list in 1969, 1984, and 1997 for largest share of research expenditures being focused on biological efficiency. Cotton and fruits and vegetables were at the other extreme. Twenty percent of the cotton research focused on biological efficiency in 1969 and 1984 and 25 percent in 1997. For fruits and vegetables, 18.9 percent of the research in 1969 focused on biological efficiency. This share rose to 31.3 percent in 1984 but has backed off to 27.5 percent in 1997. Cotton, tobacco-sugar, and potatoes are commodities where a relatively large share of public agricultural research was focused on protection-maintenance.

The SAES System: Additional Information

Additional insights into the organization of SAES research are obtained by comparing the composition of research for three non-overlapping pairs of years from different subperiods: 1920 and 1935, 1951 and 1964, and 1969 and 1984.

Comparing 1920 and 1935

For 1920 and 1935, the only available data on composition of agricultural research activities are the number of SAES system research projects (see Table 4.10). Although these data are not perfect for making comparisons with expenditures in later years, we believe they are insightful. In 1920, the SAES system had 4,123 different research projects or an average of 86 per state. The number of projects was 48 percent larger in 1935.

More than half of SAES research was plant science projects(66.8 percent in 1920 and 59.8 percent in 1935. Animal science projects composed slightly over one-fifth of the total projects in both years. There were only 102 social science projects—an average of slightly over two per state—in 1920, but the number was 7.7 times larger in 1935, reflecting the rapid development of social sciences, public and private responses to the Great Depression, and growth of the federal government in farm income enhancement. In 1935, the number of research projects focusing on post-harvest research was only one-half that of 1920.

Table 4.9 Distribution of agricultural research expenditures by major research foci for federal and state institutions, 1969, 1984, and 1997

Research Commodities	1969				
	Biological Efficiency	Mechanization	Management	Protection-Maintenance	Post-Harvest
Crops:					
Corn	36.4	1.9	2.0	28.6	25.7
Wheat	31.3	0.5	5.4	21.6	30.2
Other Cereals	49.5	1.1	3.4	24.1	14.8
Soybeans	39.1	1.3	1.7	27.2	20.0
Other Oil Seeds	26.0	5.3	3.8	13.4	28.6
Cotton	20.0	10.3	5.5	32.2	26.6
Forage	71.2	2.5	1.9	17.8	2.9
Tobacco-Sugar	39.9	4.1	2.6	33.3	4.0
Potatoes	25.1	2.6	8.6	29.7	27.1
Fruit-Vegetables	18.9	2.7	11.3	17.7	10.3
New & Misc. Crops	0.0	0.0	0.0	0.0	0.0
Livestock:					
Beef Cattle	43.1	0.0	14.5	25.3	14.6
Swine	40.7	0.0	9.6	33.3	11.6
Sheep & Wool	41.9	0.0	7.5	29.4	16.4
Dairy	40.7	0.0	11.7	24.2	18.1
Poultry	40.2	0.0	10.1	33.6	11.4
Other:					
Soils	55.1	0.0	22.9	14.2	0.0
Water	3.0	0.0	0.0	11.3	47.3
Forestry	32.6	2.3	16.1	27.3	19.2
Recre-Wildlife	0.0	0.0	23.0	4.5	0.0

Source: U.S. Department of Agriculture, Current Research Information System.
[a] Major research foci were defined by grouping CRIS Research Program Areas (RPAs). Proportions do not sum to 1.0 because a residual category is excluded.

	1984					1997				
	Biological Efficiency	Mechanization	Management	Protection-Maintenance	Post-Harvest	Biological Efficiency	Mechanization	Management	Protection-Maintenance	Post-Harvest
	(percentage of total expenditures on a commodity)[a]									
Crops:										
	43.5	1.7	6.7	33.8	6.3	36.1	0.9	7.4	27.5	18.1
	48.7	0.5	10.0	28.0	8.5	42.6	1.1	9.3	27.7	15.4
	49.7	0.7	8.1	29.0	8.5	49.0	0.7	7.1	28.8	10.5
	43.7	1.9	7.2	35.4	7.5	42.5	1.3	7.6	28.5	15.7
	33.3	3.0	7.6	32.5	16.9	39.6	0.2	8.7	21.1	21.4
	20.3	4.2	7.8	42.5	13.1	24.6	3.1	12.9	42.5	12.5
	59.0	1.6	7.6	23.6	2.0	56.3	0.4	12.1	20.8	1.7
	38.2	5.0	4.0	29.4	7.3	36.9	2.4	5.9	43.7	9.2
	37.8	0.3	3.8	40.8	15.0	29.9	0.2	4.0	54.1	8.8
	31.3	2.6	18.9	32.7	10.0	27.5	0.9	19.0	37.8	9.4
	51.9	1.0	8.6	14.2	2.5	25.2	0.7	7.2	31.8	2.6
Livestock:										
	51.5	0.1	8.0	28.7	6.9	42.2	0.0	9.5	31.4	7.4
	50.3	0.7	3.7	32.9	5.4	43.7	0.1	4.0	35.3	8.0
	43.8	0.1	4.5	37.6	3.4	40.0	0.0	2.7	47.6	2.5
	49.8	0.4	7.9	27.5	9.8	41.3	0.1	5.9	32.2	12.1
	45.0	1.2	4.3	35.4	6.8	33.2	0.2	2.2	37.4	8.1
Other:										
	44.0	0.0	14.3	12.1	8.6	35.5	0.0	10.8	25.9	11.6
	6.8	0.0	0.0	18.6	42.1	11.5	0.0	0.0	34.6	34.4
	33.0	2.6	20.7	23.9	15.8	38.6	1.4	20.0	15.0	15.9
	0.0	0.0	10.9	2.0	0.0	0.0	0.0	7.2	0.8	0.0

Table 4.10   Numbers and percentage distribution of research projects by major subject matter category, all SAES, 1920 and 1935

Research Category	1920		1935	
	No.	Percent	No.	Percent
Plant Science	2,755	66.8	3,979	59.8
Animal Science	981	23.8	1,398	21.0
Post-Harvest Technology[a]	131	3.2	61	0.9
Social Science	102	2.5	782	11.8
General Science and Engineering[b]	154	3.7	434	6.5
TOTAL	4,123	100.0	6,654	100.0

Source: Roland R. Robinson. U.S. Department of Agriculture, Cooperative States Research Service, unpublished data.
[a]Includes food technology and human nutrition.
[b]Primarily genetics and agricultural engineering.

Comparing 1951 and 1964

For 1951 and 1964, SAES research expenditures data are available for subject-matter categories from unpublished USDA-CSRS worksheets. The 62 subject-matter categories were aggregated into 32 categories. These categories were constructed to be similar, where possible, to later CRIS research commodities.

Each SAES project during the 1951–1964 period was placed in the subject-matter category that was the primary focus of the research. Under the CRIS system, multiple RPA, research commodity, and field of science classifications are possible, with a fraction of total project expenditures being assigned to each. In the national aggregate, the pre-CRIS commodity-discipline classification scheme and CRIS commodity classification scheme are expected to paint similar, but not exactly the same, pictures.

In 1951, total SAES expenditures on agricultural research were $353 million (1984 dollars). This magnitude was 71 percent larger in 1964. In both years, approximately 65.5 percent went to crop-livestock research. Sixty-two percent of the crop-livestock research was allocated to research on crops.

The distribution of SAES research expenditures was remarkably similar in 1951 and 1964 (see Table 4.11). Fruits and vegetables accounted for the largest share, excluding administrative and other costs—9.2 percent in 1951 and a slightly smaller 8.6 percent in 1964. The plant nutrition, soils, and land development category was second, with 7.5 percent in 1951 and 8.6 percent in 1964. Research on beef cattle was third, accounting for 5.3 percent in 1951 and 5.9 percent in 1964. Outside the crop-livestock category, marketing research was the single largest component, comprising 4.2 percent in 1951. Its share increased to 5.9 percent in 1964, reflecting perhaps the general emphasis on better marketing of agricultural commodities as a means of reducing the large government-held grain stocks. Foods and nutrition research, however, lost 1 percent between 1951 and 1964.

General and pre-invention biological science research appear to account for a very small share of the total SAES research in 1951 and 1964. The evidence is the small share—less than 3 percent—accounted for by genetics, general botany, and microbiology (see Table 4.11). Of course, given the classification scheme used, some of this biological science research was included in projects that had primarily other emphases.

Table 4.11 Absolute and relative support for agricultural research by subject matter area, all SAES, 1951 and 1964 (thousands of 1984 dollars)

Subject Matter	1951 Total	1951 Percent	1964 Total	1964 Percent
Crops:				
Cereals	16,861	5.3	21,145	3.6
Cotton	4,148	1.3	6,628	1.1
Fiber, exclud. cotton, and oil seeds	2,918	0.9	3,065	0.5
Forage pasture and range	12,601	3.9	21,222	3.6
Tobacco and sugar	3,773	1.2	5,346	0.9
Potatoes	3,046	1.0	2,500	0.4
Fruits and vegetables	29,531	9.2	50,695	8.6
Utilization-crops	4,673	1.4	10,769	1.8
Control field & hort. crop diseases	13,326	4.2	25,380	4.3
Control crop insects	13,621	4.3	28,856	4.9
Weed control	3,513	1.1	12,658	2.2
Plant nutrition, soils and land development	23,992	7.5	50,762	8.6
Subtotal, crops	(132,003)	(41.3)	(239,026)	(40.6)
Livestock:				
Beef cattle	17,071	5.3	34,600	5.9
Swine	8,426	2.6	11,552	2.0
Sheep and goats	4,614	1.4	9,445	1.6
Dairy	15,498	4.9	26,902	4.6
Poultry	13,764	4.3	23,732	4.0
Control animal diseases and parasites	12,007	3.8	27,207	4.6
Control insects of animals	1,916	0.6	4,635	0.8
Utilization-animal products	4,449	1.4	7,814	1.3
Subtotal, livestock	(77,745)	(24.3)	(145,887)	(24.8)
Subtotal, crops & live	($209,748)	(65.6)	($384,913)	(65.4)
Other:				
Forestry, incl. utilization and diseases	4,409	1.3	16,213	2.8
Fish and wildlife	1,561	0.5	4,412	0.7
Ag. Engineering, exclud. Land Development	9,530	3.0	14,985	2.5
Marketing	13,346	4.2	34,980	5.9
Farm Management	4,902	1.5	7,209	1.2
Other Ag. Economics and Rural Sociology	6,511	2.0	13,020	2.2
Foods and Nutrition	9,022	2.8	10,168	1.7
Textiles, clothing, housing and equipment	2,058	0.6	1,905	0.3
Home Mgt., Family Econ., Family Life & Child Dev.	736	0.2	1,764	0.3
Genetics	3,354	1.0	10,083	1.7
General Botany and Microbiology	2,618	0.8	5,788	1.0
Administration, General Expense and Other	51,666	16.2	83,059	14.1
Subtotal, other	(119,713)	(37.5)	(203,639)	(34.6)
TOTAL, Classified	319,462	100.0	588,552	100.0
Other Unclassified	33,470		16,497	
Grand Total	352,932		605,049	

Source: U.S. Department of Agriculture, Cooperative States Research Service, unpublished data from worksheets (CSESS Form 13).

Comparing 1969, 1984, and 1997

Expenditure data for these years were available from CRIS. Total SAES research expenditures were $649.6 million (1984 dollars) in 1969, $1,057 million or 63 percent larger in 1984, and $1,929 million or 60 percent larger in 1997. The field of science and funding source classifications are of interest here.

SAES research clearly has been dominated by the biological sciences.[10] The CRIS basic plus applied biological science fields accounted for about 66 percent of total SAES expenditures in 1969, 1984, and 1997 (see Table 4.7). In the physical sciences, chemistry accounted for slightly more than 5 percent. In the social sciences, economics accounted for about 6 percent. Other fields were of lesser importance.

In 1969, 30.6 percent of SAES research expenditures, or $199 million (1984 dollars), were for basic biological science research. In 1984, expenditures in these fields were 62 percent larger, and in 1997 a further 68 percent increase occurred. However, the share of total SAES research spent on basic biologic sciences was only 2.5 percentage points higher in 1997 than in 1969! In 1969 and 1984, very little molecular biology research was undertaken in the state agricultural experiment stations, but between 1984 and 1997 a dramatic 6.5 percentage point increase occurred (Table 4.7). This is evidence that the biotech revolution did not hit the SAES system until after the mid-1980s. Evenson (1983) used this as evidence that the SAES system has been slow to realize the research potential of biotechnology.

A large share of SAES research has been applied biological science research (see Table 4.7). The production systems and disease categories of applied biology accounted for about 46 percent of the total in 1969 but a significantly lower 42 percent in 1997. The bio-environmental systems area was the largest component, followed by nutrition and metabolism.

The Private Research System: Additional Information

Aggregate data on private agricultural research expenditures do not exist before 1956. To gain some perspective on composition, we have derived an estimate of private agricultural production and post-harvest research using data on assigned patents. The procedure for carrying out this estimation is similar to that described in the first section of this chapter. The allocation of private research expenditures between the two categories was according to the share of assigned patents for production and post-harvest technology fields.

We conclude that a large share (greater than 80 percent) of the early (1850–1900) private sector agricultural research was allocated to the development of farm-production-related technologies (see Table 4.12). This was the period when the U.S. farm machinery industry was developing rapidly. By 1900, it was the single largest manufacturing industry in the United States. The production share peaked in the decade of the 1860s at 89 percent and has trended downward since then. During the 1970s and 1980s, post-harvest and production-related private research were of approximately equal magnitudes.

Starting in 1956, private sector agricultural R&D expenditures were classified into five technology fields: utilization-nutrition; clothing and housing; insect and weed control and plant nutrition; plant and animal diseases; and engineering and management (see Table 4.13). In 1956, 50 percent of private agricultural R&D was in the technology field of engineering and management, 27 percent was in utilization-nutrition, and 12 percent was in insect and weed control and plant nutrition. The

Table 4.12 Private sector agricultural research expenditures in constant 1984 dollars, 1840–1985 (millions of 1984 dollars)

Years	Total	Production	Percent Production
1840–49			57.1
1850–59	0.4	0.3	84.4
1860–69	3.1	2.8	88.9
1870–79	6.6	5.7	86.3
1880–89	29.0	24.9	85.9
1890–99	32.4	27.8	85.9
1900–09	53.1	44.1	83.2
1910–19	101.6	78.6	77.3
1920–29	134.0	99.9	71.4
1930–39	405.4	256.9	63.4
1940–49	346.3	202.4	58.5
1950–59	890.6	575.0	64.6
1960–69	1367.8	848.0	62.0
1970–79	1569.6	884.2	56.3
1980–85	2444.7	1429.5	58.5

Note: Also see Huffman and Evenson 1994, Appendix 3.B.

clothing-housing field accounted for 8.6 percent and disease control for only 3.0 percent. Since 1956, the research share for the engineering-management field has trended steadily downward because total expenditures in this field have changed very little as expenditures in other fields grew. In 1984, the engineering-management expenditure share was one-half the 1956 level (or 24 percent).

The research expenditure share for the insect and weed control and plant nutrition field has shown a steady rise since 1956; and in 1984, it was 26 percent or more than double the 1956 share. The relative importance of utilization-nutrition research increased until 1970, when it was 37 percent; thereafter it has declined slowly. The share of private agricultural research expenditures for disease control increased from 3 percent in 1956 to 8.2 percent in 1984.

In 1984, the private R&D expenditures on utilization-nutrition were the largest—$780 million. Expenditures for insect and weed control and plant nutrition R&D were $637 million and for engineering and management R&D $592 million. Diseases and clothing-housing R&D each accounted for approximately $200 million.

Thus, we conclude that a major shift occurred in the distribution of private agricultural R&D during 1956–1984, and it resulted in a smaller share for engineering-management and larger share for insect and weed control and plant nutrition.

Public Extension

The distribution of total U.S. extension resources among subject-matter components for selected years, during the period 1920–1983, is presented. From 1920 to 1983, extension staff time was accounted for by subject matter (e.g., crops, livestock, management). In 1920, 26 percent of professional staff time was allocated to crops, including soils, and 24 percent to livestock (see Table 4.14). Agricultural economics, including marketing, and agricultural engineering received much less emphasis (4.0 percent and 3.3 percent, respectively) of staff time. Thus, 57 percent of extension staff time was allocated to agricultural subjects.

Table 4.13   U.S. private sector expenditures on agricultural research, total (millions of 1984 dollars) and by technology fields, 1956–1984

Year	Utilization/ Nutrition	Clothing/ Housing	Insect & Weed Control/Plant Nutrition	Diseases (Plant & Animal)	Engineering & Management	Total
1956	242.2	76.4	104.0	26.3	444.7	890.6
1957	266.2	63.2	115.6	27.4	521.7	994.1
1958	302.4	110.8	126.9	34.5	615.3	1189.9
1959	312.1	97.7	132.6	36.7	507.6	1086.7
1960	377.3	103.4	99.4	43.1	552.1	1175.3
1961	367.6	104.7	136.3	45.9	466.3	1120.8
1962	377.1	104.9	145.0	49.0	483.4	1159.4
1963	382.5	113.1	151.5	51.6	511.8	1210.5
1964	425.3	116.4	155.2	50.8	510.9	1258.6
1965	450.9	68.9	197.8	56.9	593.3	1367.8
1966	427.1	130.6	225.4	63.2	585.5	1431.8
1967	418.2	163.6	257.7	65.3	571.4	1476.2
1968	453.9	151.9	244.5	62.2	474.2	1386.7
1969	508.9	175.7	266.1	69.3	497.1	1517.1
1970	543.4	139.6	301.2	76.3	425.5	1486.0
1971	532.8	147.2	300.4	83.2	415.9	1479.5
1972	569.8	195.0	239.0	79.0	411.6	1494.4
1973	559.6	204.1	235.7	84.1	496.2	1579.7
1974	596.5	164.6	259.2	86.8	495.6	1602.7
1975	548.3	137.1	311.3	93.0	488.1	1577.8
1976	579.5	145.1	346.2	100.2	398.6	1569.6
1977	623.5	163.0	377.4	102.7	706.9	1973.5
1978	643.0	174.7	396.6	103.2	775.2	2092.7
1979	646.3	180.4	403.3	101.1	814.9	2146.0
1980	677.4	205.4	520.4	131.3	765.6	2300.1
1981	669.8	216.1	595.1	150.8	679.5	2311.3
1982	720.7	225.2	616.1	171.1	651.7	2348.8
1983	744.4	226.6	616.7	183.7	609.4	2380.8
1984	780.0	235.2	637.0	200.5	592.0	2444.7

Source: Derived by Robert E. Evenson from variety of published sources of data, including science indicators.

By 1935, agricultural subjects accounted for an even larger share of staff time (65 percent), and within agricultural subjects it shifted toward crops and agricultural economics. The share of staff time allocated to crops was 13 percent higher in 1935 than in 1920 and to agricultural economics was 3.4 percent higher. Livestock subjects, however, received less emphasis—7.8 percent less in 1935 than in 1920.

After 1950, less than 45 percent of total extension staff time was allocated to agricultural subjects. The reduction was centered primarily on crop production. In 1950, only 18 percent of extension staff time was allocated to crops. The shares allocated to livestock, agricultural economics, and agricultural engineering were similar to 1935. The distributions of extension staff time in 1962 and 1972 were also similar to 1950 (see Table 4.14).

In 1983, the relative emphasis on crops was larger and on livestock smaller than for other post-1950 years (see Table 4.14). Twenty-one percent was allocated to crops, 12 percent to livestock, 7.6 percent to agricultural economics, and 1.2 percent to agricultural engineering.

Over 1973 to 1992, data are available on the allocation of professional staff time among extension program areas (i.e., agriculture and natural resources, home economics, and community and

Table 4.14 Absolute and relative distribution of extension professional staff time, United States, 1920–1983

Total or subject	1920	1935	1950	1962	1972	1983
Professional Staff:						
Total[a]	4,265	6,454	11,788	13,960.0	13,841	15,808
State Specialists as % of total	21.1	18.1	17.9	19.8	26.9	27.0[c]
	Distribution of Agent and Specialist Time (%)					
Crops, incl. soils	25.6[c]	38.4	18.1	16.6	15.5	21.4
Livestock	23.9[b]	16.1	15.9	13.1	18.6	12.0
Agr. Econ., incl. Marketing	4.0[b]	7.4	7.5	6.4	6.9	7.6
Agr. Engineering	3.3[b]	3.0	2.1	3.1	1.8	1.2
Other	43.2[b]	35.1	56.4	60.8	57.2	57.8
Total	100.0	100.0	100.0	100.0	100.0	100.0

Source: U.S. Department of Agriculture, Extension Service.
[a]Professional staff includes total state staff, area agents, and county extension agents.
[b]An estimate. The distribution is for 1924.
[c]An estimate.

rural development). See Table 4.15. The total full-time equivalent number of professional staff years was 16,300 in 1973. It was slightly larger in 1978 but returned to the 1973 value in 1984 and was 4 percent lower in 1992, the last year in which data are available, than in 1973 or 1984. In 1973, the data on distribution of staff time by programs showed that 38.2 percent of the days were allocated to agriculture and natural resources. This share was rising to 45 percent over 1973 to 1984. The share allocated to community and rural development programs decreased about 2 percentage points from 1973 to 1984 and was then unchanged in 1992, but most of the decrease was in the 4-H youth program. This program lost 9 percentage points between 1973 and 1992. The relative size of home economics extension increased over 1973 to 1992. The bottom line on these numbers is that full-time equivalent extension staff allocated to agricultural and natural resource extension areas grew over 1973 to 1984 and then declined.

The Extension Service has prepared tabulations for the agriculture and natural resource extension area showing allocations of staff time among subject matter components for only 1978–1983.[11] Table 4.16 shows that in 1978, 44.6 percent of professional extension resources was allocated to crop production. Furthermore, the share allocated to crop production was 3.3 percent larger in 1983.

Table 4.15 Absolute and relative distribution of extension professional resources by program areas, United States, 1973, 1978, 1984 and 1992

Program Area	1973	1978	1984	1992
Total Staff, FTE years	16,291	16,873	16,183	15,497
	Distribution(%)			
Agriculture and National Resources	38.2	41.3	45.4	44.9
Home Economics	20.9	22.8	22.6	26.0
4-H Youth	32.1	27.9	25.9	23.1
Community and Rural Development	8.8	8.0	6.1	6.0
Total	100.0	100.0	100.0	100.0

Source: U.S. Department of Agriculture, Extension Service, annual reports EXT/PDEMS/MS, 1984, 1987, 2003.

Table 4.16   Absolute and relative distribution of extension professional resources, agriculture and natural resource program area, United States, 1978 and 1983

Components	1978	1983
Total Staff, FTE years	7,140.1	7,418.50
	Distribution (%)	
Crop Production	44.6	47.9
Livestock Production	22.3	21.1
Management, Marketing and Agr. Economics	18.7	15.9
Mechanization and Agr. Engineering	1.2	1.3
Total	100.0	100.0

Source: U.S. Department of Agriculture, Extension Service, annual reports EXT/PDEMS/MS, 1984.

The share of professional staff resources allocated to livestock production was about 22 percent of the total in both years and to mechanization and agricultural engineering was about 1.2 percent. Management, marketing, and agricultural economics research accounted for 18.7 percent of total professional resources in 1978 but decreased 2.8 percent between 1978 and 1983. The shift of agriculture and natural resource extension staff time away from the marketing, management, and agricultural economics components toward crop production during 1978–1983 was surprising, given the serious economic problems of farmers that became apparent during the early 1980s.

Detail by State

Additional insights about the public research and extension system can be obtained by examining the distribution of expenditures across states. Comparable data are presented for 1950, 1984, and 2000 for research and for 1950, 1984, and 1992 for extension.

SAES

The previous subsection focused on the characteristics of public agricultural research at the national level. The natural resources, geoclimatic conditions, size, and nonfarm resources differ considerably from state to state and are expected to be reflected in the intensity of their agricultural research and extension investments. Table 4.17 presents selected information on agricultural research expenditures by state for 1950, 1984, and 2000. Aggregate SAES expenditures were $311 million in 1969, $1,059 million in 1984, and $1,222 million in 1997 (in constant 1984 dollars). These numbers give an annual average growth rate over 1950 to 1984 of 3.6 percent and for 1984 to 2000 of 0.9 percent per annum. Comparisons using these dates will make long-term changes apparent. The research expenditures were limited to those of the SAES, i.e., they do not include the agricultural research conducted in each state directly by the research agencies of the USDA or business-industry. State government decisions on funding of agricultural experiment stations are examined in greater detail in Chapter 7.

Over the long term, California has made large investments in agricultural research. In 1950, $25.3 million (at 1984 price level) were invested by that state. By 1984, the investment (in 1984 prices) had grown by a factor of four to $97.2 million, but in 2000 expenditures were only 16 per-

Table 4.17   Selected measures of the intensity of investment in SAES research by state, 1950, 1984 and 2000

Production Region/States	SAES Res. Exp. (in mil. 1984 dol.)			State Gov. Share SAES Exp. (%)			Agr. Res. Exp. per $1,000 of Agr. Sales		
	1950	1984	2000	1950	1984	2000	1950	1984	2000
Lake States									
Wisconsin	12.0	36.7	52.7	50.4	49.1	34.9	2.85	7.14	6.43
Minnesota	10.2	32.4	39.1	62.9	68.7	55.9	1.93	5.19	5.31
Michigan	6.5	27.5	43.0	73.0	52.1	40.4	2.50	9.89	9.12
Corn Belt									
Illinois	10.7	21.1	28.9	63.3	46.8	50.7	1.43	3.12	3.80
Iowa	10.5	25.2	46.6	43.3	41.0	42.9	1.16	2.70	3.38
Indiana	11.6	31.3	32.6	39.4	40.4	45.6	2.88	7.97	5.92
Ohio	8.6	22.0	25.6	67.5	74.1	65.4	2.21	6.09	6.96
Missouri	4.7	18.5	24.2	23.3	36.4	35.9	1.18	4.97	3.47
Delta States									
Louisiana	8.5	29.4	23.9	58.8	69.4	60.0	6.29	19.23	14.38
Mississippi	8.1	26.0	25.9	45.3	54.9	46.6	4.34	12.01	7.54
Arkansas	4.5	19.4	25.7	46.8	61.0	63.6	2.09	5.82	6.10
Southern Plains									
Texas	14.3	50.5	70.0	43.4	65.5	55.1	1.48	5.21	5.27
Oklahoma	6.6	16.0	22.7	57.9	62.8	62.1	2.57	6.26	6.10
Northern Plains									
Kansas	5.1	26.5	30.1	47.1	48.2	50.5	1.21	4.44	3.50
Nebraska	5.9	29.6	36.3	40.6	38.6	46.3	1.37	4.18	3.43
N. Dakota	4.6	18.8	17.9	65.3	66.3	49.1	2.07	7.38	5.98
S. Dakota	2.5	7.2	8.6	40.5	58.1	46.9	1.04	2.48	1.95
Southeast									
Florida	12.3	53.4	62.7	76.1	72.1	62.7	6.60	11.64	10.31
Georgia	3.5	35.8	42.3	22.0	77.0	58.2	1.71	9.97	8.90
Alabama	7.1	19.6	20.5	41.5	42.9	59.9	2.34	8.94	6.86
S. Carolina	3.9	13.7	11.5	52.0	79.0	52.4	3.33	12.10	7.12
Appalachian									
N. Carolina	8.9	44.0	40.5	64.9	60.8	61.9	2.90	10.67	6.18
Virginia	5.3	24.1	28.2	60.8	56.4	50.0	3.10	13.42	11.30
Tennessee	4.8	16.5	17.3	37.2	47.5	61.3	2.56	8.30	9.56
Kentucky	5.3	14.9	18.4	32.6	65.9	73.4	2.31	5.61	6.82
W. Virginia	3.4	5.5	5.3	53.9	35.3	40.3	7.55	24.62	9.94
Pacific									
California	25.3	97.2	114.1	88.8	68.9	51.2	2.65	6.86	4.18
Oregon	8.3	26.1	27.0	62.4	43.9	52.0	5.06	14.58	8.41
Washington	9.2	22.6	21.4	72.3	42.3	34.3	4.59	7.70	2.65
Alaska	0.8	4.1	2.6	49.7	74.5	54.6	—	33.15	—
Hawaii	3.4	10.7	11.9	67.8	63.9	51.5	—	17.36	21.22
Mountain									
Colorado	3.7	30.1	19.8	40.9	22.8	25.0	1.57	8.98	1.98
Arizona	2.7	18.2	28.4	57.7	60.6	47.8	2.41	11.99	10.83

(*Continues*)

Table 4.17   *(Continued)*

Production Region/States	SAES Res. Exp. (in mil. 1984 dol.)			State Gov. Share SAES Exp. (%)			Agr. Res. Exp. per $1,000 of Agr. Sales		
	1950	1984	2000	1950	1984	2000	1950	1984	2000
Utah	2.8	9.5	12.4	55.8	54.8	46.4	3.98	16.33	10.38
Wyoming	2.7	5.2	4.3	52.3	58.0	69.9	3.97	9.05	5.71
Idaho	3.2	10.8	13.9	75.1	52.9	49.8	2.10	4.71	3.72
Montana	4.5	12.0	13.5	52.0	48.9	34.9	2.91	8.46	4.75
N. Mexico	2.2	8.7	5.2	46.7	58.8	56.7	2.55	8.79	2.56
Nevada	0.9	4.9	6.6	12.3	53.0	48.5	4.77	19.48	15.19
Northeast									
New York	18.6	48.4	48.3	77.3	39.4	29.0	5.37	17.91	8.18
New Jersey	6.6	11.8	17.6	56.8	57.2	65.4	5.59	23.36	25.81
Pennsylvania	7.0	17.7	26.1	37.6	47.3	52.1	2.35	5.59	6.14
Maryland	3.1	9.4	6.0	49.5	67.1	29.8	3.30	8.11	2.23
Mass.	3.4	6.2	5.8	69.9	44.4	23.5	4.55	16.18	6.32
Conn.	5.5	7.5	9.5	83.0	62.9	59.0	8.22	20.92	20.35
Vermont	1.2	3.3	3.0	25.1	40.5	39.9	2.43	8.28	4.35
N. Hampshire	1.0	3.5	2.7	23.5	40.8	64.5	3.98	31.78	20.45
R. Island	1.2	2.7	2.0	22.3	40.7	38.8	13.27	42.76	29.93
Delaware	1.9	4.3	3.3	30.1	50.1	32.3	4.51	8.33	2.61
Maine	2.4	7.7	8.1	48.4	36.8	47.9	3.52	16.84	13.93
National Average	311.0	1059.3	1221.8	—	55.8	50.1	—	4.14	5.77

Source: Data on research expenditures are from U.S. Department of Agriculture(1951,1985b). Data on sales are from U.S. Department of Agriculture(1985a). CRIS funding summaries.

cent larger than in 1984 ($114 million). New York and Texas ranked second and third for total SAES expenditures in 1950—$18.6 and $14.3 million, respectively. The expenditures in Florida, Wisconsin, Indiana, Illinois, Minnesota, and Iowa were similar, being $10.2 to $12.3 million in 1950. At the other extreme, eight states invested less than $2.5 million. Except for New Mexico, Alaska, and Nevada, they were located in the Northeast region, i.e., Vermont, New Hampshire, Rhode Island, Delaware, and Maine. In 1984, Florida and Texas ranked second and third with total SAES investments in agricultural research of $53.4 and $50.5 million. Expenditures in North Carolina were $44 million; in Wisconsin, $36.7 million; and in Georgia, $36 million. Georgia had only $3.5 million of SAES agricultural research expenditures in 1950, but by 1984, expenditures were 10 times larger. The whole Southeast region showed an increase by 4.6-fold for 1984 relative to 1950. Two of the Corn Belt states, which ranked in the top eight for agricultural research investments in 1950, showed very little increase by 1984. Illinois's investment increase in 1984 relative to 1950 was only 1.97 and Iowa's was 2.4. Other states showing small increases in 1984 over 1950 were West Virginia (1.6), Wyoming (2.0), Washington (2.5), and some of the northeastern states (New Jersey, Massachusetts, Connecticut, Rhode Island). Although the growth of total SAES research expenditures was at a low rate for 1984 to 2000, some states had significant growth. These states include Iowa (61 percent), Wisconsin (36 percent), Texas (33 percent), and Florida (16 percent). Florida relinquished its second place size to Texas in 2000.

The cost of SAES research has been shared by the state governments and federal government and other groups (see Table 4.3). Across the states, the share paid by the state governments varied immensely, and shows no simple pattern over time or by size of states. In 1950 and 1984, the state government's share of total SAES receipts was 55 percent, and the states' share declined to 50 percent in 2000. In 1950, the states' share of SAES revenue exceeded 75 percent in California, New York, and Connecticut. Other states that had large state government shares were Michigan, Florida, Washington, and Idaho. In contrast, state government shares were small in Nevada (12.3 percent), Georgia (22.0 percent), Rhode Island (22.3 percent), New Hampshire (23.5 percent), Missouri (23.3 percent), and Vermont (25.1 percent).

In 1984, the shares were bounded by a smaller range than for 1950. In 1984, the largest state government shares were for South Carolina (79 percent) and Georgia (77 percent), which had one of the smallest shares in 1950. In California, the share was 20 percentage points lower in 1984 than in 1950. At the other extreme, Colorado had the smallest state government share in 1984, 23 percent. In all the other states, the share exceeded 35 percent in 1984 (Table 4.17).

Although the state government's share of total SAES revenue was lower in 2000 than in 1984, 15 states experienced an increase: New Hampshire, New Jersey, and Pennsylvania in the Northeast; Illinois, Iowa, Indiana, Nebraska, and Kansas in the Midwest; Alabama and Arkansas in the South; and Colorado and Oregon in the West. Other work has shown that agricultural research expenditures are related to the size of current and past agricultural production (Huffman and Miranowski 1981; and Chapter 7). Furthermore, deflating SAES research investments by farm sales is one, but not the only, method of standardizing research investments before comparison across states. Table 4.17 also presents information on SAES expenditures per $1,000 of agricultural sales for 1950, 1984, and 2000.

Although much of the cross-state variation in research intensity was removed, this research intensity measure ranges in 1950 from a low of 1.0 for South Dakota, and 1.2 for Kansas and Iowa, to a high of 13.3 for Rhode Island, 8.2 for Connecticut, 7.5 for West Virginia, and 6.6 for Florida. In fact, the Northern Plains and Corn Belt regions had small values for this research intensity measure, and the Northeast region had larger values.

SAES research expenditures per thousand dollars of agricultural sales were uniformly larger in 1984 than in 1950. States that had small values were Illinois (3.l), South Dakota (2.5), and Iowa (2.7). States with large intensities were Rhode Island (42.8), Alaska (33.l), and New Hampshire (31.8). These patterns suggest that other characteristics of agriculture (such as number of different agricultural commodities produced), of the size of a state's forestry and fisheries industries, and of the organization of agricultural research—especially economies of size—were also important sources of differences in relative intensity of agricultural research investments. Although SAES research expenditures relative to cash receipts of farmers were higher in 2000 than in 1984, two-thirds of the states had a lower intensity in 2000 than in 1984.

Extension

This section presents a comparison of state investments in extension for 1950 and 1984. The 1950 investments have been inflated to the 1984 price level using the price index for research inputs (Table 4.1). The state of Texas, which has the largest number of farmers, has the largest investment in extension in both years—$21.6 million in 1950 and $59.4 million in 1984. In 1950, Texas's extension investment was, however, only slightly larger than those for North Carolina and New York. California was a distant fourth. Seventeen other states invested between $10 million and $14 million. Except for New York and Pennsylvania, the other Northeastern and all of the Mountain states invested very little in extension in 1950.

All states invested more in extension during 1984 than during 1950. The average increase (in real terms) was 232 percent. In 1984, California ranked second and North Carolina third behind Texas with extension investments of $46.9 and $41.7 million. States that had large increases between 1950 and 1984 were Minnesota (5.2 times), Florida (4.6), Arizona (3.7), Georgia (3.2), Colorado (3.2), Nevada (3.2), New Hampshire (17.5), Rhode Island (9.5), and Alaska (9.2).

The cost of extension has been shared by the federal, state, and local governments and other organizations. In 1950, the share of the cost borne by state and local governments ranged from 78.3 percent in New York and 77.4 percent in California to 36.6 percent in Kentucky and 37.9 percent in Delaware. In 1984, the state and local governments' share of the extension investment was larger than for 1950 in all states outside of the Northeast, except for Idaho. However, in the Northeast region, the state and local government share was lower in all states in 1984 than in 1950, except for Maryland, Maine, and Vermont.

The choice of the appropriate standardization for extension investments across states is not clear-cut. Extension information generally has private- and public-good aspects. If these aspects are dominated by private information, then deflating by the number of farms would be appropriate. If extension is a pure public good, then deflation by the number of different commodities produced in the state, or not deflating at all, would be appropriate. The willingness of a state's taxpayers to pay for extension may also be related to the size of its agriculture. We have chosen to report state comparisons of extension expenditures per $1,000 of gross farm sales. These expenditures are one simple indicator of differences across states. (See the two right-hand columns of Table 4.18.)

In 1950, Nebraska, Iowa, Minnesota, California, South Dakota, Arizona, Illinois, and Maine invested between $1.40 and $1.90 in extension per $1,000 of gross farm sales. States investing relatively large amounts were West Virginia, Alabama, Louisiana, Rhode Island, and New Hampshire—$8–$13 per $1,000 of gross sales. Furthermore, in general, the South (Delta, Appalachian, and Southeastern states) had relatively large extension investment intensities while the Northern Plains and Corn Belt states were relatively low.

Extension expenditures relative to gross farm sales were larger in all states in 1984 than in 1950. In 1984, states having small investment intensities were Nebraska, Iowa, South Dakota, and California, with $2.40–$3.50 of extension investment per $1,000 of gross farm sales in 1984. In addition to Alaska and most states in the Northeastern region, West Virginia, Virginia, South Carolina, Nevada, Louisiana, and Utah had large extension investment intensities relative to gross farm sales in 1984. As a region, the Northern Plains states had the lowest extension investment intensity that year.

These state extension data showed that the intensity of extension investment—gross and relative to sales—increased significantly between 1950 and 1984. Although the range of variation in extension investment by state was reduced when expenditures were deflated by sales, remaining differences in regional patterns suggest that other factors (e.g., average farm size and diversity of geoclimatic conditions) are also important in explaining differences across states and over time (Chapter 7).

Conclusions

Over the 100-year period 1900–2000, the rate of growth of U.S. real total public agricultural research expenditures was 4.0 percent per annum, which is somewhat lower than the 4.5 percent per annum for U.S. real private agricultural R&D expenditures for the same period. Up to about 1906, U.S.

Table 4.18. Selected measures of the intensity of investment in extension by state, 1950 and 1984

Production Region/States	Extension Exp. (in mil. 1984 dol.[a])		State and Local Gov. Share (%)		Ext. Exp. Per $1,000 of Agr. Sales	
	1950	1984	1950	1984	1950	1984
Lake States						
Wisconsin	10.0	30.0	53.6	71.7	$2.75	$5.84
Minnesota	4.8	25.1	46.5	66.5	1.67	3.70
Michigan	11.5	23.2	58.3	59.9	4.41	8.36
Corn Belt						
Illinois	14.2	26.7	62.3	59.9	1.89	3.97
Iowa	13.8	25.5	63.8	65.9	1.53	2.73
Indiana	10.4	24.2	58.3	64.9	2.58	6.16
Ohio	10.3	26.7	43.9	58.8	2.56	6.29
Missouri	10.8	24.3	46.1	58.0	2.39	6.53
Delta States						
Louisiana	11.1	24.3	59.9	67.5	8.24	15.93
Mississippi	12.2	25.1	43.3	63.3	6.55	11.59
Arkansas	9.3	18.0	40.1	58.3	4.31	5.38
Southern Plains						
Texas	21.6	59.4	48.6	70.0	2.25	6.13
Oklahoma	10.0	20.3	48.3	67.0	3.86	7.93
Northern Plains						
Kansas	12.1	23.9	70.3	76.2	2.88	4.01
Nebraska	6.3	17.3	52.6	71.7	1.47	2.44
N. Dakota	4.5	10.1	48.7	67.3	2.05	3.99
S. Dakota	4.3	7.8	47.0	57.0	1.83	2.74
Southeast						
Florida	6.5	30.2	70.0	57.0	3.49	6.58
Georgia	11.9	38.3	41.2	71.8	5.77	10.66
Alabama	12.5	23.0	46.2	55.7	8.34	10.49
S. Carolina	8.4	19.7	43.6	61.4	7.12	19.78
Appalachian						
N. Carolina	20.3	41.7	59.4	64.7	6.65	10.10
Virginia	11.5	32.7	55.4	71.6	6.77	18.21
Tennessee	11.1	23.0	42.4	52.6	5.94	11.60
Kentucky	10.0	26.2	36.6	59.2	4.37	9.89
W. Virginia	5.7	9.1	45.6	50.5	12.60	40.25
Pacific						
California	17.3	46.9	77.4	78.9	1.81	3.31
Oregon	8.1	14.1	77.2	73.8	4.97	7.88
Washington	7.0	16.3	67.6	73.6	3.46	5.56
Alaska	0.6	5.5	67.6	78.2	—	218.64
Hawaii	3.0	4.2	63.8	66.7	—	6.87

(Continues)

Table 4.18.   (*Continued*)

Production Region/States	Extension Exp. (in mil. 1984 dol.[a])		State and Local Gov. Share (%)		Ext. Exp. Per $1,000 of Agr. Sales	
	1950	1984	1950	1984	1950	1984
Mountain						
Colorado	4.7	15.1	57.4	78.8	2.01	4.51
Arizona	2.1	7.7	49.8	68.8	1.83	5.06
Utah	2.6	7.0	55.0	74.3	3.69	12.10
Wyoming	2.7	5.6	61.6	71.4	4.00	9.76
Idaho	3.6	7.2	57.5	55.6	2.33	3.15
Montana	4.4	7.8	64.2	67.2	2.88	5.50
N. Mexico	4.4	7.8	65.7	69.1	5.14	7.88
Nevada	1.4	4.5	51.3	72.6	7.23	17.86
Northeast						
New York	20.1	38.5	78.3	71.6	5.80	14.22
New Jersey	4.8	11.2	73.6	67.1	4.05	22.16
Pennsylvania	11.0	21.2	49.9	47.2	3.67	6.69
Maryland	4.9	13.2	64.6	65.2	5.19	11.41
Mass.	4.8	8.1	76.3	58.0	6.48	21.14
Conn.	2.9	4.7	65.4	53.2	4.32	13.03
Vermont	1.9	4.3	51.3	60.6	4.03	10.74
N. Hampshire	0.2	3.5	64.4	55.2	7.92	32.28
R. Island	0.2	1.9	50.8	36.8	8.85	30.16
Delaware	1.4	2.4	37.9	33.9	2.09	4.59
Maine	2.3	4.7	43.0	50.3	1.90	10.29

Source: Data on Extension expenditures are taken from Fred Woods 1992(also, see Appendix A4.3). Data on farm sales are from U.S. Department of Agriculture(1985a).

[a]Expenditures for 1950 were inflated to the 1984 price level by using the price index for research inputs.

private agricultural R&D expenditures exceeded U.S. public expenditures, but for about the next 44 years, the real expenditures of the public sector were larger. Since about 1950, U.S. private agricultural research expenditures have exceeded U.S. public expenditures on agricultural research. In 2000, we conclude that the private sector expenditures were 2.3 times higher than the public expenditures.

From 1888 to 1918, the expenditures of the SAES system exceeded those of the USDA research agencies. During 1919–1948, the USDA research agencies had significantly larger expenditures on agricultural research than the whole SAES system. The relative positions were dramatically reversed in 1949, and after that date, the SAES system has had expenditures on agricultural research that are much larger than research expenditures of the USDA research agencies.

Although the Hatch Act provided a large share of SAES funding during the 1890s, it actually has accounted for a relatively small share of SAES funding over the long term. The reasons for this decline were the growth of funds from non-Office of Experiment Stations sources, largely state government appropriations, and periodic passage of new federal legislation that provided funding for SAES. In 1955, federal funding of the SAES system was consolidated in the Amended Hatch Act, and in 1956, it accounted for 23.5 percent of total SAES funding. The share due to the Amended Hatch Act has declined to only 9.2 percent of total SAES funds in 2000. Furthermore, after 1941, the long-term trend in the share of SAES funding obtained from regular federal source (Office of Experiment Stations or Cooperative State Research Service) has been strongly negative.

The composition of the research activities undertaken with U.S. agricultural research funds has changed significantly over time. We provide some data for 1920 and 1935, but the relatively good data started in 1961. In 1961, the SAES system allocated 51 percent of its resources to crop production-related research and 23 percent to livestock production-related research activities. The USDA and private industry allocated significantly less of their research resources to production areas. Private industry allocated 53 percent of its research resources to post-harvest activities, and the USDA allocated 33 percent to this area. All three sectors allocated about 12 percent of their resources to economics and statistics.

Although the SAES vet-med systems have conducted significantly more research in total than the USDA after 1969, the USDA has dominated research on some crops (e.g., cotton) and natural resource commodity (e.g., water) categories and economics. Public agricultural research is largely biological science research. The share of the total expenditures for upstream biological science research has shown a significant increase since 1969.

With the exceptions of cotton and fruits and vegetables, public agricultural research is concentrated most heavily on enhancing biological efficiency of crops and livestock. For cotton, fruits, and vegetables, a larger share of the research is on protection-maintenance. A very small share of the research is on mechanization. Between 1969 and 1984, there was a decline of the share of funds allocated to post-harvest research. In the private sector, post-harvest research has become relatively more important after 1960, and after 1980 agricultural production-oriented research has become relatively less important, especially as the private sector made large investments in agricultural biotechnology.

Across the states, large variations exist in the amount invested in SAES research and in cooperative extension. The order of magnitude of the differences was reduced when expenditures were expressed relative to farm sales.

Appendix 4A. Expenditure Data for Public Research and Extension

Table A4.1. Total USDA funds for chemical, biological, and physical science research and for economics and for economics and statistics research, excluding administration, fiscal 1888–1997(thousands of dollars)

Year	Chem., Biol. & Phys. Sci. Research Total (net)[a](1)	Economics & Stat. Research Total (net) (2)	Total (3)	Year	Chem., Biol. & Phys. Sci. Research Total (net)[a] (1)	Economics & Stat. Research Total (net) (2)	Total (3)
1888	$ 146	—	146	1945	30,391	2,110	32,501
1889	143	—	143	1946	31,921	2,493	34,414
1890	225	—	225	1947	54,744	2,089	56,833
1891	208	—	208	1948	69,809	2,180	71,989
1892	193	—	193	1949	46,062	2,282	48,344
1893	194	—	194	1950	27,377	2,600	29,977
1894	195	—	195	1951	29,044	2,150	31,194
1895	244	—	244	1952	32,059	2,259	34,318
1896	202	—	202	1953	31,470	2,246	33,716
1897	206	—	206	1954	35,353	5,404	40,757

(Continues)

Table A4.1. (*Continued*)

Year	Chem., Biol. & Phys. Sci. Research Total (net)[a] (1)	Economics & Stat. Research Total (net) (2)	Total (3)	Year	Chem., Biol. & Phys. Sci. Research Total (net)[a] (1)	Economics & Stat. Research Total (net) (2)	Total (3)
1898	207	—	207	1955	36,165	6,557	42,722
1899	249	—	249	1956	39,416	7,147	46,563
1900	264	—	264	1957	52,303	7,866	60,169
1901	466	—	466	1958	58,946	8,767	67,713
1902	634	15	649	1959	64,631	9,049	73,680
1903	721	16	737	1960	65,671	8,790	74,461
1904	817	7	824	1961	73,182	8,996	82,178
1905	834	5	839	1962	77,337	8,190	85,527
1906	1,101	5	1,106	1963	82,080	9,742	91,822
1907	1,523	5	1,528	1964	96,286	10,016	106,302
1908	1,580	—	1,580	1965	114,376	10,138	124,514
1909	2,424	—	2,424	1966[b]	116,082	12,448	128,530
1910	2,248	—	2,248	1967[c]	117,822	15,252	133,074
1911	2,756	—	2,756	1968	117,612	11,685	129,297
1912	3,201	—	3,201	1969	122,243	15,252	137,495
1913	3,165	—	3,165	1970	133,013	13,480	146,493
1914	4,177	—	4,177	1971	143,078	14,901	157,979
1915	3,766	270	4,036	1972	188,491	19,172	207,663
1916	4,933	—	4,933	1973	195,438	20,282	215,720
1917	5,588	—	5,588	1974	202,758	21,585	224,343
1918	6,274	304	6,578	1975	224,095	24,693	248,788
1919	7,837	302	8,139	1976	324,902	34,966	359,868
1920	7,365	375	7,740	1977	288,623	28,817	317,440
1921	8,734	414	9,148	1978	315,700	33,301	349,001
1922	14,972	891	15,863	1979	321,642	36,824	358,466
1923	15,230	889	16,119	1980	339,692	42,553	382,245
1924	15,472	1,037	16,509	1981	396,414	46,546	442,960
1925	20,997	1,135	22,132	1982	397,816	45,487	443,303
1926	22,204	1,175	23,379	1983	431,428	45,155	476,583
1927	20,212	1,194	21,406	1984	435,346	47,146	482,492
1928	21,885	1,166	23,051	1985	480,920	48,476	529,396
1929	27,865	1,192	29,057	1986	469,920	45,806	515,726
1930	35,022	1,370	36,392	1987	502,935	46,743	549,678
1931	33,505	1,476	34,981	1988	534,866	50,259	585,125
1932	29,717	1,303	31,020	1989	545,541	50,614	596,155
1933	27,635	1,156	28,791	1990	562,764	51,316	614,080
1934	26,803	947	27,750	1991	610,382	53,445	663,827
1935	27,221	1,063	28,284	1992	643,781	59,515	703,296
1936	27,373	1,126	28,499	1993	642,984	59,039	702,023
1937	26,318	1,156	27,471	1994	666,831	55,931	722,762
1938	27,746	831	28,577	1995	677,859	56,411	734,270
1939	34,147	861	35,008	1996	658,280	55,052	713,332
1940	31,844	1,111	32,955	1997	681,675	53,307	734,982
1941	31,812	1,327	33,139				
1942	30,143	3,535	33,678				

1943	31,386	3,186	34,572
1944	28,880	2,429	31,309

Source: U.S. Congress, Statutes at Large, 1888–1922, containing the Laws and Concurrent Resolutions, Enacted U.S. Congress; U.S. Bureau of the Budget, Budget of the U.S. Government, 1922–1967, Inventory of Agricultural Research, 1967–1997.

[a]Excludes forestry or Forestry Service Research.

[b]The year 1966 is a transition between the older and new Current Research Information System. The totals reported for this year are an interpolation of magnitudes in 1965 and 1967.

[c]For the period 1967–1990, column (1) is "net" research expenditure of the Agricultural Research Service(or Research Service) and Nutrition Information Service (1984–89) and for column (2) is "net" research expenditures of the Economic Research Service (or Econ. Coop. And Statist. Serv.), Ag (or Farmer's Cooperative) Service, and Statistical Reporting Service. The net is defined as total expenditures less amounts for USDA contracts, grants, and cooperative agreements with SAES.

Table A4.2. Total funds in current dollars for schools of forestry and of veterinary medicine, fiscal 1969–2000 (thousands of dollars)

Year	Schools of Forestry Total	Schools of Veterinary Medicine[a] All Federal Funds	Total
1969	4,158		
1970	4,537		
1971	5,279		
1972	6,044		
1973	6,401		
1974	6,358		
1975	9,372		
1976	17,671		
1977	16,388		
1978	18,307		
1979	18,922		
1980	24,084		
1981	26,435		
1982	24,970	29,308	51,997
1983	26,255	20,808	46,458
1984	27,843	21,048	53,800
1985	28,533	20,157	56,411
1986	31,559	28,052	69,024
1987	32,095	28,978	82,476
1988	42,145	36,462	87,546
1989	46,735	36,928	103,549
1990	50,577	40,577	114,956
1991	50,591	39,648	116,435
1992	63,143	43,363	124,230
1993	79,850	49,589	134,530
1994	93,031	52,961	147,582
1995	98,094	61,211	158,136
1996	101,650	59,349	156,854
1997	109,276	63,364	163,128
1998	112,482	79,546	197,592
1999	118,497	116,710	239,282
2000	124,006	121,275	245,714

Source: U.S. Department of Agriculture, Cooperative States Research Service, Inventory of Agricultural Research, various years. CRIS funding summaries, various years.

[a]The Schools of Veterinary Medicine were conducting research before 1982, but the data for these earlier years were not reported to the Cooperative States Research Service.

Table A4.3. Total funds in current dollars for cooperative extension by major funding sources, fiscal 1915–1990 (thousands of current dollars)

| Year | Federal Total | From within State | | | | Grand Total |
		Total	State Appropriation	County Appropriation	Non-tax	
1915	1,486	2,111	1,044	780	287	3,597
1916	2,143	2,721	1,471	973	277	4,864
1917	2,719	3,431	1,928	1,258	245	6,150
1918	6,476	4,827	2,469	1,864	494	11,303
1919	9,039	5,623	2,961	2,291	371	14,662
1920	5,891	8,767	5,229	2,866	672	14,658
1921	6,434	10,358	6,044	3,294	1,021	16,792
1922	6,727	10,455	6,528	2,973	954	17,182
1923	7,101	11,384	7,054	3,420	910	18,485
1924	6,924	11,955	7,040	4,259	656	18,879
1925	6,862	12,388	7,203	3,858	1,326	19,250
1926	6,891	12,526	7,327	4,055	1,144	19,417
1927	6,916	12,832	7,423	4,363	1,046	19,748
1928	6,929	13,469	7,761	4,673	1,029	20,398
1929	8,573	13,940	6,406	6,282	1,252	22,513
1930	8,798	15,006	6,865	7,036	1,105	23,804
1931	9,705	15,876	7,243	7,523	1,110	25,581
1932	9,716	15,683	7,189	7,365	1,129	25,399
1933	9,653	13,752	6,390	6,394	967	23,405
1934	9,376	10,520	4,889	4,844	787	19,896
1935	9,000	11,042	5,045	5,152	844	20,042
1936	16,936	11,844	5,465	5,569	811	28,780
1937	17,256	12,508	5,795	5,888	825	29,764
1938	17,541	13,486	6,467	6,241	778	31,027
1939	17,968	14,148	6,582	6,676	890	32,116
1940	18,585	14,179	6,427	6,665	1,087	32,764
1941	18,591	14,603	6,707	6,807	1,089	33,194
1942	18,956	15,155	7,141	6,960	1,054	34,111
1943	18,957	15,908	7,312	7,442	1,154	34,865
1944	18,997	17,743	8,466	8,168	1,110	36,740
1945	18,997	18,839	9,158	8,480	1,201	37,836
1946	23,407	21,141	10,738	9,059	1,345	44,548
1947	27,323	25,670	12,855	11,076	1,739	52,993
1948	27,457	31,006	17,174	12,268	1,564	58,163
1949	30,531	35,202	18,867	14,214	2,121	65,733
1950	32,160	41,234	23,464	15,528	2,242	73,394
1951	32,174	43,809	24,942	16,534	2,333	75,983
1952	32,091	47,908	27,693	17,859	2,356	79,999
1953	32,150	52,443	30,544	19,644	2,255	84,593
1954	32,163	57,368	33,875	21,166	2,327	89,531
1955	39,675	60,942	35,998	22,403	2,541	100,617
1956	45,475	64,437	37,840	24,282	2,315	109,912
1957	49,865	69,330	40,516	26,502	2,312	119,195
1958	50,715	77,345	46,993	28,358	1,994	128,060
1959	53,715	81,121	49,517	30,102	1,502	134,836

1960	53,715	86,356	53,583	31,231	1,542	140,071
1961	56,715	93,383	57,895	32,782	2,706	150,098
1962	59,590	99,637	62,226	34,530	2,881	159,227
1963	63,430	105,191	65,704	36,402	3,085	168,621
1964	67,108	110,812	69,907	37,804	3,101	177,920
1965	71,684	117,200	74,341	39,776	3,083	188,884
1966	75,184	126,039	80,345	41,941	3,753	201,223
1967	78,256	135,413	87,461	44,096	3,856	213,669
1968	77,882	147,595	96,752	46,600	4,243	225,477
1969	80,762	161,190	106,326	50,288	4,576	241,952
1970	112,719	177,969	119,115	53,485	5,369	290,688
1971	138,191	193,706	129,562	58,613	5,531	331,897
1972	148,520	205,839	136,090	63,582	6,167	354,359
1973	163,104	221,987	148,218	66,387	7,382	385,091
1974	165,605	241,847	161,897	71,744	8,206	407,452
1975	178,821	269,513	181,848	79,126	8,539	448,334
1976	190,954	307,498	206,854	91,805	8,839	498,452
1977	199,232	326,130	220,906	93,612	11,612	525,362
1978	215,300	371,444	245,638	111,019	14,787	586,744
1979	221,076	403,847	270,047	119,193	14,607	624,923
1980	230,820	451,878	304,883	130,630	16,365	682,698
1981	248,935	497,580	335,723	142,390	19,467	746,515
1982	302,920	550,988	368,846	157,671	24,471	853,908
1983	316,197	581,106	389,423	164,706	26,977	897,303
1984	322,219	615,604	415,521	171,335	28,748	937,823
1985	330,939	665,690	452,866	182,253	30,571	996,629
1986	316,460	722,569	489,424	200,912	32,233	1,039,029
1987	322,309	729,717	500,601	194,693	34,423	1,052,026
1988	343,219	801,940	552,933	210,688	38,319	1,145,159
1989	347,238	860,519	587,801	229,070	43,648	1,207,757
1990	354,557	909,507	620,471	235,742	53,294	1,264,046

Source: Fred Woods 1992.

Notes

1. Griliches (1990) reached a favorable conclusion about the use of historical patent statistics as science indicators for the private sector for periods before R&D expenditures were available.
2. Griliches (1990) shows that there is a linear trend (over 1953 to 1985) in the relationship between the natural logarithm of patent applications and natural logarithm of real private R&D expenditures and a quadratic trend (over 1925 to 1985) in the relationship between patent applications and patents granted by the U.S. Patent Office.
3. Our adjustment factor is .7353 (per decade).
4. Marcus (1987, p. 22) points out that the USDA's sponsorship and funding of independent research dates back to the 1890s. At that time, the USDA started granting monies to scientists to pursue investigations outside the agency's normal research activities.
5. During 1888–1906, the Office of Experiment Stations had responsibility for publishing research results, and during 1896–1940, the Office of Experiment Stations administrated the agricultural experiment station funds for the U.S. territories.
6. Historically, most of the states have had enough nonfederal funding so as to more than meet their matching obligations.

7. Seventy percent of the agricultural research price index is comprised of wages and salaries, and it seems to be a reasonable inflator for extension expenditures.
8. CRIS has a primary classification scheme that is multi-dimensional: Research Program Areas, Activities, Commodities, and Field of Science. The classification by Commodities (or Commodity, Resource, or Technology) indicates the subject of the research, e.g., soil and land, corn, beef cattle, the farm as a business, and biological cell systems. Although researchers frequently complain about the rigid nature of the CRIS classification system, it has several very good features in the way it attempts to account for research expenditures.
9. Federal agencies located outside the USDA also perform a small amount of agricultural research. The 1977 Farm Bill made the USDA responsible and accountable for coordinating all federal agricultural research (Office of Technology Assessment 1981). The only federal agricultural research projects recorded in CRIS are for the USDA research agencies.
10. Each project leader determines the field of science classification of his or her project, and more than one field can be designated along with the relative weight of each. (See Huffman and Evenson 1994 , Appendix Table 15, for the translation of CRIS fields of science into the field of science categories used in this chapter.)
11. In the program of work starting in 1984, the Extension Service discontinued their "issue" emphasis. This means that it is virtually impossible to make meaningful comparisons of pre- and post-1984 extension service resource allocation patterns in greater detail than at the program area level.

References

Committee on Research Advisory to the U.S. Department of Agriculture. 1972. "Report Submitted to the National Academy of Science—National Research Council," Unpublished "Pound Report."

Evenson, Robert E. 1983. "Intellectual Property Rights and Agribusiness Research and Development: Implications for the Public Agricultural Research System." *American Journal of Agricultural Economics* 65: 967–975.

Flatt, William P., H.O. Graumann, and A.W. Cooper. 1982. "Agricultural Research Classification for Management Purposes." In *An Assessment of the United States Food and Agricultural Research System*, Vol. II, Part A. Office of Technology Assessment. Washington, D.C.: U.S. Government Printing Office.

General Accounting Office. 1977. *Management of Agricultural Research: Need and Opportunities for Improvement*. GAO Report to the Joint Economic Committee. Washington, D.C.: U.S. Government Printing Office.

Griliches, Zvi. 1960. "Hybrid Corn and the Economics of Innovation." *Science* 132: 275–280.

Griliches, Zvi. 1990. "Patent Statistics as Economic Indicators: A Survey." *The Journal of Economic Literature* 28(4): 1661–1707.

Heimlich, Ralph. 2003. *Agricultural Resources and Environmental Indicators, 2003*. Agriculture Handbook No. 722(Feb). Washington, D.C.: USDA, Economic Research Service.

Hightower, J. 1973. *Hard Tomatoes, Hard Times*. Cambridge, MA: Schenkman Publishing.

Huffman, Wallace E. 1996. "Labor Markets, Human Capital, and the Human Agent's Share of Production." In *Essays in Honor of D. Gale Johnson,* ed. J. Antle and D. Sumner. Chicago, IL: The University of Chicago Press.

Huffman, Wallace E. and R. E. Evenson. 1994. "The Development of U.S. Agricultural Research and Education: An Economic Perspective." Department of Economics, Iowa State University, Ames, IA. (Available through the Parks Library at Iowa State University.)

Huffman, Wallace E. and John A. Miranowski. 1981. "An Economic Analysis of State Expenditures on Experiment Station Research." *American Journal of Agricultural Economics* 63:104–118.

Klotz, Cassandra, K. Fuglie, and C. Pray. 1995. *Private-Sector Agricultural Research Expenditures in the United States, 1960–92.Staff Paper No. AGES-9525*. Washington, D.C.: USDA, Economic Research Service.

Marcus, Alan I. 1987. "Constituents and Constituencies: An Overview of the History of Public Agricultural Research Institutions in America." In *Public Policy and Agricultural Technology: Adversity Despite Achievement,* ed. D.F. Hadwiger and W.P. Browne. London, England: Macmillan.

National Science Foundation. Various years. *Science Indicators*. Washington, D.C.: U.S. Government Printing Office.

Office of Technology Assessment. 1981. *An Assessment of the United States Food and Agricultural Research System*. Vol. I. Washington, D.C.: U.S. Government Printing Office.

Robinson, Roland. 1956–1968. "Revised Reports of Expenditures of Agricultural Experiment Stations, OD-1044." Washington, D.C.: U.S. Department of Agriculture, Cooperative States Research Service.

Robinson, Roland. 1985. "Evolution of Research Program Content." Washington, D.C.: U.S. Department of Agriculture, Cooperative States Research Service. Unpublished report.

Rockefeller Foundation. 1982. "Science for Agriculture." New York, NY: Rockefeller Foundation ("Winrock Report").

Schmitz, A. and D. Seckler. 1970. "Mechanized Agriculture and Social Welfare: The Case of the Tomato Harvester." *American Journal of Agricultural Economics* 52: 569–577.

U.S. Bureau of the Budget. 1922–1967. *Budget of the U.S. Government*. Washington, D.C.: U.S. Government Printing Office.

U.S. Congress. 1888–1922. *Statutes at Large*. Washington, D.C.

U.S. Department of Agriculture. 1889–1894. "Report of the Director of the Office of the Experiment Stations." Department of Agriculture.

U.S. Department of Agriculture. 1895–1897. "Statistics of Agricultural Colleges and Experiment Stations." Office of Experiment Stations *Circular* Nos. 27, 29, 35. Washington, D.C.: U.S. Government Printing Office.

U.S. Department of Agriculture. 1898–1902. "Statistics of Land-Grant Colleges and Agricultural Experiment Stations in the United States." Office of Experiment Stations *Bulletin* Nos. 51, 64, 78, 97, 114. Washington, D.C.: U.S. Government Printing Office.

U.S. Department of Agriculture. 1903–1942. *Annual Report, Office of Experiment Stations*. Washington, D.C.: U.S. Government Printing Office.

U.S. Department of Agriculture. 1920–1987. "Annual Reports." Washington, D.C.: U.S. Department of Agriculture, Extension Service.

U.S. Department of Agriculture. 1944–1954. *Report on the Agricultural Experiment Stations*. Washington, D.C.: U.S. Government Printing Office.

U.S. Department of Agriculture. 1951. *Report on the Agricultural Experiment Stations*. Washington, D.C.: U.S. Government Printing Office.

U.S. Department of Agriculture. 1955–1961. *Report on State Agricultural Experimental Stations*. Washington, D.C.: U.S. Government Printing Office.

U.S. Department of Agriculture. 1962–1969. *Funds for Research at State Agricultural Experiment Stations*. Washington, D.C.: U.S. Government Printing Office.

U.S. Department of Agriculture. January 1962. *Food and Agriculture: A Program of Research*. A report of a U.S. Department of Agriculture study in cooperation with the State Agricultural Experiment Stations. (This study was commissioned by House Agriculture Committee Report 448, 87th Congress, First Session, June 2, 1961. The publication is not generally available because it was never distributed due to political opposition to projections for future research.)

U.S. Department of Agriculture. 1969–1991. *Inventory of Agricultural Research*. Washington, D.C.: U.S. Department of Agriculture, Cooperative States Research Service.

U.S. Department of Agriculture. 1985a. *Economic Indicators of the Farm Sector: State Financial Summary, 1984*. Washington, D.C.: U.S. Department of Agriculture, Economic Research Service.

U.S. Department of Agriculture. 1985b. *Inventory of Agricultural Research FY1984*. Washington, D.C.: U.S. Department of Agriculture, Cooperative States Research Service.

U.S. Department of Agriculture. 2002. *Inventory of Agricultural Research FY2000*. Washington, D.C.: U.S. Department of Agriculture, Cooperative States Research Service.

U.S. Department of Agriculture. 1986. *Compilation of Statutes Related to Agriculture and Forestry Research and Extension Activities and Related Matters*. Washington, DC: U.S. Department of Agriculture, Agricultural Research Service.

U.S. Department of Agriculture and the Association of State University and Land-Grant Colleges. 1966. *A National Program of Research for Agriculture*. Washington, DC: U.S. Government Printing Office.

U.S. Department of Agriculture, Cooperative State Research Service. 1993. *Current Research Information System: Manual of Classification of Agricultural and Forestry Research.* Version V, Beltsville, MD.

U.S. Department of Agriculture, Cooperative State Research, Education, and Extension Service. 2003. *Current Research Information System: Manual of Classification of Agricultural and Forestry Research.* Version VI, Beltsville, MD.

U.S. General Accounting Office. 1977. *Management of Agricultural Research: Need and Opportunities for Improvement.* "GAO Report to the Joint Economic Committee." U.S. Department of Agriculture. Washington, DC: U.S. Government Printing Office.

Woods, Fred. 1992. "Amount and Percent of Cooperative Extension Funds Available, by Source, 1915–1990." Washington, D.C.: U.S. Department of Agriculture, Extension Service.

5

The Private Sector, Biotechnology, and R&D for Agriculture

Private, for profit, firms require incentives to engage in invention and innovation. These incentives are typically in three forms: market power, intellectual property rights (IPRs), and trade secrecy. These forms are often in conflict (Rosenberg 1976; Scherer 1982). Intellectual property rights require that an invention be removed from secrecy. A firm must decide whether to obtain an IPR or to maintain a trade secret. A firm with market power has an incentive to invent and innovate because these firms have some price-setting power. By bringing a new improved product to the market, the firm can set a higher price and thus earn a return on its invention/innovation investment. When other firms are engaged in invention and innovation, competitive pressures to invest in new product development are created. Firms without market power (as is the case of most farms) do not have strong incentives to invest in significant invention and innovation activities unless they have some means to prevent others from copying their inventions. R&D also contributes to economic growth (see Jorgenson and Griliches 1967; Jorgenson, Gollop, and Fraumeni 1987; Jones 2002).

As indicated in Chapter 1, intellectual property rights (IPRs) create legal incentives to invent and, indirectly, to innovate. IPRs take a number of forms. The patent right is the chief IPR for mechanical, chemical, and electrical inventions used in the agricultural sector. Inventions in plant and animal improvement were effectively without IPRs until 1930 in the U.S., when the Plant Patent Act provided IPRs for asexually reproduced plants. In 1960, the Plant Variety Protection Act extended "breeders' rights" protection to sexually reproduced plants. It was not until "case law" rulings extended patent IPRs to plants and animals after 1980 that genetic inventions were given IPRs on a par with mechanical, chemical, and electric inventions.

The absence of real IPR protection for plant and animal improvement programs until recent decades presented a serious challenge to the USDA-SAES system. This challenge was met by public investment in plant and animal breeding programs. The USDA-SAES research system was built because of the limited scope of IPRs for agricultural inventions, particularly for genetic improvement. The agricultural experiment station model proved to be compatible with plant and animal breeding programs. Plant and animal breeding programs were arguably the "core" programs in the SAES-USDA system until the 1960s and even into the 1980s. Investment in public sector R&D for agriculture was greater than private sector investment until after World War II. (See chapter 4.)

As IPRs have been strengthened, private sector investment has increased. This is particularly notable with the emergence of the agricultural biotech research sector. The public sector has found that its role, and the "optimal" form of support for private sector invention and innovation, has changed as both technology and IPRs have changed.

In this chapter, we begin in part I with a review of IPR legislation and its implementation. Part II summarizes crop varieties protected by Breeders' Rights laws in the U.S. Part III provides summaries of patented inventions used in the agricultural sector. This section uses the Yale Technology Concordance to "assign" patented inventions to industries-of-manufacture (IOM) and sectors-of-use (SOU). Part IV discusses the changes brought about by investments in public and private sector biotechnology research for the agricultural and food sector. These events have dramatically changed the nature of private sector R&D. Because of relatively strong IPRs in the mechanical, chemical, and electrical fields of invention, the private sector has long been the major developer of innovations in these fields. With the development of stronger IPRs for genetic inventions, particularly for biotech-based inventions, the private sector is now dominating all fields of invention for agriculture. It is critical that the public sector USDA-SAES system develop effective programs to support private sector invention and innovation. Part V reports summaries of the National Plant Breeding Study conducted in the United States in 1994. Part VI reports a summary of R&D investment by commodities, and Part VII presents conclusions.

Intellectual Property Rights (IPRs)

In this section, we lay out the taxonomy of IPRs for agriculture, the economics of IPRs, and the role played by international conventions.

IPRs Relevant to Agriculture: A Taxonomy

A taxonomy for 11 types of IPRs is presented in Figure 5.1. Some IPRs are in conflict with others. The most important rights for inventions and innovations are patent rights. A distinction is made in Figure 5.1 between traditional patent rights and expanded patent rights. Traditional (original) patent protection has been provided to inventors in the chemical, electrical, and mechanical fields of invention for many years. The "expanded" patent protection now covers genetic inventions. The expansion in question occurred primarily in the United States and was achieved through "case law". That is, the expanded coverage of patent protection was the result of court decisions, not of federal legislation. In the case Diamond vs. Chakrabarty (447US 303[1980]), the court ruled that multicellular living plants and animal were not excluded from patent protection.[1] Further, court rulings in ex parte Hibberd for plants (227 USPQ 443(1985)) and ex parte Allen for animals (2 USPQ 2d 1425) reaffirmed this. This opened the door to patenting of plants and of genes and gene constructs.[2]

To obtain a patent right, the right holder must demonstrate that the invention:

(a)   Is novel, i.e., new to the world, the first of its kind;
(b)   Is useful in the sense that it can be incorporated into a useful device;
(c)   Entails an "inventive step." Courts test this by requiring that the invention be "unobvious to a practitioner skilled in the art."

In addition, the patent document must provide an "enabling disclosure" of the invention. This must be in adequate detail to allow the replication of the invention. This is part of the IPR "bargain". In return for IPR protection, the invention must be removed from secrecy (see below).

International conventions affect patent rights. The Paris Convention enacted in 1887 allowed for diversity in patent laws in different countries but required that each member country provide "national treatment" to inventors from another member country. The Paris Convention (and its amendments) allows member countries the right to obtain patent protection in another country within one year of the original application and to maintain the original date of filing. This is important because most countries (except the U.S.) operate on a "first to file" basis to establish novelty.[3]

IPRs	Scope	Period of Protection	Conditions for Obtaining IPR			International Convention	Disclosure Requirement	Research Example	Reproduction Rights
			Novelty	Usefulness	Inventive Step				
Patent (traditional)	Inventions: chemical, electrical, mechanical	17–20 years	Yes (global)	Yes	Strong	Paris, TRIPS	Enabling	None (conceptual)	Negotiable
Patent (expanded)	Inventions: genes, plants, and animals	17–20 years	Yes (global)	Yes	Strong	Paris, TRIPS	Enabling	None (germplasmic)	Negotiable
Utility model	Minor inventions	5–15 years	Yes (national)	Weak	Weak	None	Enabling	None (conceptual)	Negotiable
Industrial design	Designs	Permanent	Yes	None	None	None	None	None (conceptual)	None
Breeders' rights	Plant varieties	5–16 years	Yes	Yes	Weaker	UPOV	None, deposit	Allowed	Limited (farmers' rights)
Appellation of origin	Food products	Permanent	Regional	None	None	Lisbon	Location of productions	Allowed	None
Folkloric Rights	Indigenous products	Permanent	Yes	None	None	FAO/UNESO		Allowed	Negotiable
Farmers' rights (CBD)	Genetic resources	Permanent	Yes	None	None	CBD undertaking	Location of genetic resource	None (use rights)	(use rights)
Copyrights	Written works	Life + 50 years	Yes	None	None	Berne, TRIPS	None	Allowed	None (limited)
Trademarks	Brand names	Permanent	Yes	None	None	TRIPS	None	Allowed	None
Trade	Trade secrets	Permanent	None	None	None	None	None	None	None

Fig. 5.1. Intellectual property rights: a taxonomy. Source: Adapted from Evenson (2000).

The utility model is used in many countries in Asia and Europe. It is often referred to as a "petty patent" because it protects minor inventions. In some countries novelty is judged against a national standard. The utility model can be used to protect "adaptive invention." Industrial designs protect shapes and designs and are important for marketing, and breeders' rights (BR) protect plant varieties that meet uniformity and stability standards. Protection is weakened by a research exemption. The research exemption allows a researcher (plant breeder) to utilize a BR protected variety as a parent variety. The farmers' exemption allows a farmer to save seed from his crop. A recent U.S. Supreme Court ruling allows a plant variety to be protected by a patent or a breeders' right or both. The World Trade Organization (WTO)-Trade Related Intellectual Property Issues (TRIPS) *sui generis* system for plant variety is widely expected to be a breeders' rights system.[4]

Other rights are as follows: Appellation-of-origin rights are largely used for labeling and identification purposes. These rights are important for "niche" markets in wine, cheese, and similar products. Folkloric rights, including farmers' rights in the Convention on Biodiversity (CBD), are relatively new and untested in courts. They are often seen as "developing country" rights, because of the perception that developing countries actually produced most "farmers' varieties" or landraces of major cultivated crops. Copyrights protect written works. They also protect the "copying" of such items as computer programs. Trademarks are important in most food markets. They "identify" brand names and prevent other companies from benefiting from brand loyalty. Finally, trade secrets are protected in cases where an employee may reveal secrets. Fundamentally, a company must make a choice between holding an invention in secrecy and obtaining patent rights.

The Economics of Patent Rights

When an inventor obtains a patent right, this right has three features:

1. It is a "right to exclude" others from making or using the invention. It is not a right to actually make and use the invention.
2. The right to exclude is limited in time. Under current WTO-based rules, the patent right expires after 20 years from the date of application.
3. The right to exclude is granted in return for the "removal from secrecy" of the invention.

Patents provide important incentives for invention, because the patent right can be licensed. Many efforts to develop an invention fail, but the patent right does provide incentives for invention effort. Many large industrial companies invest millions of dollars in R&D programs.[5]

But patents are also important to protect investments made in innovation—the commercialization of inventions. Most inventions are made by R&D employees in a firm and assigned to the firm. The firm then has to evaluate the merit of investing to commercialize the invention. Fewer than ten percent of the patent grants made by the U.S. Patent and Trademark Office actually become innovations. Patent protection to protect innovation is important because investments in pilot production, testing, and marketing can be quite large and unless the investing firm can exclude other firms from taking advantage of these investments, innovations will not take place. This does not mean, however, that the inventor and the innovator have to be the same party. An inventor can provide an "exclusive license" to an innovator, thus preserving the incentive.

Prior to the Bayh-Dole Act of 1980 (PL.96-517), research undertaken in U.S. universities with federal government funding (NSF, NIH, USDA, etc.) were required to share patent licensing revenues with the federal government. They were also required to offer "non-exclusive" licenses to prospective licensees. Under these conditions, few university inventions were patented. Approximately 300 university and government patents were granted annually in the 1970s.

The Bayh-Dole Act changed both requirements. Universities were no longer required to share licensing revenues with the federal government. And they were no longer required to offer "non-exclusive" licenses. The Bayh-Dole Act has had a profound effect on university research. It created incentives to orient more applied research to the development of inventions. The granting of exclusive licenses also enabled universities to overcome the problem of innovation incentives. By 2002, the number of university and government generated patents had increased to more than 3,000 (Massing 2003; Henderson, Jaffe, and Trajtenberg 1998).

The dominant IPR for encouraging invention and innovation is the utility patent (usually referred to as a letters patent or simply a patent). The logic for the patent right is shown in Figure 5.2. It depicts two potential inventions. The upper panel depicts a major invention; the lower panel, a "run of the mill" invention. For each invention, the first period after the patent is granted, the last period of the patent grant, and the periods after the patent is no longer valid are shown.

For both inventions in period 1, there is a demand for the invention depicted by the curve DD. This demand is expressed as the quantity or units of use demanded at different royalty rates, $r(u)$. In period 1, the holder of the patent right has a monopoly right over the products in which this invention is embedded. That is, the IPR holder has the "right to exclude" others from making or using the product. In period 1, the monopoly rate is $r^*(u)$ for both inventions.[6] The area PR represents payments to the IPR holder. These may be licensing revenues or in the form of price premiums for the product. The area CS is the "consumer surplus" associated with the monopoly IPR grant. The area UCS is the unrealized consumer surplus associated with the monopoly IPR grant.

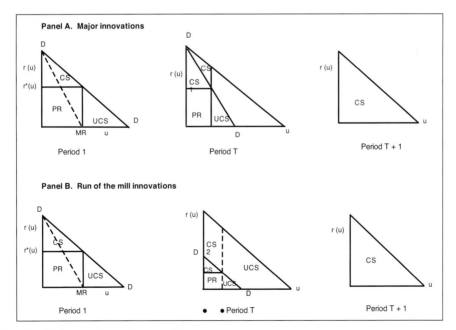

Fig. 5.2.   Benefits from major and run-of-mill innovations.

By period T, the final period of the patent monopoly, the market for both inventions will have been eroded by the development of substitute products. This erosion is a natural part of the market economy. Substitute product development is a central feature of industrial markets. The development of substitutes is accelerated by the requirement that the patent documents provide an enabling disclosure of the invention. As shown in Figure 5.2, the major invention has modest market erosion, but the run-of-the-mill invention has major market erosion.

In a patent system with significant renewal fees, the major invention is likely to be renewed to year T. The run-of-the-mill invention will not be renewed to year T.[7] In period T + 1, both inventions will be "off-patent" and will contribute consumer surplus gains that may continue for many years.[8] The run-of-the-mill invention should be converted to consumer surplus earlier than year T under an optimal renewal-fee structure.

The gains from IPR systems depend in part on alternative R&D systems. Public sector agricultural research systems produce inventions (plant varieties) without the UCS components associated with IPR systems. Ideally one would like to realize the UCS components if possible. Without the UCS gains, however, IPR systems do produce CS gains up to period T and full CS gains thereafter. These are important gains, because for many inventions, the invention lives on as a "building block" in products that have eroded the market for the original invention. As noted above the IPR is necessary to protect both the invention and the innovation. Most private sector R&D is actually D, i.e., development or commercialization expenses.[9]

Can public sector agricultural research programs really be an alternative to IPR-based private firm R&D? Almost certainly not! It would be almost unimaginable that the range of inventions used in U.S. agriculture could have been produced in a public sector agricultural research system. Literally thousands of inventions in farm machines and farm chemicals have been made by private individuals and private firms in the industries supplying products to and buying products from the agricultural sector (see below).

International Treaties—Conventions

Most countries of the world are members of at least one international agreement that attempts to protect an inventor's rights to his or her invention in foreign countries (Evenson 2000). These agreements perform a function that is similar to the way free trade agreements protect commerce from tariffs and other unilateral trade restrictions. The most widely held agreements are the International Convention for the Protection of Industrial Property. It is sometimes called the "Paris Convention" because of the location of its formulation in 1883. This agreement, as subsequently amended at The Hague (1925), London (1934), Lisbon (1958), and Stockholm (1967), provides that any country belonging to the convention should grant citizens of another convention country the same rights that it grants its own citizens.

Two other treaties have a more direct bearing on agricultural inventions: the International Convention for the Protection of New Varieties of Plants (ICPNV) and the Budapest Treaty on the International Recognition of the Deposit of Microorganisms for the Purpose of Patent Procedure. The Plant Variety Convention was amended most recently in 1978 and provides for patent or patent-like protection to breeders of new plant varieties who belong to member countries. These plants may be sexually as well as asexually reproduced (which gives protection to hybrid varieties), but member states may exclude hybrid varieties from protection at their discretion (on the grounds that the breeder retains control over the parents, rendering protection unnecessary).

The Budapest Treaty on Microorganisms was signed in 1977. It provides for an "international depository authority" in several nations that keeps samples of patented microorganisms. This special arrangement takes the place of the usual written and/or graphic description that regular patent documents employ. The treaty does not grant patent protection per se, but merely commits member countries to recognizing deposits made in other countries as equally valid with those made in their own. Thus, the treaty leaves a considerable degree of freedom in the hands of the individual countries to decide what constitutes a patentable microorganism. The treaty aims to lower the cost and reduce the inconvenience of depositing multiple samples in each country in which the inventor desires protection.

Two recent developments in the United States have dramatically expanded the scope of U.S. patent law. The first is administrative. The United States established a Federal Court of Appeals in 1980 specifically to deal with intellectual property rights. This institution has contributed both to more efficient adjudication of disputes and conflicts and to a climate for strengthened property rights.

The second and much more important development has been in case law expansion. In Diamond vs. Chakrabarty (447US303) (1980), the Court of Appeals ruled that living tissue was not excluded from patentability. This opened the door for extension of patent protection to plants and animals. In ex parte Hibberd (227USPQ443) (1985), patent protection was expanded to plants including hybrid corn and other plants. This extension was facilitated by modern biotechnology advances allowing novelty to be identified more precisely. A large number of agricultural plants now have been patented including a large number of corn hybrids. Ex parte Allen allowed for patenting of animals (2USPQ2d 1425) (PTO Bd Pat. App. & Int., 1987). The first animal patent was granted in 1988 to the "Harvard Mouse," a cancer research mouse. See Barton (1998).

Breeders' Rights Protection

Until the 1870s, all crop genetic improvement was the result of farmer selection of seed. By saving seed that performed well in a specific production environment, farmers created numerous "landraces" or "farmers' varieties." These landraces are vital parts of today's plant genetic resource collections.

Modern plant breeding programs were developed for sugarcane and cereal crops in the 1870s and 1880s. By 1900 the USDA had established breeding programs for many crops. Most SAES units were also active in plant breeding by 1900 or shortly thereafter.

With the establishment of formal plant breeding programs, most crop varietal improvements were achieved in these programs. By contrast, animal breeding programs were dominated by ranchers and dairymen until the development of the parent-grandparent system of poultry breeding in the 1930s and the development of the artificial insemination system for dairy cattle starting in the 1950s and spread to other species in the 1980s.

Hybridization and Private Firm R&D

The first major incentive system for private sector plant breeding was the development of "hybrid" maize (and later sorghum and millet) varieties in the early part of the 20th century. Hybrid maize was not successful commercially in the United States until after double crosses were developed at the New Haven, Connecticut, Agricultural Experiment Station in 1920. An additional decade was required before superior double cross varieties were generally available to farmers in the Corn Belt (Griliches 1960). Starting in the early 1930s, hybrid corn varieties rapidly replaced open-pollinated corn varieties in the Corn Belt and then spread to the rest of the nation. As shown in Figure 5.3, farmers in some states began to use hybrid corn much earlier than others. Also, once the transition to hybrids started, farmers in the center of the Corn Belt made the changeover quite quickly. For example in Iowa, farmers went from planting 10 percent to 90 percent of their total corn acreage to hybrids in only 4 years. Farmers in states on the fringe of the Corn Belt started planting hybrid corn later and did not make the transition to nearly complete usage as quickly. In almost every state, however, the adoption followed an S-shaped growth curve. Griliches (1960) was the first to show how differences in local geoclimatic conditions affect the use and diffusion of new agricultural technology.

Some of the first successful "hybrids" were obtained in corn in 1907 and 1908 by George Shull (Carnegie Institute for Experimental Evolution, Cold Springs Harbor, New York) and Edward East (Connecticut Agricultural Experiment Station, New Haven, Connecticut). Donald Jones, a Harvard graduate student working at the Connecticut Agricultural Experiment Station, New Haven, discovered how to produce low-cost hybrid seed corn varieties. He was impressed with the corn yields of some of East's single cross hybrids. He then proposed using the seeds of two of these single cross hybrids as parents for a double cross hybrid. From experiments conducted in 1916–1919, Jones obtained immediate success with his first double cross (the Burr-Leaming hybrid), which consistently yielded 20 percent better than the best open-pollinated varieties then adapted to Connecticut. The seed from double crosses could be sold at a much lower price than that from single crosses. Later, it was discovered that only a very small percentage of single cross combinations resulted in superior double cross hybrids. Hybrid corn became a wide-scale commercial success in the United States starting in the 1930s (Griliches 1957, 1960).

The SAES programs in the Corn Belt states and the USDA had corn-breeding programs during the first two decades of the 20th century, but scientists in these programs failed to piece together the

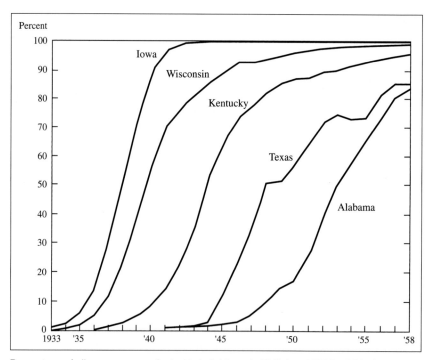

Fig. 5.3.   Percentage of all corn acreage planted to hybrid seed. (Griliches 1960 by AAAs.)

complex puzzle lying behind commercial hybrid corn. Some stations focused their corn research on improving open-pollinated varieties (e.g., Iowa), and others engaged in inbreeding, or pure line development (e.g., Illinois and Indiana). Some had tried but were not impressed with single cross hybrids. When Jones discovered double crosses in 1916–1919, C. P. Hartley, director of the USDA's corn-breeding programs, and several directors of corn research in Corn Belt agricultural experiment stations were convinced that hybrid corn had no practical importance. This attitude delayed significantly hybrid corn research in Minnesota, Indiana, and Illinois (Crabb 1947).

Between 1920 and 1963, virtually all hybrids planted by U.S. farmers were double crosses (Hallauer and Miranda 1981; Jugenheimer 1976). During the early 1960s or 30 years after the first introduction of commercially successful double-cross hybrids in the Corn Belt, commercial single-cross hybrids became available that consistently outperformed existing double-cross hybrids and that were only slightly more expensive. Single crosses have an advantage in concentrating genes for superior traits. Inbred lines were developed that were respectable seed producers, which kept the seed costs from being prohibitive. Consequently, during the 1960s, farmers made a transition from planting almost exclusively double crosses to planting almost exclusively single crosses. In 1979, 88 percent of the commercial hybrid seed was single crosses (Zuber and Darrah, 1980). The corn-breeding programs at the University of Wisconsin and the University of Minnesota began unexpectedly early, given the corn acreage in these states in 1930, and the programs have been unusually successful over time in pushing the Corn Belt north (Griliches 1960). During the 1980s, the corn-breeding program at the University of Nebraska became effective in aiding the movement of the Corn Belt west.

In the Midwest, hybrid varietal development shifted rapidly away from the public sector (USDA and SAES) to private commercial hybrid-seed companies; but up to the late 1980s, the public sector was an important source of inbred lines used in commercially successful hybrids. The public sector has shifted its research emphasis toward pre-invention science, e.g., plant physiology. In the South, public research has been more heavily involved in research and development of hybrids. Public researchers in general have become the custodians of corn germplasm stocks and watchdogs over problems in the commercial seed corn industry (Wallace and Brown, 1956).

The public sector corn-breeding research program has been and is a joint USDA and SAES activity. Although the research is located physically at SAES programs, the corn-breeding programs in major corn-producing states generally are conducted jointly by USDA-ARS and SAES researchers.

Although the early hybrid seed-corn companies engaged primarily in reproduction and marketing of hybrids developed by the public sector, Pioneer, DeKalb, and other companies successful over the long term quickly developed their own corn-breeding programs. Over time more companies have developed breeding programs; and in 1982, 66 private companies had corn-breeding programs (Kalton and Richardson 1983). They employed 155 (full-time equivalent) Ph.D. scientists and 300 (full-time equivalent) M.S. and B.S.-level scientists. Some companies are engaged in fairly sophisticated research, but others, especially small companies, continue to rely heavily upon public-sector inbred lines for developing commercial hybrids (Duvick 1996). During the 1990s, the U.S. seed industry underwent major reorganization and consolidation with related industries.

Specific IPRs for Plants: Plant Breeders' Rights

The first specific IPRs for plants were provided by the Plant Patent Act of 1930. This act allowed for patenting of asexually reproduced plants (except tubers). This form of protection prevented buyers of protected plants from simply reproducing plants through cuttings, grafts, or other asexual methods. These early "breeders' rights" were used by private breeders of trees (citrus, fruit, nuts), shrubs (azalea, rubirosum), ornamentals (rose, chrysanthemum), and fruits (blueberry, grape, raspberry, and strawberry).

By the 1960s, several European countries had enacted plant breeders' right (PBR) laws. The 1961 UPOV Convention governed their rights. The UPOV Convention governs the coverage of genera and species and regulates DUS standards. DUS refers to the fact that eligible varieties must be "distinct, uniform, and stable." PBR offices in several countries test for DUS standards.

The United States passed a PBR law, the Plant Variety Protection Act (PVPA), in 1970. This act covered plants reproduced from seed. In 1980, an amended act included protection for okra, celery, peppers, tomatoes, carrots, and cucumbers. The U.S. joined UPOV in 1980 under the terms of the 1978 UPOV Convention. The PVPA provides a 20-year period of protection. Farmers can save seed for their own use, but not for resale. The Plant Variety Protection Office in the USDA administers the PVPA (The Patent and Trademark Office administers most other IPRs).

A Supreme Court ruling in 2001 (J.E.M. Supply vs. Pioneer Hi-Bred) made plant varieties eligible for both utility patent protection and PVPA protection. Studies of the breeders' rights impact on private sector breeding programs find that the PVPA did stimulate private firm breeding programs (Alston and Venner 1998).

The number of PVPCs issued by crop for three year subperiods, starting in 1991 and continuing to 2002, is reported in Table 5.1. The USDA-SAES public sector system (with Iowa State University as the leader) obtained 21.2 percent of field crops PVPCs, 14.8 percent of grasses PVPCs, and 10.3 percent of vegetable PVPCs. Foreign inventors obtained 1.8 percent of field crop PVPCs, 7.1 percent of grasses PVPCs, and 16.4 percent of vegetable PVPCs. Two thirds of all PVPCs were granted for field crops. The total volume of PVPCs granted declined in the 1995–98 period and then rose again in the 1999–2000 period. Almost all foreign PVPCs were granted to the Netherlands, United Kingdom, and West Germany.

The different roles of the public and private sectors in soybean breeding are particularly instructive. Before 1965 only 1 or 2 private-sector soybean breeding programs existed. Fifteen or so SAES-USDA programs existed. They were producing a steady but slow rate of varietal release (three to four per year). By 1970, 15 or so new private-sector soybean breeding programs were initiated, and by 1980 this had grown to 26. The flow of new varieties increased markedly, including an increase in output by the SAES-USDA units.

After the 1970 PVPC Act, the U.S. Patent Office made major changes in patenting of living organisms. In a 1980 court case ruling (the Diamond vs. Chakrabarty case), the door was opened to patenting living organisms. As of 1987, a number of organisms, mostly bacteria, had been patented. In 1987, the Patent Office announced much broader rules for patenting living materials, including animals. These changes have altered the setting and the incentive structure for invention in both the private and public sector. They have also altered the optimal scale of SAES-USDA research programs.

Inventions Used in Agriculture

The wealth of data in the U.S. patent system can be transformed into more useful indicators by using the Yale Technology Concordance (YTC) system (Evenson 1991). This system is based on approximately 300,000 patent assignments by the Canadian Patent Office. As patent applications are received in the Canadian Patent Office, they are given an international patent class (IPC) desig-

Table 5.1 Plant variety protection certificates issued by the U.S. Plant Variety Protection Office, 1991–2002

	Total Issued	% Public	% Foreign	Total Issued by Sub-Period		
				91–94	95–98	99–2002
Field Crops:						
Barley	76	34.2	6.6	35	18	23
Beans (fld)	73	32.9	0.0	29	9	35
Corn (fld)	504	2.4	1.0	162	122	220
Cotton	115	15.7	0.9	39	17	59
Oats	34	82.4	0.0	8	13	13
Rice	28	39.3	25.0	15	10	3
Safflower	17	35.3	0.0	6	6	5
Soybeans	575	19.3	0.3	162	156	257
Wheat	344	40.4	3.8	95	75	174
Total Field Crops	1766	21.2	1.8	551	426	789
Grasses and hay crops:						
Alfalfa	46	4.3	0.0	11	9	26
Bluegrass	47	19.1	23.4	23	10	14
Clover	13	46.2	0.0	1	7	5
Fescue	87	2.3	12.6	30	16	41
Ryegrass	59	6.8	1.7	14	7	38
Other	73	34.2	0.0	17	18	38
Total Grasses and hay	325	14.8	7.1	96	67	162
Vegetables:						
Beans, Garden	132	0.0	9.8	69	18	45
Beans, Lima	2	0.0	0.0	1	1	0
Cauliflower	5	0.0	40.0	5	0	0
Lettuce	121	0.0	13.2	81	5	35
Onion	37	48.6	5.4	10	18	9
Peas	169	0.0	21.3	48	36	85
Tomatoes	27	59.3	3.7	20	2	5
Watermelons	23	21.7	0.0	6	0	17
Other Vegetables	154	19.5	26.0	27	18	109
Total Vegetables	670	10.3	16.4	267	98	300
Flowers	77	9.1	16.9	29	25	23
Total All Crops	2838	17.6	6.3	943	616	1274

nation and two industry designations. The first industry designation is to a four digit industry-of-manufacture (IOM). The second designation is to a four digit sector-of-use (SOU). Utilizing the YTC, it is possible to distribute, on a probability basis, inventions in a four-to-six digit IPC to IOMs and SOUs. It is also possible to produce an IOM-SOU matrix based on Canadian assignments.

The IOM-SOU matrix for seven agricultural sectors of patented inventions in Canada from 1978 to 1993 is reported in Table 5.2. The seven agricultural SOU sectors are livestock production, crop production, fruit and vegetable production, horticultural crop production, services to the livestock sector, services to the crop sector, and services to a residual sector. As the table indicates, the agricultural sector is the using sector for inventions manufactured in many industries. The most important

Table 5.2 Patents used in agriculture by major industries-of-manufacture, Canada 1978–1993

IOM (Industry of Manufacture)	Livestock Production	Crop Product.	Fruit & Veg. Production	Horticulture Production	Services			Total	Share of IOM
					Livestock	Crops	Other		
10—Food industry	179	2	2	0	0	0	0	197	0.11
1053—Feed industry	174	1	1	0	14	0	0	190	0.70
12—Tobacco	0	8	0	0	0	0	0	8	0.02
15—Rubber products	3	0	1	1	1	0	0	6	0.00
16—Plastic products	3	11	15	54	20	13	4	155	0.02
1610—Foamed plastic	24	2	5	0	11	0	0	42	0.03
1699—Other plastic	11	7	7	44	9	5	3	86	0.02
17—Leather products	13	0	2	0	11	0	0	26	0.02
1719—Other leather	12	0	1	0	5	0	0	18	0.08
18—Primary textiles	0	1	0	1	2	1	0	5	0.30
19—Textile products	4	3	0	4	0	0	1	12	0.00
24—Clothing industry	2	1	0	0	1	0	0	4	0.01
25—Wood industry	15	1	0	2	2	0	0	20	0.02
26—Furniture industry	4	1	0	0	3	0	0	8	0.02
27—Paper industry	8	2	10	6	4	1	1	32	0.00
2731—Folding cartons, boxes	1	0	8	5	3	0	0	17	0.01
29—Primary metals	0	1	0	0	0	0	0	1	0.02
30—Fabricated metals	44	17	11	14	34	11	3	134	0.00
3032—Prefab metal buildings	4	1	0	2	4	0	0	11	0.01
3063—Hand tools	4	0	2	2	1	6	0	15	0.03
3092—Metal valves	4	0	3	0	3	0	0	10	0.04
3099—Other fabricated metal	9	1	1	2	2	1	1	17	0.00
31—Machinery (non-electric)	400	1477	106	130	162	700	29	3004	0.01
311—Agricultural implements	251	1266	81	66	99	606	18	2387	0.04
3190—Other machinery	49	79	5	0	31	31	1	196	0.79
3192—Mach. for handling mat.	12	32	3	8	1	8	1	65	0.01
3199—Other machinery nec	84	72	16	56	30	50	8	316	0.01

32—Transportation equipment	1	15	0	0	0	0	1	19	0.00
33—Electrical products	40	22	4	9	10	8	7	95	0.00
3359—Other electronic equip	11	3	0	5	1	2	0	22	0.00
3379—Other electrical equip	13	9	2	0	2	1	1	28	0.00
35—Nonmetallic minerals	3	1	0	1	0	0	0	7	0.00
37—Chemical products	191	830	59	33	291	311	34	1749	0.03
3712—Industrial organic chem.	2	8	0	0	0	15	12	37	0.00
3722—Mixed fertilizer	11	105	4	13	0	18	3	154	0.84
3729—Other agricultural chem	34	494	24	5	12	209	6	784	0.74
3741—Pharm & medical chem	99	29	4	0	257	23	7	419	0.01
3799—Other chemicals	43	184	27	10	18	36	6	324	0.00
39—Other manufacturing	63	31	6	39	106	10	14	269	0.00
3910—Scient & profess equip	2	0	0	0	16	0	0	18	0.03
3911—Controlling instruments	7	9	1	1	20	2	2	42	0.21
3912—Other instruments	5	3	0	2	46	3	4	63	0.50
3999—Other manuf, nec	41	19	5	34	21	3	8	131	0.44
40—Building and contracting	9	3	0	17	11	4	0	43	0.21
4020—Nonresidential	3	0	0	0	7	1	0	11	0.21
4021—Light Industrial	4	0	0	4	4	0	0	12	0.44
4022—Commercial	0	2	0	12	0	1	0	15	0.21
41—Industrial construction	0	1	0	0	0	2	0	3	0.01
42—Trade contracting	1	0	0	2	0	0	0	3	0.02
49—Other utilities	5	1	0	0	0	1	2	9	0.13
4999—Other utilities	5	1	0	0	0	1	2	9	0.26
59—Other wholesale products	1	0	0	0	0	0	0	1	0.20
All industries of manufacture	1943	4755	416	585	1290	2086	179	11249	

Note: Elements are the number of patents from each IOM used in each agricultural SOU; The last column indicates the share of all patents in the IOM used in an agricultural SOU.

manufacturing sectors are the agricultural implements and the agricultural chemical industries. But these are not the only suppliers of inventions to agriculture. The feed, plastic products, fabricated metals, other machinery, pharmaceutical, and other manufacturing industries are also important contributors.

The Yale Technology Concordance System can be applied to IPC-based patent counts. These are available for some countries beginning in 1920 or so and for all countries after 1960. For more historical studies, a concordance to U.S. Patent Classes (USPCs) is required. Daniel K.H. Johnson at Wellesley College has developed the Wellesley Technology Concordance (WTC) System for this purpose.

The application of the YTC to USPCs for the entire history of the U.S. Patent Office produces numbers reported in Table 5.3. Numbers of patents granted by decade are reported where the agricultural sector is the SOU. Almost all of these patents were granted to inventors based in the United States prior to the 1950s (see Table 5.4.) As Table 5.3 indicates, the late 19th century was a period of high level of invention for U.S. agriculture. The Great Depression and World War II periods witnessed a decline in inventions. But recent decades have seen a flowering of inventions for agriculture,

Table 5.4 reports total foreign-origin inventions affecting U.S. agriculture for 1995. Twenty-nine percent of U.S. patents in agriculture were granted to foreign inventors. Japan was the leading supplier of inventions but Germany was second. Foreign invention was important in all invention fields for agriculture. In Table 5.5, patent grants for the food industries as the sector of use are reported by decade.

Table 5.6 reports foreign origin invention in the food industries. Foreign inventions are lowest in the beverages industry and highest in the tobacco industry.

Table 5.3   U.S. patents granted by decade: Agriculture sector-of-use, 1840–1999

| Year | Production Sectors | | | | Service Sectors | | | Total |
	Livestock	Crops and Combo	Fruits and Vegetables	Horticulture	Livestock	Crops	Other	
1840–1849	72	174	10	9	34	94	2	396
1850–1859	219	859	68	46	88	467	8	1753
1860–1869	1120	2848	211	157	475	1582	35	6428
1870–1879	1836	3535	263	220	777	1960	53	8644
1880–1889	2828	4602	343	273	1140	2528	72	11786
1890–1899	2587	3832	300	271	1060	2123	85	10259
1900–1909	3068	4533	363	354	1377	2527	112	12335
1910–1919	3159	5089	411	406	1638	2877	129	13709
1920–1929	2272	3919	340	379	1388	2246	105	10649
1930–1939	1886	3500	347	422	1144	1947	118	9365
1940–1949	1192	2215	214	217	749	1228	80	5895
1950–1959	2159	3953	387	422	1429	2148	135	10633
1960–1969	2212	4446	451	460	1443	2205	166	11384
1970–1979	2642	5339	537	567	1846	2478	217	13627
1980–1989	2815	5528	580	597	2217	2406	242	14385
1990–1999 (Est.)	3936	6650	752	892	3332	2833	484	18879
Total	34003	61022	5579	5692	20139	31651	2044	160129

Table 5.4   Patents granted 1995: U.S. agricultural sector-of-use by country-of-origin

	U.S. Production Sectors				U.S. Service Sectors			
	Livestock	Crops	Fruits-Vegetables	Horticultural	Livestock	Crops	Other	Total
Country of Invention								
USA	256	557	62	72	245	235	44	1471
Great Britain	14	29	3	5	11	14	3	79
Germany	19	45	4	6	14	21	3	112
France	12	29	3	6	8	13	2	71
Other European	7	53	6	2	24	11	2	105
Canada	5	1	1	2	4	4	1	27
Oceania	3	6	1	2	4	4	1	21
Japan	34	72	8	6	32	27	5	184
Total Foreign	94	244	26	27	97	94	17	599
Percent Foreign	27	30	30	27	28	29	28	29

The Agricultural Biotechnology Sector and Private Sector R&D

Little doubt exists that the biological sciences have undergone a second major "revolution" in the second half of the 20th century. And there is little doubt that private firms' R&D strategies have dramatically altered the public/private balance in R&D used in the agricultural sector. This part of the chapter first discusses the underlying science that enabled the "Gene Revolution". Then data for

Table 5.5   U.S. patents granted by decade: Food industries of use, 1840–1999

Year	Meat, Poultry and Fish	Fruit and Vegetables	Dairy Products	Cereals and Feed	Beverages	Tobacco	Other	Total
1840–1849	21	12	11	16	12	13	45	129
1850–1859	84	41	43	41	54	37	131	432
1860–1869	320	145	193	121	325	262	611	1977
1870–1879	548	239	343	177	647	502	990	3446
1880–1889	686	320	467	241	931	701	1351	4698
1890–1899	777	339	528	222	1092	844	1334	5137
1900–1909	1184	491	805	312	1626	1083	2020	7520
1910–1919	1510	628	1011	399	1959	1054	2686	9246
1920–1929	1695	717	1077	435	2128	1213	3018	10283
1930–1939	1937	789	1237	528	2323	1559	4072	12445
1940–1949	1332	541	760	377	1369	931	2943	8254
1950–1959	2120	760	1144	540	2096	1135	4033	11830
1960–1969	2517	907	1503	690	2735	1523	4978	14852
1970–1979	3016	965	1716	913	2826	1527	6179	17142
1980–1989	2930	882	1675	866	2803	1617	5952	16723
1990–1999	3731	1130	2346	1166	3989	1810	8024	22195
Total	24408	8906	14858	7044	26913	15812	48368	146308

Table 5.6   Patents granted 1995: U.S. food industries of use by country of origin

	Meats	Fruits-Vegetables	Dairy	Cereals-Food	Beverages	Tobacco	Other	Total
				U.S. Food Industries				
Country of Invention								
USA	240	85	170	96	284	101	608	1584
Great Britain	15	4	8	5	8	12	34	86
Germany	29	8	14	10	25	17	51	154
France	15	4	7	5	11	7	26	75
Other European	16	7	11	7	14	16	30	101
Canada	5	2	2	2	4	2	9	24
Oceania	4	1	2	2	2	1	7	19
Japan	32	10	23	15	29	15	95	219
Total Foreign	118	38	68	48	94	72	255	693
Percent Foreign	33	31	28	33	25	42	29	30

field trials by both public and private enterprises are summarized. Industrial development is discussed next. Finally, the adoption of agricultural biotech products is addressed.

The Science Underlying the Ag Biotech Industry

The major contribution to understanding the structure of DNA was contributed by James Watson and Francis Crick in a publication in *Nature* on April 25, 1953. This was the "Double Helix" paper (see Watson 1968). Five weeks later Watson and Crick published a second paper describing how DNA reproduces itself. In 1961, Crick suggested the "code letter" system specifying an amino acid as a sequence of three bases.

It was not until 1973, when Paul Berg at Stanford (Berg 1974) took a gene from SV40 (Simion Virus 40) and attached it to a short piece of single stranded DNA, that genetic engineering techniques were initiated. At roughly the same time, another Stanford scientist, Stanley Cohen in collaboration with Herbert Boyer at the University of California at San Francisco, developed genetic engineering techniques more fully. Cohen had invented a technique for removing plasmids from a bacterial cell and inserting them in another cell. Boyer was studying enzymes called restriction endonucleoses, which "snip" DNA into pieces. Cohen and Boyer then developed methods for inserting "foreign" pieces of DNA into a cell using the snipping technology. Cohen and Boyer later obtained a patent on this technique but granted all rights to their universities (Cohen et al. 1973).

The realization that recombinant DNA (rDNA) techniques (genetic engineering) could be used to insert "alien" DNA into living organisms without recourse to sexual reproduction led to two major developments. The first was the convening of meetings to consider the possible "dangers" of genetic engineering, particularly the Asilomar Conference in 1975 (See Berg et al. 1974). This conference set up guidelines for rDNA experiments. By 1980, scientists in this field were successful in easing government fears (today, these concerns would be termed bioethics concerns).

The early developers of rDNA technology were generally not agricultural scientists, but by 1980, agricultural scientists were employing these techniques. Steven Lindiun, a plant pathologist at UC Berkeley, developed one of the first rDNA products: the "ice-minus" bacteria. Advanced Genetic Sciences, a company formed to develop the ice-minus bacteria commercially, initiated the battle to field test the product.

Jeremy Rifkin, Director of the Foundation of Economic Trends, and an ardent foe of genetic engineering, did delay the ice-minus field trials for 4 years, but in 1987 these trials were conducted. No significant effect on ecology and climate was noted. But the tradition of opposition to biotechnology was established. The ice-minus product, however, was a commercial failure.[10]

The second major development following the demonstration of rDNA technology was the establishment of commercial companies to exploit the new technology. Many of the early companies were in the form of "University-Industry Alliances". A number of smaller companies were formed by scientists, including agricultural scientists. Literally hundreds of such companies were formed by the late 1980's. Many specialized in medical biotechnology, but many were devoted to specific agricultural problems.

Early in this phase, however, several major agricultural chemical companies entered the field with major investments. These included Monsanto, DuPont, Ciba-Giegy, and W.R. Grace. By 1990 or so the R&D budgets of a small number of agro-biotech firms were on the order of several hundred million dollars. At this point, most SAES-USDA units were rapidly changing their research portfolios to emphasize the new science-enabled fields of biotech invention, but they generally lagged the initiatives of the commercial firms. In developing countries the lag was even more pronounced. The International Agricultural Research Centers (IARCs), which provided initiatives for the Green Revolution, were relatively late in introducing the Gene Revolution (Chapter 6). Only a few National Agricultural Research Systems (NARS) were developing rDNA programs as of the early 1990s.

The rDNA products were not confined to crops. Transgenic animals have existed since 1985. Many have been engineered to produce pharmaceuticals, but progress has been slow. The major commercial product on the market is the "bovine growth hormone," or bovine somatatrophin (BST). Product development began in 1985; FDA approval was obtained in 1993 and the product is now widely used.

Many of the early transgenic crop products, notably the ice-minus product and Calgene's "flavor saver tomato", were not commercially successful. By mid-1999 more than 50 genetically altered crops, including 24 food crops, had been approved for sale in the U.S. Several thousand field tests had been conducted. Figure 5.4 depicts biotech products under development. These are so-called second-generation products, which have been shown to be feasible scientifically but are awaiting further development and commercialization.

Private and Public Field Trails

The Agricultural Plant Health Inspection Service (APHIS) has regulatory responsibility for field testing permits. Table 5.7 lists the 149 private companies granted field testing permits for genetically modified crops from 1985 to 2003. This is an impressive list of companies in terms of breadth of importance in the plant biotech field. Table 5.8 lists the public-sector programs granted field testing permits.

The Ag Biotech Industry

The agricultural biotech industry has evolved over time and is presently dominated by seven major multinational corporations (MNCs). The first phase of industrial development featured a number of academic "spin-off" companies and university-industry alliances. The "life science" company model

Corn (Maize)
    Agronomic traits, cold and drought tolerance, stalk strength, amino acid levels
    Fungal resistance
    Herbicide tolerance
    Insect resistance
    Plant-made pharmaceuticals
    Starch alternation, nutritional quality
    Phytotic reduction
    Altered coloration

Soybeans
    Agronomic properties
    Fungal resistance
    Insect resistance
    Virus resistance
    Marker genes
    Phytotic reduction

Cotton
    Agronomic properties (high oleic acid, fiber quality)
    Fungal resistance
    Herbicide resistance
    Insect resistance

Rape/Canola
    Agronomic (altered nitrogen metabolism, male sterility, cold tolerance)
    Fungal resistance
    Insect resistance
    Marker genes

Wheat
    Agronomic (drought tolerance, starch content)
    Fungal resistance
    Herbicide tolerance
    Virus resistance
    Digestibility

Sugar Beets
    Herbicide tolerance
    Virus resistance

Rice
    Agronomic (Male sterility)
    Bacterial resistance
    Fungal resistance
    Herbicide tolerance
    Insect resistance

Potatoes
    Bacterial resistance
    Fungal resistance
    Insect resistance
    Virus resistance

Squash
    Virus resistance (leading to low pesticide residual)

Sweet corn
    Insect resistance (leading to low pesticide residual)

Fig. 5.4. Biotech products under development.

Table 5.7 Private companies filing in U.S. for plant biotech field testing permits, 1985–2003[1,2]

Abbot and Cobb	Barham Seeds	Crows
Advanced Genetic Science	BASF	Dairyland Seeds
Agracetus	Bayer Crop Science	DeKalb
AgraTech Seeds	Becks Superior Hybrids	Delta and Pine Land
AgReliant Genetics	Bejo	Demegen
AgrEvo	Betaseed	DNA Plant Tech
Agrigenetics	BHN Research	Dow
AgriPro	Biogemma	Dry Creek
Agritope	Biosource	Du Pont
AgriVitis	Bio Technica	Dunn
All-Tex Seed	Boswell	Emlay and Associates
Amer Crystal Sugar	Brownfield Seed	Exelixis
American Cyanamid	Cal West Seeds	ExSeed Genetics
American Takii	Calgene	FFR Cooperative
Amoco	Cameron Nursery	Forage Genetics International
Anton Caratan & Son	Campbell	Frito Lay
Applied PhytoGenetics, Inc.	Canners Seed	Gargiulo
Applied Phytologics	Cargill	Garst
Applied Starch Tech	Chembred	gENaPPS
ArborGen	Chlorogen, Inc.	Genetics
Arcadia Biosciences	Ciba-Geigy	Goertzen Seed Research
Asgrow	Coors Brewing	Gold Harvest Seeds
Aventis	Crop Genetics	Great Lakes Hybrids
Ball Helix	Crop Tech	Harris Moran
Heinz	NC+ Hybrids	Seminis Vegetable Seeds
Hilleshog	Nestle	Shoffner FarmResearch, Inc.
Hoechst-Rousell	Northrup King	Stine Biotechnology
Holdens	Novartis Seeds	Stine Seeds
Horan Brox. Agri. Enterprises	PanAmericn Seed	Sunseeds
Hunt-Weson	Pebble Ridge Vineyards	Syngenta
ICI	PetoSeed	Targeted Growth, Inc.
ICI Garst	Pioneer	Thermo Trilogy
Integrated Plant Genetics	Plant Genetic Systems	Tilak raj Sawheny
InterMountain Canola	Plant Genetics	United States Sugar
International Paper	Plant Science Research	United Agri Products
Interstate	Plant Sciences	United States Sugar
Interstat Payco Seed	ProdiGene	Upjohn
J.R. Simplot Company	Pure Seed Testing	Van den Bergh Foods
Jacob Hartz	R J Reynolds	VanderHaven
Land O Lakes	Research for Hire	Vector Tobacco
Large Scale Biology	Rhone-Poulenc	Ventria Bioscience
Limagrain	Rogers	Western Ag Research
Lipton	Rogers NK	Westvaco
Mendel Biotechnology	Rohm and Haas	Weyerhauser
Meristem Therapeutics	Sandoz	Williams Seeds
Midwest Oil Seeds	Sanford Scientific	Wilson Genetics
Miles	Scotts	W-L Research
Monsanto	Seedco	WyFels Hybrids
Mycogen	SemBioSys Genetis	Yoder Brothers
National Starch & Chemical		Zeneca

Source: USDA, APHIS (see Appendix 1).

1. See note 9, McKelvey (1997); and Fulton and G. Konstantions (2001).

2. Company names are reported as they appear in the APHIS permit application; as a result some may appear in the tables more than once.

Table 5.8   Public institutions filing for plant biotech field testing permits, 1985–2003

ARS	New York State U/Albany	U of Connecticut
Auburn U	New York State U/Geneseo	U of Florida
Bowdoin C	Noble Foundation	U of Georgia
Boyce Thompson Institute	North Carolina Dept of Agr	U of Hawaii
Clemson U	North Carolina State U	U of Hawaii/Manoa
Cold Spring Harbor Lab	North Dakota State U	U of Idaho
Colorado State U	Ohio State U	U of Illinois
Connecticut Ag Exp Stn	Oregon State U	U of Kentucky
Cook C Rutgers U	Pennsylvania State U	U of Minnesota
Cornell U	Purdue U	U of Missouri
Duke U	Rutgers U	U of Nebraska
Fort Valley State University	Southern Illinois U	U of Nebraska/Lincoln
Hawaii Ag. Research Center	Southern Piedmont AREC	U of North Carolina
Illinois U	Stanford U	U of Rhode Island
Iowa State U	Texas A&M	U of South Carolina
Kansas State U	Texas Agricultural Exp Stn	U of Tennessee
Louisiana State U	Texas Tech U	U of Virgin Islands
Max Planck Inst Chem Ecology	Tuskegee U	U of Washington
Michigan State U	U of Arizona	U of Wisconsin
Michigan Tech U	U of California	U of Wisconsin/Madison
Mississippi State U	U of California/Berkeley	Virginia Tech
Montana State U	U of California/Davis	Washington State U
New Mexico State U	U of California/Kearney	Washington U
New York State Exp Stn	U of California/San Diego	West Virginia U
New York State U	U of Chicago	Wright State U

Source: USDA, APHIS (see Appendix 1).

emerged from this phase. Most of the originating life science companies were actually in the agricultural chemical industry marketing herbicides and insecticides (and some fertilizer). These companies recognized that they needed plant breeding expertise to succeed. Their strategy was to acquire seed firms mostly through purchase. Figure 5.5 lists these organizations and acquisitions.

Many of the purchases and mergers were not profitable, and the industry has had several years of turmoil. Part of this has been the result of exceptionally high levels of consumer resistance and political hostility to genetically modified organisms (GMOs). After several years of high levels of mergers and acquisitions, the agricultural biotech industry is now dominated by seven MNC firms and many smaller firms. Syngenta is headquartered in Basel, Switzerland but has research facilities in the U.S. (at Northrup King and at North Carolina Research Triangle Park) and is the largest agricultural biotech firm in terms of sales. The firm has products for eight crops and is also developing plant-made pharmaceuticals. Monsanto, based in St. Louis, Missouri, is the leading seller of agricultural biotech products, notably its Roundup™ herbicide tolerance products. Monsanto also has insect resistant (Bt) products. DuPont, based in Delaware has recently announced that it is becoming almost exclusively an ag biotech firm, with the sale of its polymer (nylon) division. With Pioneer Hi-Bred it offers a broad range of crop varieties. Bayer Crop Sciences, headquartered in Europe, has several research locations in the U.S. and Canada. Bayer produces herbicide tolerant (liberty) products. BASF, also European based, has research facilities in the Research Triangle Park in North Carolina, and Dow Agro Sciences is based in Indianapolis and has research programs in Brazil and other countries. Savia is headquartered in Mexico.

Leading Company	Agricultural Chemicals	Biotech	Seeds	Food/Feed Industry
Monsanto (Merged with Pharmacia March, 2000 spun off entirely Aug, 2002)		Agracetus (1995) Calgene (1996) Ecogen (13%) Millenium Pharmaceutical (Joint venture for crops genes) Paradigm (2000, $50 million contract)	DeKalb (1996) Asgrow (1997) (corn and soybeans) Holden's Foundation Seeds (1997) Cargill International Seeds, Planting Breeding Intl. (1998) Delta & Pineland (Alliance not purchase 1994)	Renassen a joint venture for feed and food with Cargill (1998) Monsanto sold brands like Nutrasweet in 2000
Bayer (Bought Aventis Crop Sciences 2001)	Hoechst & Schering create Agrevo (1994) Hoechst (Agrevo & Rhone-Poulenc 1999) Merger to create Aventis Bayer buys Aventis Crop Sciences Aug. 2001 for $6.6 billion	Plant Genetic Systems (1996)	Nunhems, Vanderhave, Plant Genetic Systems, Pioneer Vegetable Genetics, Sunseeds (1997) Nunza (Vegetables) Proagro (India) & 2 Brazilian seed companies 1999	Solavista & Novance (alliances for starch & non-food oils).
Syngenta (Novartis + Astra-Zeneca Ag.)	Formed by merger of Novartis agriculture division and Astra-Zeneca's Ag. Chemicals Dec. 1999 Novartis buys Merck's pesticide Business for $910 mil. (1997) Novartis formed by Ciba-Geigy and Sandoz (1996) Merger	Zeneca Ag. Bought Mogen International N.V. (1997) Alliance with Japan Tobacco on Rice (1999) Alliance with Diversa (2002)	1996 merger brings together Northrup-King, S&G Seeds Hilleshog, Ciba Seeds, Rogers Seed Co.	Owns Gerber Foods Novartis formed Altus a joint venture with Quaker Oats on nutraceuticals 2000
Dow Chemicals	Dow purchases Eli Lilly's 40% share of Dow Elanco for $900 million. (1997) Rohm and Haas Ag. Chem (2001)	Mycogen (1996) Ribozyme Pharmaceuticals Inc. Proiteome Systems Limited (1999 contract).	Mycogen buys Agrigenetics (1992) United AgriSeeds becomes part of Mycogen (1996) Danisco Seeds (JV) (1999) Illinois Foundation Seed Agreement (1999) Cargill Hybrid Seeds U.S. (2000)	Agreement with Cargill on plastics from corn

Fig. 5.5. Mergers and acquisitions in the U.S. and European agricultural, chemical, biotechnology, seed, and food/feed industries, 1994–2003. Source: Pray and Naseem 2003.

(Continues)

173

Leading Company	Agricultural Chemicals	Biotech	Seeds	Food/Feed Industry
DuPont		Alliances with Human genome Sciences (1996) Curagen (1997)	Pioneer (1997) (20%) Hybrinova (1999) (France) Bought other 80% of Pioneer in 1999???	Quality Grain (1998) Joint venture with Pioneer, Protein Technologies (food), Cereal Innovation Centre UK JV with General Mills on soy protein. Working on fiber from starch. JV with Bunge on soy products (2003).
BASF	Bought Sandoz N. American Herbicide business 1996 American Cyanamid From AHP for $3.8 Billion (2000)	SunGene (JV with Institute of Plant Genetics & Crop Plant Research) Metanomics (JV with Max Planck Institute) Plans to Invest $700 Million in plant Biotech over 10 Years starting in 2000	Bought 40% of Svalöf Weibell (1999).	
Savia (was Empresas La Moderna)		DNA Plant Technology (1996) is part of Bionova	Seminis (SAVIA's seed division) is made up of Asgrow (1994) (Sold corn & soybeans To Monsanto in 1997) Petoseed (1994) Royal Sluis	Bionova (fresh fruits and vegetables

Fig. 5.5. (Continued)

The R&D data indicate total R&D expenditures of more than 3 billion dollars for these firms. From 50 to 80 percent of this is on ag biotech products. Each of these companies has several crop biotech plant products on the market. Most have animal biotech products on the market as well. Major agricultural biotech products in the pipeline are shown in Figure 5.4.

Biotech Invention

A recent study by Zohrabian and Evenson (2000) of biotechnology patents utilized a "biotech" keyword to identify international patent classes (IPCs) with high biotech content. Patent counts from these IPCs and the biotech preparations of these counts were used to form a biotechnology invention database.

Table 5.9 provides the IOM-SOU matrix for U.S. biotech inventions for the period 1976–1998. Biotech inventions in the U.S. grew rapidly over this period from 814 patents in 1990 to 7000 in 1998. In 2002, this was more than 10,000. Table 5.9 shows that only 3.3 percent of all biotech inventions were used in production agriculture. The food sector was the sector of use for 4.7 percent of biotech inventions. Many of the biotech inventions were "process" inventions used in the chemical and drug industries. More than 40 percent of biotech inventions were used in the health sector.

Table 5.10 shows details for the six U.S. production agricultural sectors, seven food sectors, and the health sector from 1990 to 1999. Table 5.11 illustrates the international dimension of U.S. inventions for 1995. Total patents in the using sectors of agriculture, food, and heath are shown. The portion of inventions by country of origin (domestic, U.S., and Japan) shows that 40 percent of all biotech inventions are from U.S. based firms (Frey, 2000).

The National Plant Breeding Study 1994

For plant breeding programs, the National Plant Breeding Study of 1994, conducted by Iowa State University provides a comparative perspective on public and private resources devoted to plant breeding. A survey of public and private firms was conducted (see Frey 1997, 1998, 2000).

Table 5.12 provides details by crop on the number of scientist years employed by the SAES system, the ARS/USDA system, and by private industry. For many commodities the number of private companies with plant breeding programs is also reported. For all crops, the SAES system accounted for 24 percent of total breeder scientist years (SYs), the ARS/USDA system for 8 percent, and private industry for 68 percent. These proportions are reflecting a major shift in public-private sector

Table 5.9   IOM-SOU matrix of U.S. biotechnology patents, 1976–1998

Industry of Manufacture	Sector of Use (%)					
	Agriculture	Chemicals	Drugs	Food	Health	Total
Agriculture	0.5	0.0	0.0	0.0	0.0	0.6
Chemicals	1.0	3.8	10.9	1.2	1.5	20.2
Drugs	1.4	1.6	22.7	2.1	35.7	64.7
Food	0.1	0.0	0.0	1.1	0.0	1.2
Total	3.3	6.5	34.3	4.7	40.4	100.0

Table 5.10 Biotechnology patents by sector-of-use in the USA, 1990–1999

Sector of use	1990	1991	1992	1993	1994	1995	1996	1997	1998–1999
Agriculture livestock	3.2	3.4	5.1	5.5	5.9	6.6	10.0	16.0	36.3
Crops and combo farms	6.2	6.3	7.7	11.5	11.9	14.3	22.3	33.3	80.7
Fruits and vegetables	0.6	0.6	0.8	1.1	1.2	1.4	2.2	3.3	8.0
Horticulture	0.5	1.3	0.7	1.0	2.9	1.7	6.8	5.3	12.7
Services to livestock	6.9	7.7	11.2	12.1	13.4	14.2	20.1	34.7	77.4
Services to crops	3.2	3.3	3.8	6.1	6.4	7.8	12.5	13.3	31.0
Total Agriculture	22.0	26.1	33.1	40.6	48.5	50.3	89.4	125.0	291.5
Food									
meat, poultry and fish	1.4	1.6	1.8	1.9	2.2	? 7	4.1	5.0	11.5
Fruit and vegetables	1.1	1.2	1.3	1.6	1.8	2.2	3.5	4.8	11.1
Dairy products	3.6	3.7	4.6	6.5	6.9	8.2	13.1	19.6	46.7
Cereals and feed	2.8	4.9	4.9	4.4	7.5	14.3	19.0	15.4	50.7
Beverages	4.1	4.7	8.3	7.7	7.7	8.6	12.8	18.4	43.6
Tobacco	0.9	1.7	1.3	1.9	3.8	2.7	8.6	8.0	19.1
Other	24.1	25.4	34.5	41.0	41.5	49.5	75.6	109.2	259.9
Total Food	37.9	43.2	56.7	64.9	71.4	88.1	136.8	180.3	442.6
Health	316.8	357.2	515.3	611.8	640.8	687.9	1055.9	1683.1	3884.6
Total number of patents used in all sectors	814	894	1311	1497	1589	1735	2662	4044	9505

Table 5.11 International biotech inventions, 1995

	Patents Granted			Proportion by Origin		
	Agriculture	Food	Health	Domestic	U.S.	Japan
Australia	13.6	17.8	168.2	6	53	8
Belgium	15.5	19.3	198.6	—	42	12
Canada	5.5	9.7	81.7	9	54	17
Denmark	9.1	10.6	124.9	8	32	16
France	23.3	29.5	307.6	24	32	16
Germany	19.0	24.9	246.9	15	38	19
Greece	10.0	11.3	127.7	—	38	13
Ireland	6.9	7.6	89.2	3	44	4
Italy	16.9	21.2	220.2	3	41	15
Japan	60.0	85.7	779.7	32	28	32
Netherlands	16.4	20.6	211.3	2	41	14
New Zealand	5.1	5.0	69.0	8	45	5
Switzerland	16.3	20.4	213.3	2	41	14
U.K.	18.9	24.4	244.6	11	39	19
USA	50.3	88.1	688.0	82	82	3

activities (Ruttan 1983). The ARS/USDA role in plant breeding appears to have been in decline for some time. The SAES role has also been declining. The private industry role reflects the effects of the biotechnology revolution.

Across commodities, we see that private industry is active in virtually all crops. The ARS/USDA proportion of SYs in breeding is less than the private industry role in virtually all crops. The SAES proportion is less than the private sector role in most crops as well.

Table 5.13 shows the allocation of plant breeding SYs to plant breeding (PBR), germplasm enhancement (GE), and cultivar development (CD). This table is instructive. Private industry SYs are predominately devoted to cultivar development (varieties for commercialization), and ARS/USDA resources are dominated by genetic enhancement activities.

Table 5.12 U.S. Scientist Years in Plant Breeding: 1994

	Scientist years employed by					Shares	
	SAES	ARS/USDA	Private Industry	(No. Companies)	Total	R&D	Cash Receipts
Cereal Crops							
Corn							
Field	27.1	8.2	509.8	(91)	545	24.7	18.0
Sweet	5.4	.4	27.0	(12)	32.8	1.5	.6
Sorghum	11.8	2.5	40.8	(19)	55.1	2.5	1.4
Rice	13.8	6.3	21.9	(8)	42.0	1.9	1.5
Wheat	64.5	12.0	54.0	(27)	130.5	5.9	7.8
Other	32.8	4.9	49.5		87.2	4.0	4.9
Fiber Crops							
Cotton	19.2	11.7	103.5	(35)	134.3	6.1	5.7
Other	.5	1.0	0.0		1.5	.1	.1
Forage Crops							
Grasses	13.5	14.0		(60.5)	36.0	2.7	
Legumes							4.1
Alfalfa	15.2	11.9	41.0	(12)	68.1	3.1	
Other	9.1	7.0	2.2	(21.5)	18.3	1.0	
Grain Legume Crops							
Bean	13.8	2.0	8.8		24.1	1.1	.5
Soybean	45.0	9.6	101.4	(38)	156.0	7.1	16.2
Other	8.5	2.6	15.9		27.0	1.2	1.4
Oilseed Crops							
Canola	5.7	1.0	28.0	(4)	34.7	1.6	.1
Peanut	14.0	2.5	3.2		19.7	.9	.9
Sunflower	.6	2.6	31.5	(14)	34.6	1.6	.5
Vegetable Crops							
Fruit Vegetables							
Pepper	5.5	.5	37.6	(27)	43.6	2.0	.5
Tomato	20.7	4.6	59.5	(24)	84.7	3.8	1.5

(Continues)

Table 5.12   (*Continued*)

| | Scientist years employed by | | | | | Shares | |
	SAES	ARS/USDA	Private Industry	(No. Companies)	Total	R&D	Cash Receipts
Muskmelon	.8	1.6	20.5	(15)	22.9	1.0	.5
Other Leafy	11.0	1.4	49.8	62.2	62.9	2.8	
Bulbs-Stem	16.0	2.0	77.0	95.0	4.3		0.8
Lettuce			20.0	(18)			1.5
Temporary Fruit & Nut Crops	50	23	32		105	4.8	
Tropical Fruit & Nut Crops	10	6	—		16	.7	11.8
Root & Tuber Crops	45	12	24		81	3.7	2.1
Stimulant Crops	13	2	5		20	.9	
Sugar Crops	4	15	25		44	2.0	1.7
Beets			24	(7)			1.1
Ornamental Crops	18	5	64		87	3.9	3.5
Medicinal Spice Specialty	7	4	7		18	.8	—
All Crops	529.0	177.0	1499.0		2205.0	100.0	

Source: Frey 1997, 1998, 2000.

Research Shares and "Congruity"

Suppose two different commodity research programs had the same "discovery" parameters in their invention probabilities. Then the optimal allocation of research expenditures between the two programs would be to allocate research in proportion to the economic value of the commodities.

This was first demonstrated in models of research developed by Evenson and Kislev (1975, p. 140–150). Evenson and Kislev applied models of search to applied research programs. A specific

Table 5.13   Numbers and percentages (in parentheses) of U.S. plant breeding SYs devoted to plant breeding research (PBR), germplasm enhancement (GE), and cultivar development (CD) arranged by employer, 1994

Category	SAES	ARS/USDA	Private Industry	Activity Totals
CD	217 (41%)	22(12%)	1,191 (80%)	1,430
GE	153 (29%)	85 (48%)	165 (11%)	403
PBR	159 (30%)	70 (40%)	143 (9%)	372
TOTALS	529	177	1,499	2,205

Source: Frey 1996, p. 8.

case for an experimental function was developed. In this case the following invention function was developed:

$$E_{z,n} = \theta + \lambda \sum_{1}^{n} \frac{i}{1+i}$$
(5.1)

where $E_{z,n}$ is the expected maximum value, in a sample of size n drawn from an exponential distribution. This expression increases in n but with diminishing returns. Kortum (1994) showed that this expression is well approximated by:

$$\ell n\, V = \alpha + \beta\, \ell n(N)$$
(5.2)

When V is the number of inventions and N is the resources devoted to search. It is easily seen that the marginal product of search is $V \delta \ell nV / \delta N = \beta V/N$ where V is the value of the commodity.

Now suppose two commodity research programs with equal discovery parameters α and β. Suppose they have the same marginal costs of undertaking research. Then setting marginal products equal gives:

$$\frac{\beta}{N_1} V_1 = \frac{\beta_2}{N_2} . V_2$$
(5.3)

If $\beta_1 = \beta_2$ the optimal allocation of resources will be proportional to commodity value, i.e.:

$$\frac{N_1}{N_2} = \frac{V_1}{V_2}$$
(5.4)

Table 5.14 reports a crude comparison between research system shares (including private sector R&D for crops) and cash receipt shares. Note that 32.8 percent of crops research is commodity oriented, and 20.9 percent of livestock research is commodity oriented. About 9 percent of SAES/USDA research is allocated to forestry research. Thus, 37 percent of all research cannot be allocated to commodities.

If we use the same private/public ratios for livestock as for crops (1499/2205 =.68), then livestock research would be higher. Adjusting for this, the crops research share would be roughly 38 percent. The livestock research share would be roughly 40 percent and the forestry share would be roughly 8 percent. These shares do not suggest congruity although for most crops it appears that research shares are not too far from cash receipts shares.

There are two clear cases of what appears to be "overinvestment" in research—sheep-wool research and forestry research. Both are puzzling. This topic is reconsidered in Chapter 9.

Conclusion

Over the past two decades, new legislation, new court decisions, new IPRs, and new discoveries in biotechnology and information science have greatly changed the incentives and opportunities available to both the private and public sectors for R&D. The strengthening and broadening of IPRs and advances in scientific opportunities have increased the domain of R&D for profit in the United States. The private sector has responded by dramatically increasing its investments in agricultural R&D, number of new products developed, and contributions to productivity grown and environmental quality.

Table 5.14   Economic importance and SAES-USDA research shares by commodity, 2000

Commodity	Share of Public Ag. Research Exp.	Cash Receipt Share In Farm Income
Corn	9.4	10.0
Wheat	2.1	4.2
Other Cereals	3.0	4.2
Soybeans	2.6	8.8
Other Oil	1.5	.8
Cotton	2.6	3.1
Forage	3.3	2.2
Tobacco-Sugar	1.1	2.4
Potatoes	1.3	1.1
Fruits-Vegetables	5.5	12.4
New & Misc. Crops	1.7	7.3
All Crops	32.8	53.4
Livestock Research		
Beef Cattle	5.6	16.6
Swine	3.3	6.0
Sheep-Wool	4.3	.4
Dairy	4.6	9.7
Poultry	3.2	10.3
All	20.9	46.3
Forestry	9.3	.6

Source: U. S. Department of Agriculture.

Not all areas of agricultural R&D are expected to yield profits, and these are areas where the public sector must continue its R&D efforts. They include research on self-pollinated crops, on minority crops, on resource and environmental quality, on food safety, and on human health. However, resources for these activities are being challenged by the Bayh-Dole Act, which provided new opportunities for income and revenue generation by universities, but this challenges the production of knowledge to support the "Intellectual Commons." The economic issues continue to be sorted out about what the long-run new relationships between the public and private sectors will be in agricultural R&D over the 21st century.

Notes

1. Other IPR systems have not fully adapted U.S. practice in this regard, but the WTO-TRIPs agreement puts pressure on many countries to follow the U.S. lead.
2. The Board of Patent Appeals and Interferences of the US Patent and Trademark Office has interpreted *Diamond v. Chakrabaty* to mean that any plant can be patented provided it satisfies the basic standards for patentability. The U.S. Supreme Court in *JEM. Ag Supply vs. Pioneer Hi-Bred Int. Inc.* (534US124 (2001)) agreed with this interpretation and ruled that the availability of plant variety protection was not in conflict with patent regulations for plants (Barton, 2004).
3. The U.S. uses a "first to invent" principle. This is a costly principle because of disputes over which inventor was first.

4. Lesser (1998) provides an evaluation of the U.S. plant varieties protection, and Braga (1996) and Lesser (2000) present discussions of Breeders' Rights under TRIPS.
5. See Nordhaus (1969), Machlup (1958), and Siebeck et al. (1991) for studies of patent systems.
6. The monopoly price is set where marginal revenue is equal to marginal cost. For simplicity we assume marginal cost to be zero in Figure 5.2.
7. Actually, the consumer surplus from the invention can continue even though the market for the invention is eroded by substitutes. This is because following an invention may have "built upon" the original invention, but the new invention can only obtain IPR rights for the added value of the new invention.
8. It is important that renewal fees be high enough to convert patented inventions into public domain inventions. The market erosion in time T can reduce PR and create CS2. But unless renewal fees are set high enough, CS1 + CS2 in period T may be less than CS_1 in period 1.
9. Many analysts ignore these gains, focusing instead on the UCS component. But unless an alternative to a patent system is in place, these gains are real.
10. Commercial failures in the field of inventions are not unusual (Rifkin, 1998).

References

Alston, Julian M., and Raymond J Venner. 1998. "The Effects of the U.S. Plant Variety Protection Act on Wheat Genetic Improvement." Paper presented at the symposium on "Intellectual Property Rights and Agricultural Research Impact," sponsored by NC208 and the CIMMYT Economics Program–El Batan, Mexico, March 5–7, 1998.

Barton, J.H. 1998. "The Impact of Contemporary Patent Law on Biotechnology Research." In *Global Genetic Resources: Access and Property Rights*, ed. S.A. Eberhart. Madison, WI: Crop Science Society of America, pp. 85–97.

Barton, John H. 2004. "Acquiring Protection for Improved Germplasm and Inbred Lines." In *Intellectual Property Rights in Agricultural Biotechnology*, ed. F.H. Erbisch and K.M. Maredia. CABI Publishing, Cambridge, MA.

Berg, P. et al. 1974. "Potential Biohazards of Recombinant DNA Molecules." *Science* 185 (July 26):303.

Berg, P., D. Baltimore, S. Brenner, R.O. Roblen, III, and M.F. Singer. 1975. "Asilomar Conference on Recombinant DNA Molecules." *Science* 188:991–994.

Braga, P. 1996. "Trade Related Intellectual Property Issues: The Uruguay Round Agreement and Its Economic Implications." In *The Uruguay Round and the Developing Countries,* ed. W. Marin and L.A. Winters. Cambridge, UK: Cambridge University Press.

Cohen. J. 1997. "The Genomics Gamble." *Science* 275 (February 7):767–772.

Cohen, S.N., A.C.Y. Chang, H. Boyer, and R.B. Helling. 1973. "Construction of Biologically Functional Bacterial Plasmids in Vitro." *Proceedings of the National Academy of Sciences U.S.A.* 70(November):3240–3244.

Crabb, A. R. 1947. *The Hybrid-Corn Makers: Prophets of Plenty*. New Brunswick, NJ: Rutgers University Press.

Diamond vs. Chakrabarty, 1980 447 U.S. 303.

Duvick, D.N. 1996. "Plant Breeding, An Evolutionary Concept." *Crop Science* 36:539–548.

Evenson, R. E. 1991. "Inventions Intended for Use in Agriculture and Related Industries: International Comparisons," *American Journal of Agricultural Economics* 73:887–891.

Evenson, R.E. 2000. "Economics of Intellectual Property Rights for Agricultural Technology." In *Agriculture and Intellectual Property Rights*, ed. V. Santanello, R.E. Evenson, D. Zilberman, and G.A. Carlson. Wallingford, UK: CABI Publishing.

Evenson, R.E. and Y. Kislev. 1975. *Agricultural Research and Productivity*. New Haven, CT: Yale University Press.

Frey, K.J. 1996. "National Plant Breeding Study—I: Human and Financial Resources Devoted to Plant Breeding Research and Development in the United States in 1944." Iowa Agriculture and Home Economics Experiment Station.

Frey, Kenneth. 1997. "National Plant Breeding Study-II." *National Plan for Promoting Breeding Programs for Minor Crops in the U.S.* Iowa State University, Iowa Agriculture and Home Economics Experiment Station, Special Report 100.

Frey, Kenneth. 1998. "National Plant Breeding Study-IIII." *National Plan for Genepool Enrichment of U.S. Crops.* Iowa State University, Iowa Agriculture and Home Economics Experiment Station, Special Report 101.

Frey, Kenneth. 2000. "National Plant Breeding Study-IV." *Future Priorities for Plant Breeding.* Iowa State University, Iowa Agriculture and Home Economics Experiment Station, Special Report 102.

Fulton, M. and G. Konstantions. 2001. "Agricultural Biotechnology and Industry Structure." *AgBioForum* 4:137–151.

Griliches, Z. 1957. "Hybrid Corn: An Exploration in the Economics of Technological Change." *Econometrica* 25:501–522.

Griliches, Z.. 1960. "Hybrid Corn and the Economics of Innovation." *Science* 132:275–280.

Griliches, Z. 1980. "Returns to Research and Development Expenditures in the Private Sector." In *New Developments in Productivity and analysis Measurement and Analysis*, ed. J.W. Kendrick and B.N. Vaccara, 419–54. Conference on Research in Income and Wealth: Studies in Income and Wealth, vol. 44. Chicago, IL: University of Chicago Press for the National Bureau of Economic Research.

Hallauer, A. and J.B. Miranda. 1981. *Quantitative Genetics in Maize Breeding.* Ames, IA: Iowa State University Press.

Henderson, R. A.B. Jaffe, and M. Trajtenberg. 1998. "Universities as a Source of Commercial Trechnology: A Detailed Analysis of University Patenting, 1965–1988." *Review of Economics and Statistics* 80(1998): 119–127.

Jones, C.I. 2002. *Introduction to Economic Growth.* New York, NY: W.W. Norton & Co.

Jorgenson, Dale and Zvi Griliches. 1967. "The Explanation of Productivity Change." *Review of Economic Studies* 34:249–283.

Jorgenson, Dale, F. Gollop, and B. Fraumeni. 1987. *Productivity and U.S. Economic Growth.* Cambridge, MA: Harvard University Press.

Jugenheimer, R.W. 1976. *Corn: Improvement, Seed Production, and Use.* New York: John Wiley & Sons, Inc.

Kalton, R.R., and P. Richardson. 1983. "Private Sector Plant Breeding Programs: A Major Thrust in U.S. Agriculture." *Diversity* 5:16–18.

Kortum, S. 1994. "A Model of Research, Patenting, and Productivity Growth." Boston University, Institute for Economic Development, Discussion Paper Series 37 (Feb.).

Lesser, W. .1998. *Sustainable Use of Genetic Resources Under the Convention on Biological Diversity: Exploring Access and Benefit Sharing Issues.* Wallingford, UK: CABI Publishing.

Lesser, W. 2000. "An Economic Approach to Identifying an 'Effective *sui* generic System' for Plant Variety Protection Under TRIPs." Department of Agricultural, Resource and Managerial Economics, 405 Warren Hall, Cornell University, Ithaca, NY.

Machlup, F. 1958. *An Economic Review of the Patent System.* Study of the Subcommittee on Patents, Trademarks and Copyright, Committee on the Judiciary. U.S. Senate, Study No. 15.

Massing, D.E., Ed. 2003. *AUTM Licensing Survey: Fiscal 2002.* Chicago, IL: The Association of University Technology Managers, Inc.

McKelvey, M.D. 1997. "Emerging Environments in Biotechnology." In *Universities and the Global Knowledge Economy*, ed. H. Etzkowitz and L. Leydesdorff. London: Cassell Academic.

Pray, Carol E. and Anwar Naseem. 2003. "The Economics of Agricultural Biotechnology Research." *The Food and Agriculture Organization of the United Nations.* Agriculture and Economic Development Analysis Division, ESA Working Paper No. 03–07.

Rifkin, J. 1998. *The Biotech Century: Harnessing the Gene and Remaking the World.* New York: Putnam.

Rosenberg, N. 1976. *Perspectives on Technology.* New York, NY: Columbia University Press.

Ruttan, V.W. 1983. "Changing Roles of Public and Private Sectors in Agricultural Research." *Science* 216 (April 2):23–29.

Scherer. F.M. 1982. "Demand Pull and Technological Inventions: Schmookler Revisited." *Journal of Industrial Economics* 30:225–237.

Siebeck, W. (ed.). 1991. "Strengthening Protection of Intellectual Property in Developing Countries." World Bank Discussion Paper 112.

Wallace, H.A. and W.L. Brown. 1956. *Corn and Its Early Fathers*. East Lansing, MI: The Michigan State University Press.

Watson, J.D. 1968. *The Double Helix: A Personal Account of the Discovery and Structure of DNA*. New York: Athenium.

Zohrabyan, A. and R.E. Evenson. 2000. "Biotechnology Inventions: Patent Data Evidence." In *Agriculture and Intellectual Property Rights*, ed. V. Santanello, R.E. Evenson, D. Zilberman, and G.A. Carlson. Wallingford, UK: CABI Publishing.

Zuber, M.S., and L.L. Darrah. 1980. "1979 U.S. Corn Germplasm Base." In *Proceedings of the Thirty-fifth Annual Corn and Sorghum Research Conference*. American Seed Trade Association. Washington, D.C. Publication No. 35.

6

International Dimensions of U.S. Agricultural Research

U.S. universities and research systems are part of an international network of higher education and knowledge creating organizations. Universities in the United States, both private and public, including the land grant universities (LGUs), host students from other countries. U.S. students also study in other countries. Networks of scientific exchange are increasing internationally, and the United States is both an importer and an exporter of agricultural technologies.

These international dimensions hold for all fields of graduate study and research. The university system in the United States has enjoyed an excellent reputation for producing scientists and engineers. This is particularly true for the training of agricultural scientists. Many agricultural scientists from other countries earned Ph.D. degrees in the U.S. land-grant university system and many of these scientists have, in turn, become producers of scientists in other countries. This is part of the broad system of international technology transfer.

Technology in the form of inventions and innovations (the commercialization of products embodying inventions) is exported from the United States and imported from other countries. Some of this international trade is directly in the form of innovation products, but much of it is in the form of licensing of intellectual property rights (IPRs) as noted in the previous chapter. IPR systems vary greatly in scope of coverage and in enforcement regimes across countries. The WTO-TRIPS agreement serves to facilitate a certain degree of "harmonization" of IPRs on member countries. Most developing countries, however, have concluded that it is in their best economic interests to resist implementing strong IPR systems.[1]

The subject of this chapter is the international dimension of the U.S. research and higher education system. Part I addresses the role of technological capital (TC) in the economic development process.[2] Two indexes of technological capital are defined and related to development indicators. The major role of the U.S. land-grant universities in creating technological capital is then documented.

Part II of this chapter addresses the exchange of crop varieties between countries. It is well known by plant and soil scientists and farmers that crop varieties are highly location specific and that limited exchange between countries occurs. This chapter utilizes data from the Plant Variety Protection Office of the USDA to document imports of crop varieties. Green Revolution crop varieties and the export of biotech or genetically modified (GM) crop products are also examined.[3]

Part III discusses the flow of inventions between countries. The United States is one of the world's leading producers of inventions and a net exporter of inventions in almost all fields. Finally, Part IV addresses the role of scientific discoveries through the citation frequencies of three fields of U.S. agricultural science by research scientists in other countries.

Technological Capital in Developing Countries

Defining Technological Capital

Two forms of "human capital" have been recognized in developing countries for some years. The oldest form is "schooling capital." Numerous studies of investment in schooling and returns to investment in schooling have been made in developing countries. Schooling capital of an individual is typically measured by years of formal schooling completed (Huffman and Orazem 2004). More sophisticated measures adjust years of schooling completed for quality dimensions, e.g., intensity of classroom teaching, major fields of study, and peer effects.

The second form of human capital is "social capital." Social capital emphasizes membership in social and political organizations and communication networks (Becker 1996). A number of measures of social capital have been developed, but most proponents of the importance of social capital agree that measurement requires several indexes or indicators.[4]

There is widespread agreement among those who study technological change that two activities are crucial. The first activity is technological discovery or invention, followed by innovation—the commercialization of the invention in the form of a marketed product embodying the invention. It is also recognized that modern product markets have "room" for many products. This explains why we observe R&D programs in all industries and significant R&D expenditures by a large number of firms in all industries in modern market economies.[5]

The second crucial activity is "technology mastery." Virtually all economists concerned with the agricultural sector recognize that genetic inventions in the form of new crop varieties are highly location specific (Huffman and Evenson 1993; Evenson and Gollin 2003b). This is also illustrated in an international study of rice variety improvement by Evenson and Gollin (1997). Of 1,709 rice varieties released in different countries, 390 were produced outside the country of release. Of these, 294 were developed at the International Rice Research Institute (IRRI) and released in several countries. Thus, only 96 (6 percent) of rice varieties developed by National Agricultural Research Systems (NARS) programs were transferred across national boundaries. Breeding programs in the U.S. contributed very few varieties to developing countries.[6]

This phenomenon of high specificity of biological materials to soil and climate conditions is well understood by plant and soil scientists and farmers. The policy implication is that a country without a breeding program in place cannot expect plant improvement products developed in other countries to "spill-in". As a result, almost all countries in the world now have some form of public sector plant breeding program, although, as noted below, these vary greatly in strength and effectiveness.

Economists are less in agreement regarding the potential for international spill-ins of mechanical and chemical inventions. Some argue that little investment is required in the receiving country for the country to benefit from spill-in discoveries. Others argue that the TC and human capital of a particular country must reach at least a critical threshold level in producing firms before there is a capacity to absorb technology from abroad. In crop research, indirect spill-in in the form of breeding germplasm has been shown to be important, but the ability of a country to exploit that potential depends on host-country scientific capacity.[7]

Typically, developing countries build intellectual capital to first produce "domestic" goods and only later to produce "international" goods. Food is one important domestic good, and Johnson (2000) and Huffman and Orazem (2004) have shown that technical change in the agricultural sector generally precedes growth and development of the nonagricultural sector. They argue that technical change reduces the price of agricultural goods relative to nonagricultural goods. Economic growth, however, increases the demand for labor, and with restricted immigration, real wage rates rise. In the Western developed countries new inputs have been developed to substitute for local labor. Thus, wheat, barley, and rice harvesting has been undertaken mechanically in the U.S. for most of the past century. Hundreds of machines have been developed and sold by a number of farm machinery manufacturers. For the past 60 years, Brazil's farmers have realized falling costs of rice harvesting equipment because of the R&D of multinational and domestic machinery manufacturing firms. Farmers in Bangladesh and India, however, have not benefited from invention in rice or wheat harvesting technology because in those countries real wages are low and hand harvesting is still the minimum cost technique for rice harvesting (see Evenson, 2004c).

The technology mastery required for effective technology spill-in is almost certainly subject to a threshold investment effect. Below we define two indexes, one for invention innovation (II) and a second for technology mastery (TM). The indexes are defined and computed empirically for 92 developing countries for 1970 and 1990.[8]

The II index is based on two indicators—agricultural scientists per unit of cropland and research and development (R&D) as a percent of Gross Domestic Product (GDP). The first indicator is derived from several studies conducted by ISNAR (The International Service for National Agricultural Research). The second indicator is reported by UNESCO. UNESCO data may include some agricultural research, but it is interpreted here primarily as related to industrial research. Countries are given II index values of 1, 2, or 3 based on the following criteria:

Agricultural Scientists / Cropland (million ha)
Index   = 1   if indicator value is .02 or lower
           = 2   if indicator value is between .021 and .06
           = 3   if indicator value is greater than .06

R&D / GDP
Index   = 1   if indicator value is .002 or lower
           = 2   if indicator value is between .002 and .006
           = 3   if indicator value is greater than .006

The II index is obtained by adding together the scores on the above two indicators. A number is obtained for 1970 and 1990.

The TM index is also based on two indicators. The first indicator is the number of extension workers per unit of cropland. This reflects the fact that agricultural extension programs have been widely utilized to provide advice on technological practices to farmers. In poor countries, extension information is a substitute for farmers' schooling (Huffman, 2001). The second indicator is the schooling level of males over age 25 in the country. Schooling of the farm operators, and more generally of the labor force, is also a factor in technology mastery for both agricultural and industrial practices. Local industrial R&D is the most important factor facilitating technology spill-in in the nonagricultural sector.

Countries are given TM index values of 1, 2, or 3 based on the following:

Extension Workers / Cropland (million ha)
Index   = 1   if indicator value is .2 or lower
           = 2   if indicator value is .2 to .6
           = 3   if indicator value is higher than .6

Average Schooling of Males over 25
Index   = 1   if indicator value is less than 4 years
           = 2   if indicator value is 4 to 6 years
           = 3   if indicator value is greater than 6 years

The TM Index is the summation over these two indexes and is reported for 1970 and 1990.

The invention-innovation indexes reported in Table 6.1 have the following interpretation. An II index value of 22 means that the countries were in II class 2 in both 1970 and 1990. An II index value of 23 means that the country was in II class 2 in 1970 and moved to II Class 3 in 1990. TM index values are reported in parentheses in Table 6.1. (An asterisk means that the R&D/GDP component of the II index was actually reported to be zero).

Consider the countries that started with an II index of 2 in 1970. Nine of these countries were in II class 2 in both 1970 and 1990, six moved to class 3 in 1990, and seven moved to class 4 in 1990. Seven of the nine II "22" countries also had TM "22" indexes, as did two of the six "23" countries. Only two of the "22" countries had R&D/GDP indexes of 2. Fourteen of the "22" countries reported R&D ratios of zero. Two countries, Guinea Bissau and Sudan, actually lost II rank, reverting from level 3 in 1970 to level 2 status in 1990. Of the 24 countries either starting in II class 2 in 1970 or ending in that class in 1990, none would be considered to be industrially competitive by UNIDO.[9] Most are in Sub-Saharan Africa where the end of the colonial period dates from 1960. These countries inherited virtually nothing from their colonial mother countries (not all countries were in colonial relationships, however).

Next, consider the 33 countries starting in 1970 in II class 3. Two reverted to class 2 in 1990, and 10 remained in innovation class 3. Sixteen improved to II class 4, and five improved to II class 5. Of the 10 countries starting in II class 3, only two were in II class 3 because of industrial R&D. Six of these countries reported zero R&D. Eight of these countries were in II class 3 because of public sector investment in agricultural research. Of the 16 countries moving to class 4 in 1990, nine moved on the strength of R&D investments. Seven countries moved to II class 4 on the strength of public sector investment in agricultural research. Of the five countries moving to II class 5 in 1990, all had agricultural research investment indexes of 3 and all invested in industrial R&D.

Twenty-six countries had II index scores of 4 in 1970. Two, Saudi Arabia and Zimbabwe, reverted to II Class 3 in 1990. Thirteen remained in II class 4. Nine moved to II class 5, and two moved to II class 6. Of the other 13 countries remaining in II class 4, eight had agricultural research indicators of 3. Four reported zero R&D levels to UNESCO. All the countries improving to II classes 5 and 6, of course, have significant R&D capacity. Nine countries began in II class 5, and five of them moved to II class 6.

It is clear that the 25 countries in II classes 5 or 6 in 1990 have had good to excellent economic performance. Conversely, the 36 countries in II classes 2 or 3 in 1990 have had poor economic performances. The latter set of countries are all in "mass poverty." The 29 countries in II class 4 in 1990 have had mixed economic success. In general, the countries in II class 4 in 1990 with R&D capacity have tended to do a little better than the countries without R&D capacity in industry.

There are two simple questions that can be asked of these classifications. First, is movement upward in II class based on investment in agricultural research or on more general R&D? Consider the six countries improving from II class 2 in 1970 to II class 3 in 1990. All improvement was from public sector agricultural research. Six countries improved from II class 2 to 4. Four of these improvements were based on public agricultural research. Fourteen countries improved from II class 3 to 4. Twelve improved on the basis of public sector agricultural research. Thus, for II class improvement up to II class 4, 22 of 26 improvements were based on public sector agricultural investment. Among those with improvements from II class 4 to 5, 4 to 6, or 5 to 6, improvements were about equally divided between agricultural research investment and industrial R&D investment.

Second, does a higher TM class level in 1970 facilitate improvements in II capital? For countries having improvements from II class 2 in 1970 to class 3 in 1990, 2 to 4, or 3 to 4, TM capital was

Table 6.1 The classification of developing countries into invention innovation (II) classes, 1970 and 1990[a]

II Classes 2 and 3 in 1970

(1990)/22		23		24		32		33		34		35	
Afghanistan	(22)	Benin	(34)	Dominican Republic*	(24)	Guinea Bissau*	(22)	Chad	(22)	Algeria	(22)	Guatemala	(33)
Angola	(22)	Burundi*	(22)	Ecuador	(23)	Sudan	(22)	Gabon*	(32)	Cameroon	(34)	Kenya	(45)
Cambodia	(23)	Burkina Faso	(43)	Guinea*	(33)			Haiti	(33)	Guyana*	(44)	Malawi	(44)
Congo (Zaire)*	(23)	Central African Republic	(33)	Mali	(34)			Laos	(33)	Indonesia	(25)	Peru	(45)
Ethiopia	(22)	Rwanda	(44)	Nicaragua	(23)			Madagascar	(22)	Iran	(23)	Venezuela	(33)
Mongolia*	(34)	Somalia*	(22)	Togo	(23)			Mauritania*	(33)	Libya	(33)		
Mozambique*	(22)	Tunisia	(24)					Morocco	(33)	Nepal*	(34)		
Namibia*	(22)							Nigeria	(22)	Myanmar	(33)		
Niger	(22)							Paraguay	(24)	Panama	(56)		
								Zambia	(34)	Senegal	(23)		
										Swaziland*	(23)		
										Syria	(25)		
										Tanzania	(44)		
										Uganda*	(34)		
										Uruguay	(34)		
										Vietnam*	(44)		
										Yemen	(22)		

II Classes 4 and 5 in 1970

(1990)/43		44		45		46		55		56	
Saudi Arabia	(23)	Bangladesh	(33)	Argentina	(44)	Turkey	(25)	Cuba	(44)	Brazil	(24)
Zimbabwe	(45)	Bolivia	(33)	Botswana	(45)	India	(24)	Costa Rica	(44)	Chile	(35)
		Colombia	(44)	Egypt	(35)			Philippines	(46)	China	(56)
		Cote d'Ivoire	(23)	Iraq	(22)			South Africa	(46)	El Salvador	(25)
		Gambia	(22)	Malaysia	(35)					Pakistan	(24)
		Ghana	(34)	Mauritius	(56)						
		Honduras	(24)	Mexico	(35)						
		Jamaica	(45)	Sri Lanka	(56)						
		Jordan	(45)	Thailand	(45)						
		North Korea	(22)								
		Sierra Leone	(44)								
		Surinam	(22)								
		Trinidad-Tobago	(45)								

[a]The column numbers refer to invention innovation class in 1970 and 1990, and the row numbers in parentheses refer to technology mastery classes.
* = RSD ÷ GDP is zero.

189

below II capital in 20 cases, equal in 18 cases, and above in 14 cases. For II class improvements from 3 to 5, 4 to 5, 4 to 6, or 5 to 6, TM capital was below II capital in 20 cases, equal in 14 cases, and above in seven cases. Hence, II capital is more important than TM capital for the economic progress of a country.

Technological Capital and Economic Performance

A summary of economic performance measures using II and TM classes is presented in Table 6.2. The most important performance measure for agriculture is growth in multi-factor productivity (MFP). MFP growth rates are reported in Avila and Evenson (2004). Tables 6.2 through 6.5 compute indicators weighted by a country's value of agricultural product.[10]

Table 6.2 provides very limited support for the proposition that investment in agricultural extension produces higher MFP growth. MFP growth rates when both II and TM classes are lowest (level 2) averaged .775 percent per year. That was not enough to offset the real price declines in world markets for agricultural commodities. MFP growth was even lower when the TM class was 3. When the TM class was 4, there is a suggestion that MFP growth is higher. But clearly, the move from II class 2 to II class 3 or 4 was the major driver of MFP growth, and once the II class is 3 or 4, there is no indication that higher TM class levels result in higher MFP growth.

Table 6.3 shows the relationship between the adoption of Green Revolution varieties and II and TM classes. It essentially supports the interpretation for Table 6.2. Table 6.4 also reports several agricultural indicators of agricultural sector progress as they relate to II classes. They show very low cereal yields for the "22" and "23" countries and very low fertilizer use levels. Cereal yields of less than one ton per hectare are very low. Even with farms of substantial size the farmers do not produce enough products to earn more than one dollar per day per capita.

Table 6.2 MFP growth: Invention innovation (II) class vs. technology mastery (TM) class for 92 developing countries, 1970 and 1990

II Class	TM Class			
	2	3	4	5, 6
2	.775	.394	1.172	
3	2.466	1.459	.131	.955
4	2.310	1.270	1.665	−.187
5, 6	.758	.687	2.582	3.216

Table 6.3 Adoption of Green Revolution varieties (%) by invention innovation (II) class and technology mastery (TM) class for 92 developing countries

II Class	TM Class			
	2	3	4	5, 6
2	18	10	30	
3	37	50	37	
4	19	54	57	82
5, 6		62	76	80

Table 6.4   Agricultural sector indicators by invention innovation (II) class for 92 developing countries

II Class	Growth in TFP[a]	Adoption of Green Revolution Varieties (%)	Cereal Yields (kg)	Fertilizer per Hectare (kg)
22	.55	14	960	6
23	1.84	21	928	9
24	1.26	45	1733	48
33	.78	44	1393	16
34	1.33	62	2368	81
45	1.83	79	2922	91
56	3.86	81	3760	210

[a]total factor productivity.

Table 6.5 shows how the level and rate of growth of GDP per capita related to II class. Incomes of the poorest II class actually declined. Clearly II capital is important for economic growth.

U.S. Land-Grant Universities and Technological Capital

U.S. land-grant universities (LGUs) have been major producers of technological capital, particularly invention-innovation capital for developing countries. In fact, as will be discussed further below, U.S. LGUs trained a significant share of the agricultural scientists who produced the Green Revolution.

From 1950 to 1980, several foundations (the Rockefeller Foundation, the Agricultural Development Council, and the Ford Foundation), international aid agencies (FAO, UNDP, World Bank), and bilateral aid agencies (particularly USAID) were committed to building II capital in developing country universities. The basic model was simple. The most promising faculty members in the "flagship universities" of developing countries were provided financial support to complete Ph.D. programs in U.S. LGUs and in European and Japanese universities. Upon completion of their doctoral programs, these individuals returned to teach and undertake research in the national flagship university. With sufficient faculty mass, these flagship universities in developing countries introduced master's and doctoral degree-granting programs and trained the next generation of agricultural scientists. This was to some extent an exportation of the "research-university" model, in effect, the LGU model, to developing countries (see Huffman and Evenson 1993). [11]

Table 6.5   Economic growth of GDP per capita in dollars by invention innovation (II) class for 92 developing countries

II Class	Income Lever PPP ($) (1998)	Growth in Income Percent (1962–92)
22	1160	−1.08
24	3203	2.14
33	2291	.60
34	2881	2.49
45	8430	3.49
56	4156	3.67

Note: PPP (Purchasing Power Parity).

This research-university model was very successful because it provided the scientific expertise needed to continue the Green Revolution that Borlaug and others initiated in the early 1960s. An assessment of the accomplishments of this system over its twenty-year history is presented by Evenson and Gollin (2003a). But, by 1980, several factors led to the abandonment of these research-capacity building programs. First, and perhaps, most importantly, a number of studies of "returns to schooling" showed highest rates of return to primary schooling, next highest rate to secondary schooling, and the lowest rate to university education (see Psacharopoulos and Woodhall 1985). This discovery focused national policies on reducing the illiteracy rates. These results, combined with a siege of egalitarian concern in the World Bank and other NGOs, were factors in the demise of university-capacity building programs. Finally, donor fatigue in the CGIAR system was also a factor as the foundations downsized their support to research-capacity building programs.[12]

These were not valid reasons for eliminating agricultural R&D-capacity building programs in developing countries. Although most developing countries in Latin America and Asia did take advantage of these programs, most countries in Africa did not. This leaves many African countries woefully short of II and TM capital. Reference to the technological capital discussion above makes it clear that the "price of admission" to the "economic growth club" is a minimal level of technological capital.[13] There is no other option! NGO programs and food aid alone will not bring countries into the growth club.

USAID published a roster of leading agricultural scientists in the developing world in 1973 (USAID, 1973). The country coverage of the roster is incomplete in some respects. Cuba and the Peoples Republic of China are not included, but most other countries are. Four International Agricultural Research Center (IARCs)—CIAT, CIMMYT, IITA and IRRI—were included in the roster.

This 1973 report showed that the U.S. was a major source of training of scientists in the IARCs and NARs over the capacity building years 1950—1973. For the IARCs, 48 of the 71 (67 percent) scientists listed in the roster had received U.S. LGU graduate degrees. For Asian National Agricultural Research Systems (NARS), 91 of 244 (37 percent) scientists had U.S. LGU graduate degrees. For the Latin American NARS, 74 of 216 (34 percent) of listed scientists had U.S. LGU graduate degrees. For the Middle East-North African NARS, 33 of 77 (43 percent) scientists had U.S. LGU graduate degrees. For the Sub-Saharan African NARS, 28 of 87 (43%) scientists had received U.S. LGU graduate degrees.

The 1973 report also showed that the U.K. was another important source of scientists for IARCs and NARs capacity building. Approximately 4 percent of the NARS scientists had graduate degrees from the U.K. Another 5 percent had graduate degrees from other developed country universities. The contribution of the U.S. LGUs to developing country-scientific capabilities has to be regarded as impressive. The agencies supporting this capacity building exercise made a massive difference in the world during the last two decades of the 20th century.[14]

Foreign Graduate Training in the United States

The U.S. continues to train large numbers of international graduate students in the sciences (also, see Chapter 3). Tabulations of Ph.D. degrees awarded by U.S. universities by major fields of science for the periods, 1960–64, 1975–79, and 1990–1999 are reported in Table 6.6. Also, the percentage of foreigners (non-U.S. citizens) in U.S. university graduates is reported by major fields. The total number of Ph.D. degrees awarded in all fields of science in 1960–64 was 31,894. The number increased by 85 percent for the 1975–79 period, and from 1975 to 1979 it increased by 34 percent up to 104,550 for 1995–1999. For agricultural sciences, the number of Ph.D. degrees awarded in 1960–64 was 2,305, and this number grew by 60 percent for the 1975–79 period. It further increased by 19 percent to 5,057 in 1995–99 relative to 1975–79. In 1960–64, the share of foreign students in all Ph.D.s awarded by U.S. universities in the sciences was only 14.8 percent. In 1975–79, their share had risen only slightly to 16.9 percent. However, in the final period, 1995–99, the foreign students comprised a much larger 26.8 percent of all doctorates. In agricultural sciences, the percentage of foreign students receiving Ph.D. degrees has been much larger than in the other areas of science (Table 6.6). The foreign student share of agricultural science Ph.D. degrees was 25.6 percent in 1960–64, 36.0 percent in 1975–79, and 51.5 percent in 1995–1999. This increase has taken place even though aid agencies had effectively ended their capacity building programs by 1991.

International students have become increasingly competitive as entrants to U.S. doctorate awarding programs, including the U.S. LGU system. This rise in foreign student enrollment in U.S. LGUs affects all LGUs. In part it reflects the fact that a huge demand exists by foreign students for U.S. graduate training. The U.S. training is of high quality by international standards. It is also training in the English language, which is increasingly the language of science and of international commerce in the age of globalization. Student demand for U.S. degree training is also showing up in undergraduate programs.

But the U.S. LGU graduate programs also have a demand for high quality doctoral students, e.g., those with strong undergraduate training in the sciences. With the decline in the share of the U.S. labor force engaged in agricultural production, many U.S. LGUs see the domestic demand for training in the agricultural sciences dwindling. This is true at all degree levels, except for the "business" related undergraduate degrees. As a consequence, most U.S. LGU departments have an "excess demand" for doctoral students in the agricultural sciences. As a consequence, the implicit "price" of a high quality international doctoral student has risen. And U.S. LGU programs are increasingly paying this "price" in the form of assistantship income and tuition remissions.

Thus, for at least the next decade, the U.S. LGU agricultural science departments are in a position to demand more foreign doctoral students. The operative question is whether this combination of demand and supply can be used to build vitally needed technological capital (TC) in developing countries, particularly in Sub-Saharan Africa. Current graduate training programs are not particularly oriented toward doctorates returning to their native countries for work or capacity building. A large share of these students are planning to stay in some developed country after graduation.

Table 6.6   U.S. doctorate degrees awarded and percent foreign students for selected time periods

Fields	1960–1964		1975–1979		1995–1999	
	Total	Foreign (%)	Total	Foreign (%)	Total	Foreign (%)
Physical Sciences	9,525	13.2	13,943	22.2	18,852	40.3
Earth, Atmospheric & Ocean	1,380	15.9	3,242	17.8	4,071	35.9
Mathematics & Computer Science	2,082	16.8	5,052	24.0	10,303	48.5
Biological Sciences	7,099	15.8	17,716	13.1	28,317	33.2
Agricultural Sciences	2,305	25.6	4,183	36.0	5,057	51.5
Social Sciences	9,503	11.9	30,405	12.6	37,950	20.9
TOTAL	31,894	14.8	74,541	16.9	104,550	26.8

Source: Science and Engineering Doctorates: 1960–82; Science and Engineering Doctorate Awards: 2001.

Table 6.7   Doctorates awarded (number) in agricultural sciences by U.S. land grant universities, 1991–2002

Year	1991	1992	1993	1994	1995	1996	1997	1998	1999	2000	2001	2002
Total	1073	1063	968	1078	1036	1037	982	1037	965	943	844	891
Women	209	233	228	249	228	282	260	298	285	274	na	na
U.S. Citizens	564	511	457	519	492	473	468	464	440	446	401	389
Non-U.S Citizens	491	534	496	550	550	544	474	541	510	471	446	502
With Permanent Visas	56	62	49	97	106	103	83	79	51	55	28	31
With Temporary Visas	435	472	447	453	429	441	391	462	459	415	359	400

Source: National Research Council.
Note: na not available.

TC-building programs must find ways to deal with this "brain drain" issue. In the 1950–80's TC building period, brain drain was a continuing problem. Those agencies that placed visiting faculty/staff in the universities of developing countries and had a mentoring philosophy were more successful in reducing the brain drain than agencies simply offering fellowships. Many Asian and Latin American governments were effective supporters of local flagship university development and this also facilitated TC building.[15]

Can TC-building programs work in our "globalized" world? Today international mobility is high and many developed countries are facing low rates of labor force entry because of a rapidly aging population. Talented doctoral students are very likely to have employment opportunities in the U.S. or Europe upon completion of their degrees. Consider the number of U.S. Ph.D. degrees awarded to students in the agricultural sciences by gender and citizenship over 1991–2002. Table 6.7 shows a downward trend in overall doctorates and an upward trend in doctorates awarded to women. The number of Ph.D. degrees awarded by U.S. universities to non-U.S. citizens shows little trend.

Table 6.8 reports post-graduation plans for doctoral students in the agricultural sciences of U.S. universities. This table shows that a low proportion of U.S. citizens receiving doctorates plan to work outside the United States. Non-U.S. citizens who have permanent visas are generally planning employment in the United States. Also, about half of non-U.S. citizens who have a temporary visa

Table 6.8   Postgraduate plans of U.S. doctorates in agricultural science

Citizenship/year	Total	Total in US	In the U.S. (number)				Abroad
			Post Doc	Academic	Industry	Other	
US Citizens							
2000	293	282	67	99	74	44	10
2002	255	245	76	78	43	18	10
Non-US Citizens Permanent Visa							
2000	26	20	8	3	8	1	6
2002	17	15	9	2	3	1	2
Non-US Citizens Temporary Visa							
2000	249	121	84	8	24	5	126
2002	254	127	93	10	22	2	125

Source: National Research Council.

plan to stay in the United States. Most non-citizen doctorates will obtain post-doctorate training in the U.S. (one year is permitted under a J1 visa), and many will eventually return to their home countries. But 15 percent already have obtained employment in the U.S., and they are unlikely to return to their home country. Thus, the brain drain issue continues to be important to developing countries. In order for capacity building programs to be successful in developing countries students must return to the home country and university. It appears that at least 50 percent do plan to eventually return.

The International Exchange of Crop Varieties

One important form of technology transfer occurs through plant varietal exchanges between countries.

Plant Variety Protection Data

A broad range of IPRs provide incentives for private sector R&D activity (see Chapter 5). In the U.S., two important IPRs relevant to agriculture are the utility patent and "breeder's rights" in the form of plant variety protection certificates (PVPCs). The utility patent is used for biotechnology inventions, the PVPC is used for plant varieties.[16]

The PVPC is administered by the Plant Variety Protection Office in the USDA. The patent right is administered by the U.S. Patent and Trademark Office (USPTO). Both patents and breeders' rights are subject to international conventions specifying the terms under which a foreign (non-citizen) party can obtain the IPR. In the case of breeder's rights, this is the UPOV convention. In the case of utility patents, the Paris Convention and the WTO-TRIPS agreement govern.

Table 6.9 reports PVPC issuance by the USDA from 1971 to 2002. For most crops, the 1999–2002 period shows the highest rate of PVPC issuance. The 1995–98 period was a period when utility patents became available for plant varieties, and this reduced issuance of PVPCs temporarily. Foreign applications for PVPCs have risen. After 1991, 2 percent of field crop PVPCs, 7 percent of grasses PVPCs, and 15 percent of vegetables PVPCs were issued to foreigners. The SAES-USDA shares in this period were 21 percent for field crops, 15 percent for grasses, and 4 percent for vegetables.

Crop Varieties in the "Green Revolution"

The "Green Revolution" is the term applied to the development of a cluster of "modern" or "high-yielding" varieties in the early 1960s which were first adopted in developing countries in Asia and Latin America in 1964 and 1965. Industrialized developed countries experienced their Green Revolution during an earlier time period, beginning the late 1800s and in the decades before World War II (Huffman and Evenson 1993).

While it was true that most industrialized countries, including the United States, did achieve genetic improvement in most crops by 1940, it was also true that many developing countries had also made gains before the Green Revolution. This was particularly true for the "colonial" crops: bananas, coffee, tea, rubber, spices, and sugar.[17]

The major impetus for the development of Green Revolution varieties was the building of the system of IARCs. The first of these was the International Rice Research Institute (IRRI) founded in

Table 6.9 Plant variety protection certificates (PVPCs) awarded by the USDA by crop 1971–2002

	Total Issued	Percent SAES		Percent Foreign		Certificates Issued by Sub-Period							
		1971–90	91–2002	1971–90	91–2002	71–74	75–78	79–82	83–86	87–90	91–94	95–98	99–2002
Field Crops													
Barley	117	26	34	21	5	0	12	2	23	4	35	18	23
Beans (fld)	108	30	33	0	0	0	1	5	18	11	29	9	35
Corn (fld)	650	4	2	0	0	0	1	6	40	99	162	122	220
Cotton	271	10	16	0	1	22	31	38	31	34	39	17	59
Oats	60	73	82	0	0	0	11	5	1	9	8	13	13
Rice	47	23	39	0	25	0	8	4	2	5	15	10	3
Safflower	32	22	35	0	0	0	1	4	1	9	6	6	5
Soybeans	1,078	17	19	0	0	34	69	135	155	110	162	156	257
Wheat	557	36	40	0	4	13	42	60	30	68	95	75	174
Total Field Crops	2,920	19	21	1	2	69	176	259	301	349	551	426	789
Grasses													
Alfalfa	116	12	4	0	0	0	3	21	16	30	11	9	26
Bluegrass	90	14	19	23	23	0	6	9	11	17	23	10	14
Clover	23	39	46	0	0	0	0	4	2	4	1	7	5
Fescue	169	8	2	33	13	0	5	16	29	32	30	16	41
Ryegrass	128	8	7	41	2	0	1	12	30	26	14	7	38
Other	95	28	33	54	0	0	1	4	2	15	17	18	38
Total Grasses	621	14	15	30	7	0	16	66	90	124	96	67	162
Vegetables													
Beans, Garden	274	0	0	15	10	32	39	21	29	21	69	18	45
Beans, Lima	9	0	0	0	0	0	2	4	1	0	1	1	0
Cauliflower	24	0	0	80	40	0	2	6	9	2	5	0	0
Lettuce	207	0	0	0	13	14	17	14	15	26	81	5	35
Mushroom	12	0	9	0	-	3	3	3	3	0	0	0	0
Onion	63	35	0	0	5	1	3	8	2	2	10	18	9
Peas	358	0	0	0	21	20	48	46	61	14	48	36	85
Total Vegetables	947	2	4	6	15	70	114	102	120	65	214	78	174

1959.[18] IARC plant-breeding programs have now been developed for more than 15 major crops in nine IARC programs. But most Green Revolution varieties were produced in NARS crop-breeding programs. At least 500 NARS breeding programs contributed to the Green Revolution.[19]

Figure 6.1 depicts the production of Green Revolution modern varieties (MVs) for 11 crops: rice, wheat and maize, other cereals (sorghum, millets, barley), protein crops (beans, lentils, ground nuts), and root crops (cassava and potatoes). All varieties are evaluated for official "release" by release boards in each country. Release standards are generally high. This figure is based on over 8,000 Green Revolution varieties. Approximately 5,000 of these are "unique" varieties. IARC-crossed varieties were generally released in multiple countries, but most NARS-crossed varieties were released only in the "home" country, reflecting the high degree of location specificity of crop varieties.

Evenson and Gollin (2003a) summarize varietal production during the Green Revolution as follows:

1. NGO programs produced no Green Revolution varieties (except for the IARCs which are technically NGOs).
2. Private sector firms produced only "hybrid" varieties (maize, sorghum, millets and rice) and then only after significant gains were achieved in open pollinated varieties.
3. Few varieties were crossed in breeding programs in developed countries. The U.S. SAES/USDA programs did produce maize, soybean and cotton varieties that were directly transferred to Brazil and Argentina. But virtually none of the SAES/USDA varieties were released in Asia and Africa.

The Green Revolution varieties were produced using sexual crossing methods including "wide-crossing" or wide hybridization methods. IARC-crossed varieties were widely used as breeding materials in NARS breeding programs.[20]

Figure 6.2 depicts adoption rates of Green Revolution varieties. A comparison of Figures 6.1 and 6.2 shows that the production of Green Revolution varieties does not always convert directly to farmer adoption. Farmers typically evaluate new varieties in a structured way. They initially experiment by planting only a small area to a new variety, but when it is "successful" they increase acreage planted to the variety and decrease acreage planted to varieties that do not produce higher income.

Table 6.10 provides an accounting analysis of the Green Revolution crops. Data for two periods, 1961–1980 and 1981–2000, are reported. For each period, the growth in production can be allocated to growth in area and growth in yield. For this comparison, we note that in the early Green Revolution period, growth in area planted to the 11 Green Revolution crops increased in all regions. For all developing countries growth in cropped area accounted for .68 percent of production growth.

By the late Green Revolution period, cropland area expansion had effectively ended in Latin America and Asia, but not in the Middle East-North Africa region. In Sub-Saharan Africa, cropland area expansion increased and accounted for most of production growth.

Yield growth can be further decomposed into two components. The first is the MV component. The second is the other input or "intensification" component, i.e., fertilizer, labor, etc. Here we note that the MV component was important in all regions in both periods, but the size of this component varied by region. Countries in Asia clearly had the highest MV gains and received them roughly 30 years earlier than did countries in Sub-Saharan Africa. The Green Revolution in Sub-Saharan Africa was not accompanied by the intensification gains realized in the Asian Green Revolution.

The Green Revolution, as noted, was based on "conventional" breeding methods requiring a "sexual" cross. Many of these crosses entailed the incorporation of "landraces," or farmer-selected varieties, to achieve host-plant resistance to insect pests and plant diseases. In more advanced

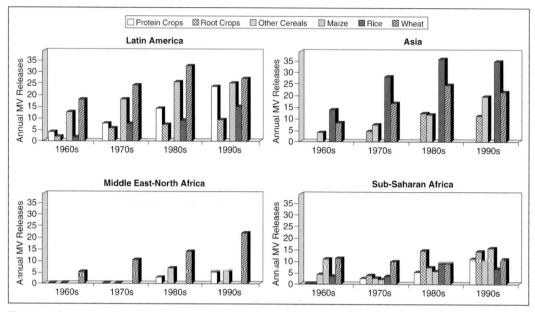

Fig. 6.1.   Green Revolution production of new varieties by decade and region of world. Source: Adapted from Evenson and Gollin 2003.

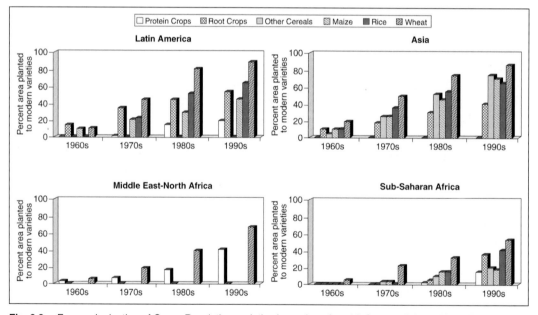

Fig. 6.2.   Farmers' adoption of Green Revolution varieties by region of world. Source: Adapted from Evenson and Gollin 2003.

systems achieving host-plant tolerance to abiotic stresses (drought, flooding, etc.) also entails the search for landrace-bred sources.

Table 6.11 reports the actual use of landraces in Green Revolution rice varieties (Gollin, 1998). Many writers on landrace development ignore the fact that U.S.-origin landraces are very important sources of the best plant traits. The table shows that the U.S. is a major exporter of landraces for

Table 6.10   Analysis of Green Revolution crops: rates of growth 1961–2000

	Early Green Revolution (1961–1980)	Late Green Revolution (1981–2000)
Latin America		
Production	3.083	1.631
Area	1.473	−0.512
Yield	1.587	2.154
MV[a] contributions to yield	0.463	0.772
Other input per hectare	1.124	1.382
Asia		
Production	3.649	2.107
Area	0.513	0.020
Yield	3.120	2.087
MV contributions to yield	0.682	0.968
Other input per hectare	2.439	1.119
Middle East-North Africa		
Production	2.529	2.121
Area	0.953	0.607
Yield	1.561	1.505
MV contributions to yield	0.173	0.783
Other input per hectare	1.389	0.722
Sub-Saharan Africa		
Production	1.697	3.189
Area	0.524	2.818
Yield	1.166	0.361
MV contributions to yield	0.097	0.471
Other input per hectare	1.069	−0.110
All developing countries		
Production	3.200	2.192
Area	0.683	0.386
Yield	2.502	1.805
MV contributions to yield	0.523	0.857
Other input per hectare	1.979	0.948

Source: Evenson and Gollin 2003.
[a]modern varieties.

Green Revolution rice varieties. But the major feature of the Green Revolution, as reported earlier in this chapter, is that U.S. LGUs trained a high proportion of the scientists who produced the Green Revolution.

Contributions of Genetically Modified (GM) Crops

The agricultural biotech industry has undergone major reorganization and restructuring over the past decades. It has emerged with six major players, so to speak, which are multinational firms. These firms were all established agricultural chemical firms that have been converted to "life science" firms by acquiring seed-producing firms (See Figure 5.5, Chapter 5). The top seven agricultural biotech companies reported annual R&D expenditures in 2002 of roughly $3 billion. These expenditures are about 10 percent of sales for these companies, which means that agricultural biotech is relatively R&D intensive (see Table 6.12).[21]

Ag biotech entails the use of "recombinant DNA" (rDNA). rDNA technologies differ from conventional breeding technologies of the Green Revolution in that they do not require a sexual cross.

Table 6.11 Summary of international flow of landraces in Green Revolution rice varieties

Country	Total Landrace Progenitors (1)	Domestic Landraces (2)	Borrowed Landraces (3)	Domestic Landraces Used in Other Countries (4)	Net Lending (2) + (4) − (3)
Bangladesh	233	4	229	10	235
Brazil	460	80	380	43	343
Burma	442	31	411	9	389
China	888	157	731	2052	2626
India	3917	1559	2358	1749	2548
Indonesia	463	43	420	420	797
Nepal	142	2	140	0	138
Nigeria	195	15	180	0	165
Pakistan	195	0	195	10	215
Philippines	518	34	484	299	749
Sri Lanka	386	64	322	57	315
Taiwan	20	3	17	669	683
Thailand	154	27	127	220	320
United States	325	219	106	2420	566
Vietnam	517	20	497	89	(0.789)

DNA from a wide variety of sources can be "installed" in crop varieties using gene-splicing technology (e.g., agrobacterium or a gene gun). Monsanto can then agree with cotton producers in a region or country to install its insect resistant products (Bollgard™ and Bollgard II™) in one or more locally popular cotton varieties in a cotton growing country (e.g., China). Typically this agreement has a restriction against farmers saving and planting their own seed. The agreement also typically specifies a "technology fee" to be paid by farmers for each hectare planted.

The total area worldwide that was planted to transgenic crops was 44 million hectares in 2000 and 68 million hectares in 2003. Of 18 countries planting transgenic crops, the U.S. has by far the largest area planted (see Table 6.13). Four crops are the center of commercial attention: soybeans and canola (for herbicide tolerance); cotton (for insect resistance); and maize (for both herbicide tolerance and insect resistance) (See Table 6.14). Herbicide-tolerant soybeans are the leading transgenic crop worldwide. Evenson (2004c) reports estimates of production-cost reductions from GM

Table 6.12 Research and development as a percentage of agriculturally-related sales in top seven ag biotech companies

Company	Annual Sales (million)	R&D (million)	R&D as % of Sales
Syngenta	$6,197	$697.2	11.2
Monsanto	$4,673	$527.8	11.3
Dupont/Pioneer Hi-Bred	$4,510	$506.2	11.2
Bayer Crop Science	E[a]4,697	E598.2	12.7
BASF	E4,924	E367.0	7.5
Dow	$2,700	na	na
Total	$27,701	$2,695	10.8

Source: Company Annual Reports, circa 2002.
[a]E = Euro.

Table 6.13   Transgenic crop area (million hectares) by country: 2000 and 2003

Country	Year 2000	Year 2003
U.S.	30.3	42.8
Argentina	10.0	4.4
Canada	3.0	4.4
Brazil	0.0	3.0
China	.5	2.8
South Africa	.2	.4
Australia	.2	.1
India	0.0	.1
Romania	<.1	<.1
Spain	<.1	<.1
Uruguay	<.1	<.1
Mexico	<.1	<.1
Bulgaria	<.1	<.1
Indonesia	0.0	<.1
Colombia	0.0	<.1
Honduras	0.0	<.1
Germany	<.1	<.1
Philippines	0.0	<.1
Total	44.2	67.7

Table 6.14   Transgenic crop area worldwide, by crop: 2003

Panel A: Transgenic trait by crop	Million Ha
Herbicide tolerant soybeans	41.4
Bt maize	9.1
Herbicide tolerant canola	3.6
Bt/herbicide tolerant maize	3.2
Herbicide tolerant maize	3.2
Bt cotton	3.1
Bt/herbicide tolerant cotton	2.6
Herbicide tolerant cotton	1.5

Panel B: Percentage of total crop area in transgenics	Percent
Soybean	55
Cotton	21
Canola	16
Maize	11
Total of these crops	25

crops and notes that the countries with advanced institutions in the form of food safety and environmental safety systems have been able to realize significant cost reductions for GM crops because GM products can be "installed" on many different varieties.[22] Many developing countries, however, are not realizing potential cost reductions. It is clear that the producers of the GM crop products have made large contributions to reducing farm costs.

Table 6.15 Domestic patenting and international-invention flows by country: 1999

Country of residence of applicant	AT	AU	BE	CA	CH	DE	DK	FI	FR
Austria	1284	86	166	51	243	453	125	45	308
Australia	101	1239	103	63	100	140	102	2	138
Belgium	167	105	1050	39	150	373	129	23	364
Canada	157	283	152	1347	151	315	137	22	309
Switzerland	839	391	750	240	1455	1509	585	98	1308
Germany	3758	822	3469	714	3980	18811	2075	496	7792
Denmark	160	112	185	50	169	280	353	37	237
Finland	218	183	184	111	185	384	191	7	360
France	1188	549	1658	637	1413	3046	964	234	11500
United Kingdom	813	959	1051	416	944	1881	894	153	1829
Italy	548	175	539	147	589	1171	415	90	1185
Japan	617	1207	903	2070	1157	8454	563	38	6734
Luxembourg	31	14	35	17	30	55	29	9	50
Netherlands	440	372	614	127	456	1108	404	77	1083
Norway	74	97	78	46	70	124	107	13	118
New Zealand	10	107	10	19	12	20	12		18
Sweden	343	366	350	149	360	707	366	27	639
USA	3163	5924	4052	7190	3488	9538	2935	211	9081
Bulgaria		1							
Czech Republic	6	2	3	1	5	8	3	—	7
Hungary	21	8	19	5	19	24	15	2	22
Romania	2		1	—	2	2	2	—	2
Russian Federation	9	14	11	9	11	18	9	—	16
Poland	5	2	3	—	5	14	3	—	7
Slovakia	4	1	1	—	2	4	1	—	3
Spain	73	42	89	28	75	137	71	7	179
Greece	8	2	8	1	9	13	7	1	15
Ireland	30	34	37	19	38	49	32	5	45
Israel	54	75	54	19	66	102	48	2	97
Republic of Korea	22	83	29	87	27	345	22	2	263
Portugal	6	3	7	1	6	14	4	2	16
Argentina	1	6	2	5	2	3	1	—	4
Brazil	4	7	4	5	3	15	2	—	15
China	2	17	2	9	1	7	2	—	9
India	1	7	3	1	3	9	1	—	8
Mexico	4	2	4	3	5	10	3	—	7
Turkey	1	1	1	1	1	1	1	1	1
Others	183	230	175	151	202	404	141	14	518
# of patents granted to residents	1284	1239	1050	1347	1455	18811	353	7	11500
# of patents granted to non-residents	13063	12289	14752	12431	13979	30737	10401	1611	32787
Total # of patents granted	14347	13528	15802	13778	15434	49548	10754	1618	44287

Note: Austria (AT), Australia (SU), Belgium (BE), Canada (CA), Switzerland (CH), Germany (DE), Denmark (DK), Finland (FI), France (FR), United Kingdom (GB), Italy (IT), Japan (JP), Luxembourg (LU), Netherlands (NL), Norway (NO), New Zealand (NZ), Sweden (SE), USA (US), Bulgaria (BG), Czech Republic (CZ), Hungary (HU), Romania (RO),

GB	IT	JP	LU	NL	NO	NZ	SE	US
266	316	76	94	191	25	9	196	479
199	126	76	87	115	14	195	112	707
354	233	163	96	337	24	45	151	648
410	236	187	100	193	22	63	199	3226
1192	1204	592	353	871	84	107	753	1279
7283	6492	2665	1170	4213	297	187	3458	9337
251	220	99	108	209	56	35	198	487
395	280	131	99	222	73	31	406	649
2903	2741	1085	800	1646	166	78	1467	3820
4465	1506	646	585	1260	135	214	1086	3572
1034	6481	283	281	603	35	34	586	1492
7900	3194	1E+05	447	2213	107	115	1166	31104
46	55	14	68	51	7	1	32	22
1058	797	551	256	2960	79	43	504	1247
147	101	36	45	100	431	12	134	224
37	16	6	8	13	2	321	12	114
673	579	287	162	454	193	91	2526	1401
10243	7004	7049	2488	5056	542	834	4177	83907
								2
8	7	2	2	3	1	1	3	24
22	16	6	10	18	2	3	20	39
2	2	—	1	2	—	—	2	4
17	13	6	4	11	2	1	13	181
7	6	3	2	5	—	2	6	19
2	2	—	1	1	—		3	5
140	141	25	44	88	13	11	82	222
14	14	2	4	11	2		12	23
157	51	12	23	37	6	4	34	94
111	91	36	36	76	4	10	63	743
707	151	1628	13	118	1	4	33	3562
14	25	—	5	7	—	—	5	5
3	7	—	—	3	1	4	1	44
18	18	6	2	6	8	—	5	91
11	5	18	—	3	—	2	2	90
11	4	6	1	4	1	1	2	112
6	7	2	2	4	—	—	3	76
1	3	—	1	1	—	—	1	4
576	332	401	92	298	29	—	196	4432
4465	6481	133960	68	2960	431	321	2526	83907
36218	25995	16099	7422	18443	1931	2142	15123	69580
40683	32476	150059	7490	21403	2362	2463	17649	153487

Russian Federation (RU), Poland (PL), Slovakia (SK), Spain (ES), Greece (GR), Ireland (IE), Israel, (IL), Republic Korea (KR), Portugal (PT), Argentina (AR), Brazil (BR), China (CN), India (IN), Mexico (MX), Turkey (TR).

(Continues)

Table 6.15 (*Continued*)

Country of residence of applicant	BG	CZ	HU	RO	RU	PL	SK	ES	GR
Austria	10	57	62	2	54	29	42	203	86
Australia	3	4	12	1	56	12	—	112	86
Belgium	11	18	20	3	77	22	13	200	98
Canada	6	14	14	4	51	20	5	176	106
Switzerland	18	97	98	7	222	69	67	819	374
Germany	77	418	433	23	586	361	182	4451	1190
Denmark	12	30	15	3	75	26	8	188	122
Finland	4	19	19	1	101	39	11	198	110
France	22	107	146	7	301	88	66	2234	726
United Kingdom	29	59	78	12	208	49	47	1261	639
Italy	7	39	53	13	154	37	28	867	391
Japan	7	23	68	9	206	18	7	1214	421
Luxembourg	4	5	3	—	16	4	7	39	23
Netherlands	4	58	69	5	106	52	46	558	275
Norway	3	7	11	1	40	15	3	92	65
New Zealand	—	1	—	—	3	1	1	13	7
Sweden	6	46	62	4	145	67	15	443	219
USA	80	201	365	46	1041	247	100	4677	2368
Bulgaria	204	—	—	—	1	—	—	—	—
Czech Republic	—	228	1	—	5	2	31	3	2
Hungary	2	10	300	4	12	4	6	21	14
Romania	—	—	1	926	—	—	—	1	1
Russian Federation	6	4	3	2	15362	2	—	9	6
Poland	—	—	1	—	6	1022	—	5	3
Slovakia	2	7	—	—	4	1	75	1	1
Spain	2	4	4	—	18	7	4	1843	65
Greece	1		2	—	3	—	1	12	7
Ireland	1	5	1	7	8	2	1	40	26
Israel	1	5	2	—	19	6	1	76	40
Republic of Korea	2	1	5	—	192	9	—	56	16
Portugal	—	—	1	—	—	—	—	11	3
Argentina	—	—	2	—	1	—	—	5	2
Brazil	—	—	—	—	3	5	—	17	1
China	—	—	—	—	13	1	—	6	—
India	—	1	—	—	2	—	—	4	—
Mexico	—	1	—	—	2	—	2	7	2
Turkey	—	—	—	—	—	—	—	1	1
Others	9	13	30	3	415	19	4	203	102
# of patents granted to Residents	204	228	300	926	15362	1022	75	1843	7
# of patents granted to Non Residents	329	1254	1581	157	4146	1214	698	18223	7591
# of patents granted Total	533	1482	1881	1083	19508	2236	773	20066	7598

Source: These data are from the World Intellectual Property Organization.

IE	IL	KR	PT	AR	BR	CN	IN	MX	TR
80	8	51	96	14	25	32	9	15	12
68	13	72	55	12	51	58	31	25	3
90	20	63	116	16	29	24	7	30	32
114	15	94	119	14	26	40	28	69	22
379	52	325	451	66	123	164	105	152	70
1154	192	1217	1563	82	379	492	268	351	231
113	22	76	116	10	9	30	28	15	10
121	3	51	148	2	32	41	10	11	7
703	73	584	939	99	202	209	71	209	77
714	167	388	700	53	125	173	78	124	74
312	44	196	432	21	119	95	29	59	31
291	72	10230	385	26	170	1465	99	134	22
23	—	16	25	1	6	10	13	—	3
289	30	416	318	31	152	120	65	70	79
62	2	23	67	5	28	14	9	5	5
11	2	4	8	—	6	4	2	3	—
206	44	138	225	15	86	118	24	64	77
2080	823	5027	2274	542	1150	1127	559	2324	285
—	—	—	—	1	—	2	—	—	—
2	1	3	4	—	—	—	3	—	1
11	3	5	15	—	2	4	1	2	5
2	—	1	1	—	—	—	—	—	—
5	1	14	5	—	2	8	10	1	2
1	1	2	2	—	—	—	—	—	—
1	—	—	1	—	—	—	—	—	—
56	6	18	129	24	22	13	1	18	4
6	—	1	6	—	1	1	4	—	2
248	—	12	24	1	2	1	1	1	6
36	419	15	42	3	7	14	17	18	8
12	5	43314	12	14	6	237	7	29	2
3	—	—	88	—	1	1	2	—	1
1	—	1	2	155	7	—	1	1	—
1	—	7	5	16	424	7	1	4	1
—	—	16	—	—	2	3097	10	3	—
—	—	4	—	—	—	—	633	1	—
3	—	—	4	—	—	—	—	120	—
1	—	—	1	1	—	—	—	—	33
89	6	251	115	17	23	36	34	36	17
248	419	43314	88	155	424	3097	633	120	33
7040	1605	19321	8405	1086	2795	4540	1527	3779	1089
7288	2024	62635	8493	1241	3219	7637	2160	3899	1122

The International Exchange of Inventions

The international patent system is regulated by the "Paris Convention" and more recently by the WTO-TRIPS agreement. Both of these agreements allow for some variation in patent laws, although the WTO-TRIPS agreement has a high degree of "harmonization" of patent law pressure built into it. Both the Paris Convention and the WTO-TRIPS agreement require that each party give "national treatment" to inventors from another member country. Thus, an inventor in the U.S. has the option to obtain patent protection in another member country. The decision to pursue this option (available within one year from the original filing date) depends on the cost of obtaining protection and on the perceived value of the invention/innovation in the second country.

As a general rule, the following observations hold for the international patent system:

1. All OECD countries support strong patent systems in terms of laws and court administration of these laws. In fact, most courts engage in "case law" in which court decisions are used to provide a basis for later decisions and to strengthen property rights, e.g., the *Diamond v Chakrabarty* court decision in the U.S. (See Chapter 5).
2. Most recently industrialized countries (RICs) are engaged in strengthening their patent systems to conform with Organization for Economic Cooperation and Development (OECD) standards. They perceive themselves as joining the invention/innovation markets of the OECD countries.
3. Most transition countries are also attempting to align their patent systems with OECD systems.
4. Most newly industrialized countries (NICs) are reluctant to strengthen patent systems. They generally offer two reasons for this. First, they are concerned that they will be committing themselves to payment of licensing fees for technologies they feel entitled to use freely. Second, they observe that trade in patents between their inventors and inventors of OECD countries is "one-sided." Few NIC inventions are marketed in OECD countries, because they are for the most part minor "adaptations" of earlier OECD inventions.
5. Developing countries that have not yet developed industrial competence have little interest in patent systems. They conduct no industrial R&D, and hence, have virtually no inventions for sale. We saw this earlier in Table 6.1. None of the countries in II classes 2 and 3 have functioning IPR systems, and about half of the countries in II class 4 do not have functioning IPR systems. All countries in II classes 5 and 6, however, have functioning IPR systems.

Table 6.15 reports patterns in domestic patenting and international-invention flow for the latest year that data are available (Johnson and Evenson 2000). Because of Paris Convention rules, inventors in an origin country can obtain protection in other countries for their inventions. In Table 6.15, patent-origin countries are the rows and destination countries the columns. Thus, Austrian inventors obtained 1,284 patents in Austria, and they patented 453 of these inventions in Germany and 308 in France. The decision to obtain patent protection in another country is based on the perceived market abroad for the invention.

Table 6.16 covers four blocks of countries organized by rows in the table. The first block is the OECD countries, or market oriented economies (from Austria to the U.S.). The second block is the transition economies, Bulgaria to Slovakia. The third group is the "recently industrialized" market economies, Spain to Portugal. And the fourth group is the newly industrialized economies, Argentina to Turkey. This table reports the number of patents obtained in the origin-group of countries and in other groups of countries. Two ratios are also reported in parentheses. First, the ratio of patents granted in the group to origin patents is presented. Second, the ratio of patents granted in the origin-group to destination patents is presented. The diagonal elements indicate the extent to which patents have markets in other countries in the group. These numbers include all origin patents plus patents granted by other countries in the group to origin-patent holders. These numbers confirm the observation that innovation markets in OECD countries are active. Many OECD countries grant patents to inventors in other OECD countries. However, for the transition (T), recently industrialized (RI), and newly industrialized (NI) blocks, the within-block markets are negligible. In these blocks, few patents are granted to inventors from other countries.

The OECD *row* in Table 6.16 shows OECD-origin patents granted in the T, RI, and NI block countries. The OECD *column* in Table 6.15 shows T, RI, and NI-origin inventions patented in OECD countries. Note that the OECD countries obtain far more patents in the T, RI, and NI countries than the T, RI, and NI countries obtain in the OECD countries. The transition economies obtain only 10 percent as many patents in OECD countries as they grant to OECD countries. For the recently industrialized countries, the grant-to-receipt ratio is a more favorable 19 percent. For newly industrialized countries, the grant-to-receipt ratio is only 6 percent.

These numbers suggest major international-invention market asymmetries. However, when patent flows are viewed relative to origin-region patents, the picture is sharply different. Transition economies grant 3.3 percent of OECD-origin inventions to OECD inventors, but 5 percent of transition-origin inventions are patented in OECD countries. RI countries grant 22.5 percent of OECD-origin inventions to OECD inventors but obtain OECD protection for 25.1 percent of RI-origin inventions. NI countries grant protection to only 5.2 percent of OECD-origin inventors, but they obtain OECD protection for 18 percent of NI-origin inventions.

Thus, there are two ways of looking at asymmetry. Most developing countries look at absolute numbers and conclude that high numbers of patent grants to foreigners are contrary to national pride and policies. But, when viewed relative to the percentages of origin inventions, the asymmetry does not hold.

Table 6.16   Invention flows from origin block to destination block, 1999

Origin Block	Origin Patents	Origin Patents Protected In:			
		OECD Economies	Transition Economies	Recently Industrialized Economies	New Industrialized Economies
OECD Economies	272,112	607,546 (223)	8,930 (3.3) (49.3)	61,216 (22.5) (133)	14,206 (5.2) (305)
Transition Economies	18,117	902 (5.0) (0.3)	18,235 (100.7)	148 (0.8) (0.3)	72 (0.4) (1.5)
Recently Industrialized Economies	45,919	11,515 (25.1) (4.2)	318 (0.7) (1.8)	46,648 (101.6)	449 (1.0) (9.6)
Newly Industrialized Economies	4,686	841 (18.0) (0.3)	33 (0.7) (0.2)	40 (0.9) (0.1)	4,715 (101.1)

Note: First number in parentheses: percent of origin patents; second number in parentheses: percent of destination patents.

Table 6.17   U.S. inventions patented in agricultural production fields and food industries as sector-of-use by country-of-origin: 1995

	Agricultural Production	Food Industries
U.S.	1471	1584
Great Britain	79	86
Germany	112	154
France	71	75
Other European Countries	105	101
Canada	27	24
Oceania	21	19
Japan	184	219
Total Foreign	599	593
Percent Foreign	29	30

Agricultural Inventions

U.S. agriculture both imports and exports inventions. Imported inventions are primarily from other OECD countries. Exported inventions are to other OECD countries and to RIC and NIC countries.

Table 6.17 reports a summary of data for patented inventions in the agricultural-production sector of use and the food-industry sector of use by country of origin (see Tables 5.4 and 5.6, Chapter 5). This provides an assessment of imported inventions. Japan and Germany are the leading suppliers of inventions to U.S. agriculture. Roughly 30 percent of U.S. agricultural technology is imported in recent years.

Table 6.18 reports the number of agricultural production inventions in 4 RIC and 10 NIC economies by origin of inventor. For the RIC economies, 49 percent were of domestic origin, 12

Table 6.18   Agricultural production inventions (patented) in RIC-NIC economies by origin of inventor: 1990 (sector-of-use)

RIC Economies	Origin				
	Domestic	U.S.	Japan	Europe	Total
Hungary	85	13	5	55	158
Korea	101	35	84	10	230
Poland	127	24	5	76	232
Turkey	3	6	2	18	29
NIC Economies					
Argentina	33	33	4	43	113
Brazil	89	63	21	118	291
China	289	41	51	81	462
Egypt	5	7	4	13	29
India	20	10	3	7	40
Kenya	2	25	5	92	124
Malaysia	1	9	10	14	34
Mexico	10	27	2	6	45
Philippines	13	48	10	28	99
Zambia	4	3	1	7	15
Total	782	344	207	568	1901
Percent	.41	.18	.11	.30	

percent of U.S. origin, 15 percent of Japanese origin, and 24 percent of European origins. For the NIC economies, 37 percent were of domestic origin, 21 percent of U.S. origin, 9 percent of Japanese origin, and 33 percent of European origin. Clearly the U.S. inventors produce important inventions for both RIC and NIC economies. Many countries without functioning patent systems have also benefited from OECD country agricultural inventions.

Biotechnology Inventions

United States firms have been dominant in the field of biotechnology. A recent study by Zohrabyan and Evenson (2000) of biotechnology inventions utilized the Yale Technology Concordance to identify biotechnology inventions. Table 6.19 reports biotechnology inventions in 1993 and 1995 by sector of use (agriculture, food, health) and by origin of invention. In 1995, inventors in the U.S. produced 82 percent of U.S. biotechnology inventions. But more impressively, U.S. inventors produced a high proportion of biotech inventions in other countries. For most other countries, U.S. inventors account for 35–55 percent of biotech inventions. This is roughly double the proportion for general agricultural inventions.

The International Exchange of Agricultural Science

The U.S. has a well-established leadership role in the production of science as well as in the production of crop varieties and agricultural inventions. By the 1990s, most LGU-SAES programs had established "pre-invention" sciences to support their applied agricultural science fields. The pre-invention sciences differed from the "basic" sciences in that they responded to demands from "inventors," including plant breeders. The pre-invention sciences use the same research design as used for basic scientific discoveries (see Chapter 2). The LGUs of agriculture were integrated with SAES programs and this demand-responsiveness was institutionalized to a considerable degree.

The development of the agricultural biotech fields of invention/innovation represented something of a disturbance to the pattern of pre-invention and basic sciences. As the biological sciences began to pursue two different paths—the molecular biology path and the more traditional evolutionary biology path—the basic sciences themselves become pre-invention sciences. To be more specific, the molecular biology sciences became pre-invention sciences (see Chapter 2), but the evolutionary biology fields did not.

The scientific studies culminating in the development of genetic engineering (rDNA) technologies were largely the product of the molecular biology sciences. They did not represent a natural development path for traditional plant breeders. Specifically, while wide-crossing or wide-hybridization techniques were developed in the agricultural sciences, they did not represent steps toward genetic engineering of crops. Wide-crossing methods were developed to extend the range of genetic resources that could be incorporated into conventional breeding programs. The rDNA techniques did not rely on sexual crosses.

These developments presented traditional agricultural scientists with some real challenges. Many LGU programs quickly developed research programs in these new fields. A number of private sector firms were very quick to see the potential of these techniques. The U.S LGU system is still responding to the challenges presented by agricultural biotech industrial development (more on this in Chapter 9).

Table 6.19 Estimates of biotechnology inventions by the agriculture, food, and health sectors for selected countries, 1993 and 1995

Sector of use	Australia 1993	Australia 1995	Belgium 1993	Belgium 1995	Canada 1993	Canada 1995	Switzerland 1993	Switzerland 1995
Agriculture—livestock	2.3	1.8	2.2	2.1	0.9	0.8	2.4	2.3
Agriculture—crops and combo farms	5.0	4.0	4.0	4.3	1.4	1.4	4.4	4.5
Agriculture—fruits and vegetables	0.5	0.4	0.4	0.4	0.0	0.1	0.4	0.5
Agriculture—horticulture	0.2	0.2	0.2	0.2	0.1	0.1	0.2	0.2
Agriculture—services to livestock	6.1	4.5	5.6	5.4	2.4	2.2	6.1	5.9
Agriculture—services to crops	2.7	2.1	2.1	2.3	0.7	0.7	2.3	2.4
Agriculture—other	0.8	0.6	0.6	0.7	0.2	0.2	0.7	0.7
Total agriculture	17.6	13.6	15.2	15.5	5.8	5.5	16.4	16.3
Food—meat, poultry and fish	0.6	0.4	0.4	0.5	0.2	0.2	0.5	0.5
Food—fruit and vegetables	0.5	0.4	0.4	0.4	0.1	0.1	0.4	0.4
Food—dairy products	2.6	2.1	2.1	2.3	0.7	0.7	2.2	2.3
Food—cereals and feed	3.2	2.1	1.3	1.2	.08	2.9	1.7	1.6
Food—beverages	3.0	2.2	2.1	2.4	0.7	0.8	2.2	2.4
Food—tobacco	0.5	0.4	0.4	0.4	0.1	0.1	0.4	0.4
Food—other	13.1	10.2	11.7	12.1	4.3	5.0	12.6	12.7
Total food	23.4	17.8	18.3	19.3	7.0	9.7	20.1	20.4
Health	232.7	168.2	195.3	198.6	75.8	81.7	212.3	213.3
Total grants	543.8	399.7	457.0	479.1	178.5	185.6	484.9	495.2
Domestic share	4%	6%	na	na	7%	9%	2%	2%
US share	55%	53%	34%	42%	48%	54%	33%	41%
Japan share	7%	8%	13%	12%	12%	17%	15%	14%

Germany		Denmark		France		UK		Greece	
1993	1995	1993	1995	1993	1995	1993	1995	1993	1995
2.6	2.6	0.5	1.3	2.8	3.3	2.6	2.6	1.1	1.4
4.9	5.4	0.7	2.2	5.1	6.4	4.9	5.3	1.7	2.6
0.5	0.5	0.1	0.2	0.5	0.6	0.5	0.5	0.2	0.3
0.2	0.3	0.0	0.1	0.3	0.3	0.2	0.3	0.1	0.1
6.7	6.5	1.6	3.7	7.3	8.5	6.7	6.6	3.1	3.8
2.5	2.8	0.4	1.2	2.6	3.3	2.5	2.8	0.9	1.4
0.7	0.8	0.1	0.3	0.8	1.0	0.7	0.8	0.3	0.4
18.1	19.0	3.4	9.1	19.4	23.3	18.2	18.9	7.3	10.0
0.5	0.6	0.1	0.3	0.6	0.7	0.5	0.6	0.2	0.3
0.5	0.5	0.1	0.2	0.5	0.6	0.5	0.5	0.2	0.3
2.5	2.8	0.4	1.1	2.6	3.3	2.5	2.8	0.9	1.4
1.6	1.6	0.3	0.7	2.0	1.9	1.6	1.6	0.6	0.7
2.5	2.98	0.5	1.2	2.6	3.5	2.5	2.9	0.8	1.4
0.5	0.5	0.1	0.2	0.5	0.6	0.5	0.5	0.2	0.3
14.7	15.8	2.7	6.8	15.5	18.9	14.7	15.5	5.1	7.0
22.7	24.9	4.0	10.6	24.3	29.5	22.7	24.4	7.9	11.3
235.3	246.9	51.7	124.9	251.9	307.6	236.1	244.6	95.9	127.7
562.6	610.9	121.6	281.8	604.2	739.9	562.9	602.3	207.2	288.4
15%	15%	5%	3%	13%	24%	12%	11%	na	na
31%	38%	30%	32%	30%	32%	32%	39%	29%	38%
20%	19%	13%	16%	19%	16%	20%	19%	10%	13%

(*Continues*)

Table 6.19 (*Continued*)

Sector of use	Ireland		Italy		Japan		Netherlands	
	1993	1995	1993	1995	1990	1995	1993	1995
Agriculture—livestock	0.5	1.0	2.3	2.4	6.8	7.8	0.1	2.3
Agriculture—crops & combo farms	0.8	1.7	4.3	4.6	15.2	18.2	0.1	4.5
Agriculture—fruits and vegetables	0.1	0.2	0.4	0.5	1.5	1.8	0.0	0.5
Agriculture—horticulture	0.0	0.1	0.2	0.2	0.8	1.0	0.0	0.2
Agriculture—services to livestock	1.3	2.7	6.0	6.1	16.3	19.0	0.2	5.8
Agriculture—services to crops	0.4	0.9	2.2	2.4	7.7	9.4	0.0	2.4
Agriculture—other	0.1	0.3	0.6	0.7	2.2	2.7	0.0	0.7
Total agriculture	3.3	6.9	16.1	16.9	50.5	60.0	0.4	17.4
Food —meat, poultry and fish	0.1	0.2	0.5	0.5	1.6	2.0	0.0	0.5
Food—fruit and vegetables	0.1	0.2	0.4	0.5	1.5	1.8	0.0	0.5
Food—dairy products	0.4	0.9	2.2	2.4	7.6	9.3	0.0	2.4
Food—cereals and feed	0.3	0.5	1.3	1.4	6.4	8.5	0.0	1.3
Food—beverages	0.4	0.9	2.2	2.5	8.3	10.4	0.0	2.5
Food—tobacco	0.1	0.2	0.4	0.5	1.5	1.7	0.0	0.5
Food—other	2.5	4.7	12.4	13.5	47.1	52.0	0.3	13.0
Total Food	3.9	7.6	19.4	21.2	73.9	85.7	0.4	20.6
Health	43.6	89.2	206.2	220.2	665.5	779.7	5.5	211.3
Total grants	92.6	193.3	486.9	530.0	1814.2 2155.6	14.0	512.2	227.6
Domestic share	na	3%	1%	3%	57%	32%	10%	2%
US share	na	44%	na	41%	30%	28%	34%	41%
Japan share	6%	4%	16%	15%	na	na	32%	14%
Agriculture—livestock	0.5	0.2	0.4	0.2	0.2	0.5	0.0	0.1
Agriculture—crops & combo farms	0.6	0.1	0.3	0.3	0.4	0.7	0.0	0.3
Agriculture—fruits and vegetables	0.1	0.0	0.0	0.0	0.0	0.1	0.0	0.0
Agriculture—horticulture	0.0	0.0	0.0	0.0	0.0	0.0	0.0	0.0

New Zealand		USA		Bulgaria		Czechoslovakia		Slovenia	
1993	1995	1990	1995	1988	1993	1991	1995	1993	1995
1.0	0.8	3.2	0.0	0.2	0.3	0.2	0.3	0.6	2.3
2.1	1.1	4.1	0.1	0.0	0.1	0.5	0.5	0.8	1.1
0.2	0.1	0.4	0.0	0.0	0.0	0.0	0.1	0.1	0.2
0.1	0.1	0.2	0.0	0.0	0.0	0.0	0.0	0.0	0.1
2.6	2.2	9.0	0.7	0.1	0.2	0.1	0.9	1.7	7.6
1.1	0.6	2.1	0.0	0.1	0.3	0.3	0.3	0.4	0.7
0.3	0.2	0.6	2.1	3.7	0.0	0.0	0.0	0.0	0.0
7.4	5.1	19.6	29.3	0.0	0.0	0.2	0.2	0.5	0.4
0.2	0.1	0.5	0.8	0.0	0.0	0.0	0.0	0.0	0.0
0.2	0.1	0.4	0.7	0.0	0.0	0.0	0.0	0.0	0.0
1.1	0.6	2.1	3.6	0.0	0.0	0.0	0.0	0.0	0.0
0.8	0.3	1.8	6.6	0.0	0.0	0.0	0.0	0.0	0.0
1.3	0.6	2.2	4.2	0.0	0.0	0.0	0.0	0.0	0.0
0.2	0.1	0.4	0.7	0.0	0.0	0.0	0.0	0.0	0.0
5.6	3.0	15.1	24.4	0.1	0.1	0.2	0.1	0.1	0.2
9.4	5.0	22.5	41.1	0.1	0.1	0.3	0.2	0.3	0.3
101.4	69.0	277.4	0.4	0.4	3.5	3.0	5.8	6.2	
		425.1							
140.9	619.7	1.2	7.3	10.3	5.4	10.1	11.4		
	958.9								
na	8%	72%	82%	na	na	na	7.3%	na	22%
50%	45%	na	na	na	34%	na	na	30%	23%
5%	5%	15%	8%	na	na	na	na	na	na
0.3	0.0	0.0	0.0	0.2	0.3	0.2	0.3	0.6	2.3
0.6	0.0	0.0	0.1	0.0	0.1	0.5	0.5	0.8	1.1
0.1	0.0	0.0	0.0	0.0	0.0	0.0	0.1	0.1	0.2
0.0	0.0	0.0	0.0	0.0	0.0	0.0	0.0	0.0	0.1

(*Continues*)

Table 6.19   (*Continued*)

Sector of use	Ireland		Italy		Japan		Netherlands	
	1993	1995	1993	1995	1990	1995	1993	1995
Agriculture— services to livestock	1.3	0.5	1.2	0.7	0.4	1.3	0.1	0.2
Agriculture— services to crops	0.3	0.0	0.2	0.2	0.2	0.4	0.0	0.1
Agriculture — other	0.1	0.0	0.1	0.0	0.1	0.1	0.0	0.0
Total agriculture	2.9	0.8	2.2	1.4	1.4	3.2	0.2	0.8
Food—meat, poultry and fish	0.1	0.0	0.0	0.0	0.1	0.1	0.0	0.0
Food—fruit and vegetables	0.1	0.0	0.0	0.0	0.0	0.1	0.0	0.0
Food—dairy products	0.3	0.0	0.2	0.1	0.2	0.4	0.0	0.1
Food—cereals and feed	0.2	0.2	0.1	0.1	0.1	0.2	0.0	0.1
Food—beverages	0.3	0.0	0.1	0.1	0.2	0.4	0.0	0.0
Food—tobacco	0.1	0.0	0.0	0.0	0.0	0.1	0.0	0.0
Food—other	2.0	1.0	0.9	1.0	1.4	2.0	0.2	0.8
Total Food	3.1	1.3	1.4	1.5	2.1	3.2	0.3	1.2
Health	39.2	15.0	28.8	19.4	15.0	39.0	3.2	9.0
Total grants	84.7	28.5	51.2	39.5	49.9	90.4	6.2	29.4
Domestic share	17%	5%	20%	51%	44%	21%	9%	22%
US share	25%	10%	30%	32%	29%	33%	56%	38%

Note: na, data not available.

One of the means by which the influence of science can be measured is through knowledge exchanges as measured by citation counts. When a scientific paper cites a prior study, this is one indication of "scientific influence." Citations are not perfect measures of influence, but it is possible to develop citation indicators with a regional focus. One measure of influence is the difference between citations received from other regions and citations given to other regions. Table 6.20 reports data for three fields of agricultural sciences—plant sciences, animal sciences, and soil sciences—based on data from the Science Citation Index database. The table shows the "number of citations per citing journal in 2002" to "science articles published in the period 1993–2002." For plant sciences, the leading ten cited journals from the United States, Europe, other developed and developing countries were identified. For animal sciences and soil sciences different numbers of source (citing) journals were identified.

Using the difference in citations received from other regions relative to citations given to other regions, the U.S. has "net scientific influence" with respect to all other regions, except for soil sciences, where U.S. citations of European journals are higher than European citations of U.S. journals. By this measure, the net scientific influence of U.S. science in plant and animal sciences is very high. The net difference in citations received minus citations given for U.S.-based journals is

New Zealand		USA		Bulgaria		Czechoslovakia		Slovenia	
1993	1995	1990	1995	1988	1993	1991	1995	1993	1995
1.7	0.1	0.2	0.1	0.7	0.9	0.5	0.9	1.7	7.6
0.3	0.0	0.0	0.0	0.0	0.1	0.3	0.3	0.4	0.7
0.1	0.0	0.0	0.0	0.0	0.0	0.1	0.1	0.1	0.2
2.0	0.1	0.2	0.3	0.9	1.4	1.7	2.2	3.8	12.2
0.1	0.0	0.0	0.0	0.0	0.0	0.1	0.1	0.1	0.2
0.1	0.0	0.0	0.0	0.0	0.0	0.1	0.1	0.1	0.1
0.3	0.0	0.0	0.0	0.0	0.1	0.3	0.3	0.4	0.6
0.2	0.0	0.0	0.0	0.0	0.1	0.2	0.2	0.4	1.7
0.3	0.0	0.0	0.0	0.0	0.1	0.2	0.2	0.4	0.6
0.1	0.0	0.0	0.0	0.0	0.0	0.1	0.0	0.1	0.1
1.7	0.0	0.0	0.3	0.1	0.5	1.9	1.9	3.4	4.6
2.6	0.1	0.1	0.5	0.2	0.7	2.8	2.7	4.8	7.9
25.2	2.0	3.3	3.9	14.5	21.6	21.1	27.8	51.6	181.2
62.9	3.4	5.0	12.4	21.6	34.6	55.0	75.0	136.8	318.5
9%	na	26%	49%	14%	9%	45%	33%	39%	62%
45%	70%	47%	43%	49%	39%	25%	16%	30%	9%

435 in the plant sciences (60 percent of U.S. cites given, 37 percent of U.S. cites received); 580 in the animal sciences (56 percent of U.S. cites given, 85 percent of U.S. cites received); and 66 in the soil sciences (59 percent of U.S. cites given, 37 percent of U.S. cites received).

Clearly U.S. agricultural sciences have been scientifically influential on an international basis. For developing countries, both U.S. and European science have been influential.

Conclusion

This chapter has reviewed and summarized the technology transfer contributions made by the USDA/SAES/LGU system to other countries. Measures of the U.S. role in creating technological capital in other countries, in developing crop varieties, in developing inventions, and in the influence of agricultural science were presented.

The LGU system had a major impact on the development of technology capital (particularly on II capital) in developing countries. A high proportion of the agricultural scientists who created the Green Revolution received graduate degrees from U.S. LGUs. They helped develop new plant germplasm that was made freely available to users in developing country NARS programs. NARS

Table 6.20   Citations per citing journal in 2002 to science journals published in 1993–2002

Citing Journal Location	Cited Journal Location			
	United States	Europe	Other Developed Country	Developing Country
Plant Sciences				
United States	1272	667	59	4
Europe	976	1566	88	8
Other Developed Country	94	63	114	4
Developing Country	95	103	16	81
Animal Science				
United States	897	63	36	3
Europe	436	208	51	12
Other Developed Country	96	19	88	7
Developing Country	150	24	14	134
Soil Science				
United States	201	53	57	2
Europe	149	373	131	4
Other Developed Country	27	27	164	0
Developing Country	2	4	2	136

programs then integrated this material into leading local varieties to obtain enhanced crop performance and yields. The actual development of Green Revolution crop varieties was achieved in developing-country locations. The IARC system and developing country NARS programs produced the Green Revolution (a few U.S.-bred Green Revolution varieties were adopted in Latin America). But to a considerable extent, U.S. LGUs produced the scientist who produced the Green Revolution.

In contrast, the Gene Revolution is being led by the private sector. Modern agricultural biotech GM products are almost entirely the product of three U.S.-based private multinational firms—Monsanto, DuPont, and Dow—and three European-based firms—Syngenta, BASF, and Bayer Crop Sciences. They have emerged as the dominant firms in the ag biotech industry, and they have large R&D budgets, both in absolute terms and relative to sales.

Developing countries have already adopted many GM products primarily from U.S. firms. This is the case because GM products are highly transportable even though crop varieties are not. An insect-resistance product produced by these companies (Bt) can be "installed" on locally developed cotton or maize varieties in a low-income or transition economy. Significant screening of new plants is then needed to see if the expression of the technology is successful, and traditional breeding methods, as in the Green Revolution, can be used to refine the new varieties. Hence, Green Revolution and Gene Revolution plant-breeding techniques are complementary.

The U.S. is also a net exporter of agricultural inventions, and many U.S.-origin inventions have contributed to productivity improvement in other developed and developing countries. The extent to which U.S.-origin inventions are valuable in other countries varies with the type of invention and with economic and geo-climatic conditions in the recipient country. Also, foreign inventions have contributed to improved productivity of U.S. agriculture (see Chapter 8).

Finally, this chapter has provided evidence of the leadership role played by U.S. agricultural science. One of the great strengths of the U.S. university system is the production of scientific discoveries. Another strength is the successful training of new scientists. The U.S. played a major role in research-capacity building in developing countries in the 1960s to the 1980s. When support for capacity building withered in the 1990s, the training continued, but the new doctorates were less likely to return to their home countries. This was a benefit to the developed countries but a negative factor for world development. The biotech industries are a clear example of science-enabled invention and innovation. Similar developments in informational technology are also science enabled. The U.S. agricultural sciences continue to enjoy a position of leadership and influence.

Notes

1. See Evenson (2004a) for an evaluation of IPRs for agriculture.
2. Evenson (2004c) and Avila and Evenson (2004) report relationships between technology capital and multifactor Productivity Growth.
3. The Green Revolution discussion is based on Evenson and Gollin (2003a).
4. See Dasgupta for a discussion of social capital.
5. Some economists develop "patent race" models and argue that patent races are wasteful. But these models ignore the fact that many products have markets, even in narrowly-defined industrial classes.
6. In fact, except for soybean varieties for the U.S. adopted in Brazil, developed-country programs did not contribute varieties to developing countries in the Green Revolution (Evenson and Gollin 2003a).
7. See Westphal (2002).
8. See Evenson (2004c) for a fuller development, including the classification data.
9. See UNIDO 2002 for industrial competition rankings.
10. FAO reports data suited to estimates of MFP grants by country. Avila and Evenson (2004) computed MFP grant rates for two periods, 1961–80 and 1981–2000 for 78 countries. Table 6.2 is based on MFP growth rate for aggregate agricultural products. MFP growth rates are weighted by the dollar volume of agricultural production in each country for each period. (See Avila and Evenson (2004) for technical details).
11. In the 1980s an alternative model, the "Development University" was pushed. This did not succeed because the Development University was perceived to be a "second rate" university.
12. Development assistance has, through its history, been driven by more than a concern for development and development strategy. It has also been driven by the political, economic, and institutional circumstances of both donors and recipients. Furthermore, in the 1990s rich countries and NGOs became "fatigued" with traditional development assistance (Kanbur, Sandler, and Morrison 1999).
13. The discussion suggests that countries with II or TM indexes below 4 are not members of the "economic growth club."
14. See Chapter 9 for a discussion of the "unraveling" of USAID in recent years.
15. The programs supporting visiting professors in developing countries are generally successful in TC-building efforts.
16. See Strachan (2004) for details of the PVPC.
17. Evenson (2004b) discusses this further.
18. Dalrymple (1986) documents MV adoption of rice and wheat varieties.
19. See Evenson and Gollin (2003a).
20. See Evenson (2004b) for a discussion of the value of IARC germplasm.
21. Runge and Ryan (2003) and Evenson (2004d) provide analysis of cost reduction gains from GM crops.
22. This means, however, that the Gene Revolution cannot replace the Green Revolution. Conventional breeding methods are still required for "dynamic" gains in crop genetic improvement.

References

Avila, Antonio Flavio Dias and R.E. Evenson. 2004. "Total Factor Productivity Growth in Agriculture: The Role of Technological Capital." In *Handbook of Agricultural Economics*. Elsevier Science/North Holland Publishing Company (forthcoming).

Barton, John H. 2004. "Acquiring Protection for Improved Germplasm and Inbred Lines." In *Intellectual Property Rights in Agricultural Biotechnology*, ed. F.H. Erbisch and K.M. Maredia. Cambridge, MA: CABI Publishing.

Becker, G.S. 1996. *Accounting for Tastes*. Cambridge, MA: Harvard University Press.

Berg, P., et al. 1974. "Potential Biohazards of Recombinant DNA Molecules." *Science* 185: 303.

Berg, P., D. Baltimore, S. Brenner, R.O. Roblen, III, and M.F. Singer. 1975. "Asilomar Conference on Recombinant DNA Molecules." *Science* 188: 991–994.

Cohen, S.N., A.C.Y. Chang, H. Boyer, and R.B. Helling. 1973. "Construction of Biologically Functional Bacterial Plasmids." *Proceedings of the National Academy of Sciences U.S.A.* 70 (November): 3240–3244.

Dalrymple, D. 1986. "Development and Spread of High Yielding Rice Varieties in Developing Countries." Washington, D.C.: Bureau for Science and Technology, Agency for International Development.

Dasgupta, P. 2000. "Economic Progress and the Idea of Social Capital." In *Social Capital: A Multifaceted Perspective,* ed. P. Dasgupta and I. Serageldin. Washington, D.C.: The World Book.

Evenson, Robert E. 1991. "Inventions Intended for Use in Agricultural and Related Industries: International Comparisons." *American Journal of Agricultural Economics* 73(3).

Evenson, R.E. 2004a. "Intellectual Property Rights and Asian Agriculture." *Asian Journal of Agriculture and Development* 1(1): 1–26.

Evenson, R.E. 2004b. "Food and Population: D. Gale Johnson and the Green Revolution." *Economic Development and Cultural Change* 52: 543–570

Evenson, R.E. 2004c. "Technological Capital in Developing Countries." New Haven, CT: Yale University, Department of Economics.

Evenson, R.E. 2004d. "Developing Country Access to Biotechnology." New Haven, CT: Yale University, Department of Economics.

Evenson, Robert E., D. Birkhaeuser, and G. Feder. 1991. "The Economic Impact of Agricultural Extension: A Review." *Economic Development and Cultural Change* 39(3).

Evenson, Robert E. and D. Gollin. 1997. "Genetic Resources, International Organizations, and Rice Varietal Improvement." *Economic Development and Cultural Change* 45(3): 471–500.

Evenson, Robert E. and D. Gollin. 2003a. "Assessing the Impact of the Green Revolution, 1960–2000." *Science* 300: 758–762.

Evenson, Robert E. and D. Gollin, eds 2003b. *Crop Variety Improvement and Its Effect on Productivity: The Impact of International Agricultural Research*. Wallingford, UK: CAB International.

Evenson, Robert E. and Daniel K. N. Johnson. 1999. "R&D Spillovers to Agriculture: Measurement and Application." *Contemporary Economic Policy* 17(4):432–456.

Evenson, Robert E. and L. Westphal. 1994. "Technological Change and Technology Strategy." In *Handbook of Development Economics*, Vol. 3, ed. T.N. Srinivasan and J. Behrman. Amsterdam: Elsevier Science/North Holland Publishing Company.

Gollin, D. 1998. "Valuing Farmers' Rights." In *Agricultural Values of Plant Genetic Resources,* R.E. Evenson, D. Gollin, and V. Santaniello, eds. Wallingford, UK: CABI Publishing.

Huffman, W.E. 2001. "Human Capital: Education and Agriculture." In *Handbook of Agricultural Economics,* Vol. IA, ed. Bruce L. Gardner and Gordon C. Rausser. Amsterdam, Netherlands: Elsevier Science/North Holland Publishing Company.

Huffman, W.E. and Robert E. Evenson. 1993. *Science for Agriculture: A Long-Term Perspective*. Ames, IA: Iowa State University Press.

Huffman, Wallace E. and Robert E. Evenson. 2004. "Agricultural Productivity, Demand for Experiment Station Resources and Impacts of Research on Productivity." Department of Economics, Ames, IA: Iowa State University.

Huffman, W.E. and P. F. Orazem. 2004. "Agriculture and Human Capital in Economic Growth: Farmers, Schooling and Nutrition." Ames, IA: Iowa State University, Dept. of Econ Working Paper #2004.

Johnson, D. Gale. 2000. "Population, Food, and Knowledge." *American Economic Review* 90: 1–14.

Johnson, Daniel K.N. and Robert E. Evenson. 2000. "How Far Away Is Africa? Technological Spillovers to Agriculture and Productivity." *American Journal of Agriculture Economics* 82: 743–749.

Kanbur, R., T. Sandler, and K.M. Morrison. 1999. *The Future of Development Assistance: Common Pools and International Public Goods*. Baltimore, MD: The Johns Hopkins University Press.

Psacharopoulos, G. and M Woodhall. 1985. *Education for Development: An Analysis of Investment Choices.* New York, NY: Oxford University Press.

United Nations Industrial Development Organization. 2002. *Industrial Development Report 2002/2003.* Vienna.

U.S. Agency for International Development. 1973. *Roster of Scientists for the Major Food Crops of the Developing World.* Washington, D.C.: Office of Agriculture Technical Assistant Bureau, Agency for International Development, Department of State.

Runge, Ford C. and Barry Ryan. 2003. "The Economic Status and Performance of Plant Biotechnology in 2003: Adoption, Research and Development in the United States." A study prepared for the Council for Biotechnology Information (CBI), Washington, D.C. St. Paul, MN: University of Minnesota.

Strachan, J.M. 2004. "Plant Variety Protection in the USA." In *Intellectual Property Rights in Agricultural Biotechnology, 2nd Edition,* ed. F.H. Erbisch and K.M. Maredia. Wallingford, UK: CABI Publishing.

Westphal, L.E. 2002. "Technology Strategies for Economic Development in a Fast Changing Global Economy." *Economics of Innovation and New Technology* 11(4–5): 277–317.

Zohrabyan, A. and R.E. Evenson. 2000. "Biotechnology Inventions: Patent Data Evidence." In *Agriculture and Intellectual Property Rights,* ed. V. Santanello, R.E. Evenson, D. Zilberman, and G.A. Carlson. Wallingford, UK: CABI Publishing.

7

Economics of the Provision of Public Agricultural Research

Real expenditures on public agricultural research have increased dramatically over the past century. The broad trends were summarized in Chapter 4. The USDA's own research activities are funded almost exclusively by the federal government. The state agricultural experiment stations are the dominant state institution engaged in agricultural research, and they have developed a more diversified funding base by obtaining funds from regular federal (USDA) sources, other federal grants and contracts, state governments, and other sources.

Chapters 1 and 4 presented a brief history of the expanding federal legislation that authorized federal support of state agricultural experiment stations. Although federal formula funding of SAES research remains an important part of agricultural research funding, this source has shrunk to less than 10 percent in 2000. Funding from the state governments have accounted for roughly one-half of all SAES funding over the past 50 years, but some weakening of support occurred in the late 1990s. Other non-USDA federal and private-sector funds have been growing in importance.

This chapter presents a brief examination of institutional factors affecting public funding of agricultural research, the changing composition of the financial support for public agricultural research, and theories of public-good provision.

Background

The public provision of agricultural research is important to the long-term welfare of the United States. The U.S. has public institutions at the federal and state level engaging in agricultural research. The research agencies of the USDA, primarily the Agricultural Research Service and the Economic Research Service, are funded by the federal government. The state agricultural experiment stations were established by the Hatch Act in 1887 with guaranteed federal support, but other types of support were also permitted. Initially all states received exactly the same federal amount— $15,000 per year (also see Chapter 1). A new federal formula was established in 1955 that allocated 20 percent to each state equally, 26 percent according to a state's percentage of the U.S. farm population, and 26 percent according to a state's percentage of the U.S. rural population. In addition, 25 percent was allocated to cooperative regional research, now called multi-state research, and 3 percent to administration.[1]

Over the last half-century, major sources of SAES funding have been state appropriations; federal formula funding; federal grants, contracts, and cooperative agreements; and private industry, commodity group, and NGO funding. The shares associated with each major source have been changing over time, and debate and discussion continues about the appropriate size of federal formula and federal competitive-grant funding for agricultural research (see Huffman and Just 1994, 1999, 2000; NRC 2000, 2003; Alston, Pardey, and Taylor 2001; Echeverria and Elliott 2001).

The demand for services from SAES programs is related to the demand for services to students in a state. The major component of this demand for SAES research is the expectation that a state's farmers will have lower production costs because of SAES research. It is further understood that a "high-quality" SAES program delivers more cost reductions than a "low-quality" program. The complementary relationship between high-quality research and high-quality teaching services reinforces the willingness of states to compete to attract high-quality scientists.[2]

Consumers have an interest in the production of high-quality food at low cost. The SAES system in the U.S. has an excellent record of delivering this product, but consumer groups are generally not important supporters of SAES programs. In fact, consumer groups support only very modest food-industry or food-technology research programs.[3] Most consumers groups see the food price question as "beyond their control", i.e., determined by national and international market forces. Thus, they do not constitute a group supporting higher production and productivity at the farm level and often are actually unaware of farm-based technology developments.

The USDA established a competitive-grants research program relatively late—in 1977—to address high-priority research areas. In 1985 it was amended to emphasize biotechnology, and in 1990 it was labeled the National Research Initiative (NRI) Competitive Grants Program (NRC 1995). The 1996 farm bill established a new grants program, the Fund for Rural America, and the 1998 Agricultural Research Extension and Education Reform Act established a new grants program called The Initiative for Future Food Systems. The latter competitive grants programs, however, have not been able to sustain federal appropriations.

During the 1980s, congressional academic earmarking of funds for agricultural research grew. This program results in direct congressional funding of particular research projects in particular states. In this process, a congressman attaches a provision to a USDA agency's research budget that a specified quantity of research funds must be "passed through" to a particular state's agricultural research institution. This process has been criticized for (1) compromising the process of making federal funds more competitive (in the scientific merit sense) and (2) diverting funds from USDA agencies' budgets and programs, including those going to the Competitive Research Grants and formula-funded programs administered by USDA-CSRS/CSREES. One possibility is that appeals to Congress directly for agricultural research funds are a direct reaction to reduced USDA emphasis on formula funds and increased emphasis on Competitive Grants Program. Hence, the net impact of all of these changes on the quality of public agricultural research in the states is difficult to judge.

The Current Funding Situation for Public Agricultural Research

In Table 4.1 the funding history of the research agencies of the USDA (column 3) and of the SAES (column 4) was presented. From 1900 to 1948, the USDA's research agencies had very rapid growth of funding: an average of 9.2 percent compounded annually in real terms. Over the same period of time, the funding of the SAES system grew at a slower 6.2 percent per annum. However, from 1948 to 1950, there was a major contraction of the USDA's research funds as post-World War II needs for food and fiber disappeared and a realignment occurred with the SAES system's research program shooting upward and exceeding that of the USDA's system. From 1950 to 1980, funds for the USDA's own research programs and for the SAES system grew at 3.8 percent per annum. From 1980 to 2000, the funds for ARS and ERS declined at an average annual rate of 0.7 percent per annum. In contrast, the funds for the SAES system grew slowly at 0.8 percent per annum.

In Chapter 4, we indicated that the SAES system receives research funds from (1) regular federal sources, i.e., OES/CSRS/CSREES, (2) contracts, grants, and cooperative agreements with USDA research agencies and non-USDA federal agencies, (3) state government appropriations, (4) endowments, commodity groups, and private industry, and (5) other sources. Between 1900 and today, a strong negative time trend exists in the share of regular federal government funds going to state agricultural experiment stations (see Figure 4.3). Although the share of regular federal funds was 65 percent in 1900, it dropped to 25 percent in 1920. In 1940 it was higher at 33 percent as the federal government helped states recover from the Great Depression, but the regular federal share thereafter gradually declined.

Table 7.1 documents the change in regular federal support of SAES research over the past two decades. In 1980, regular federal support of the agricultural experiment station system was $322 million (of 2000 dollars), but it declined to $293 million in 2000. This is a 9 percent decline. Also, in 1980, 17 percent of experiment stations' funds for research came from federal formula programs, but in 2000, this percentage had dropped to 8.4 percent and has dwindled since then.

No significant CSRS-administered competitive-grant funds were available in 1980; but in 1990, the state agricultural experiment stations obtained $27 million NRI research funds (Table 7.1). In 2000, this amount had increased to $44.7 million, which was 2.0 percent of the grand total. From 1980 to 2000, CSRS/CSREES-administered Special Grant Funds going to the SAES increased by $25 million. They comprised 1.1 percent of the grand total of SAES funds in 1980 and 2.1 percent in 2000 (see Table 7.2).

Although SAES funding through regular federal appropriations has been declining, the state agricultural experiment stations have had considerable success tapping into other federal government research funds through contracts and cooperative agreements with the USDA's research agencies. These include ARS and ERS and grants and contracts with non-USDA federal agencies such as the National Science Foundation, National Institutes of Health, Department of Energy, and Department of Defense. This source of SAES funds, not well documented before 1960, amounted to $216 million in 1980 and $360 million in 2000, which is 51 percent larger (Table 7.1). These funds were 11.4 percent of the SAES grand total in 1980 and 16.2 percent in 2000 (Table 7.2).

From 1960 to 1990, the state agricultural experiment stations obtained more than 55 percent of their funds from the state governments. In 1980, this amount was $1,052 million, and by 1990 it increased to $1,198 million. But as the states struggled with unfunded federal mandates in the 1990s, state government support of SAES was cut to $1,118 million in 2000 (Table 7.1). State government funding for SAES research accounted for 55 percent of the grand total in 1980 and 1990, but it declined to 50 percent in 2000 (Table 7.2). For some states changes have been more dramatic. States fund SAES research out of state revenue collections including transfers, and Figure 7.1 shows that the share of state government revenues allocated to SAES research was one-quarter of 1 percent of state government revenues in 1970. This share fell up to 1975 reaching a low of 0.23 percent. It then gradually returned to its 1970 level by 1980. However, since 1980 this share has steadily declined.

The private sector provides direct support for SAES research through contracts and grants, and this share has been increasing since 1960. One type of private support comes from federal and/or state marketing-orders provisions for fluid milk and many fresh fruits and vegetables. Federal marketing-order provisions were established in 1937, and they facilitated producers banding together to tax themselves in order to fund commodity promotion activities such as advertising milk and dairy research. In addition, the 1985 Farm Bill enacted provisions for a national mandatory commodity check-off program levied at the point of sale by various producer groups.

Table 7.1 Current and constant dollar revenue of U.S. state agricultural experiment stations and distribution by major sources, 1980–2003

Sources	Current $ Millions				Constant 2000 $ Millions[a]			
	1980	1990	2000	2003	1980	1990	2000	2003
Regular federal appropriations	136.9	223.6	292.6	393.0	322.1	304.6	292.6	350.3
Hatch and other formula funds	121.2	163.3	186.9	179.9	285.2	222.5	186.9	160.4
CSRS/CSREES special grants	9.6	39.7	47.0	72.2	22.6	54.1	47.0	64.3
NRI Competitive grants	—	20.0	44.7	58.7	—	27.2	44.7	52.3
Other CSRS/CSREES administered funds	6.1	0.6	14.0	82.2	14.4	0.8	14.0	73.3
Other federal government research funds	91.8	193.3	360.4	537.9	216.0	263.7	360.4	479.4
Contracts, grants, and cooperative agreements with USDA agencies	24.4	49.5	75.0	107.2	57.4	67.5	75.0	95.5
Contracts, grants, and cooperative agreements with non-USDA federal agencies	67.4	143.9	285.4	430.7	158.6	196.3	285.4	383.9
State government appropriations	446.9	877.9	1,117.8	1,124.8	1,051.5	1,197.7	1,117.8	1,002.2
Industry, commodity groups, foundations	74.0	210.0	340.9	387.1	174.1	286.5	340.9	345.0
Other funds (product sales)[b]	55.2	91.6	118.0	128.3	129.8	125.0	118.0	114.3
Grand total	804.8	1,596.5	2,229.7	2,571.0	1,893.6	2,178.0	2,229.7	2,291.4

Source: U.S. Dept. Agr. 1982, 1991, 2001, 2004.
[a]Obtained by deflating data in first three columns using the Huffman and Evenson (1993, p.95–97 and updated to 2003) agricultural research price index, with 2000 being 1.00.
[b]Amount received from industry and "other non-federal sources", excluding state appropriations and product sales or self-generated revenue.

Table 7.2 Relative distribution of U.S. state agricultural experiment station revenue by major source, 1980–2003[a]

Sources	Distribution (%)			
	1980	1990	2000	2003
Regular federal appropriations	17.0	14.0	13.1	15.3
Hatch and other formula funds	[15.1]	[10.2]	[8.4]	[7.0]
CSRS/CSREES special grants	[1.2]	[2.5]	[2.1]	[2.8]
NRI competitive grants	—	[1.2]	[2.0]	[2.3]
Other CSRS/CSREES administered funds	[0.7]	[0.1]	[0.6]	[3.2]
Other federal government research funds	11.4	12.1	16.2	20.9
Contracts, grants, and coop. agreements with USDA agencies	[3.0]	[3.1]	[3.4]	[4.2]
Contracts and grants with non-USDA federal agencies	[8.4]	[9.0]	[12.8]	[16.7]
State government appropriations	55.5	55.0	50.1	43.7
Industry, commodity groups, foundations	9.2	13.2	15.3	15.1
Other funds (product sales)	6.9	5.7	5.3	5.0
Grant total	100.0	100.0	100.0	100.0

Source: Table 7.1.

[a]Brackets are inserted around subcomponent amounts; the summation of unbracketed quantities in each column yields the grand total.

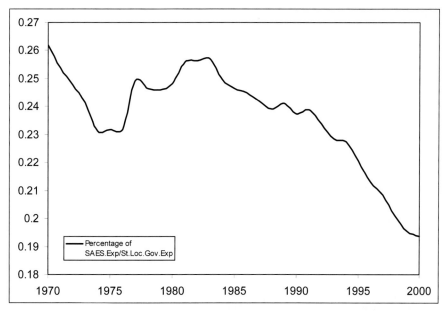

Fig. 7.1. Total SAES expenditures relative to state and local government expenditures, 1970–2000.

Private sector support of SAES research totaled $174 million in 1980 and increased to $341 million in 2000, which is a 67 percent increase (Table 7.1). The share of private sector support of SAES research rose to 9.2 percent of the grand total in 1980 and to 15.3 percent in 2000.

From 2000 to 2003, growth in the grand total of SAES funding continued up to $62 million. However, notable changes from past trends occurred. Regular federal support of SAES research increased dramatically—by $57.7 million (Table 7.1). The primary source of the increase in regular federal funds was the jump by $60 million in "other" CSREES-administered funds. This is partly due to the expenditure of funds under the Initiative for Future Food Systems competitive-grant program. Since this program received Congressional appropriations in only two years and the projects have a maximum length of 4 years, this category will soon shrink.[5] Growth of other federal government research funds continued, but the increase was an especially large $119 million, which boosted its share of the SAES grand total to 20.9 percent. However, state government appropriations to SAES continued the decline started in the 1990s and fell by $115 million, and the state's share of the SAES grand total fell to 43.7 percent in 2003 (Table 7.2).

The Cost of Public Agricultural Research

Figure 7.2 displays a graph of our research price index over the period 1970–2000 (in 1984 dollars). Note that this research price index is slightly different from the one reported in Table 4.1, because better data are available for recent years. Both research price indexes have a weight of 70 percent for personnel compensation and 30 percent for expenditures on other research inputs. The same index is used for the nonlabor inputs in the new and old research price index, but a new index is used for

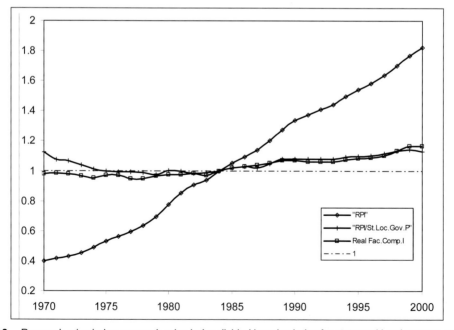

Fig. 7.2. Research price index, research-price index divided by price index for state and local government purchases, and faculty compensation relative to the implicit price deflator for personal-consumption expenditures (1984 = 1.00).

personnel compensation. It is a Tornqvist price index of faculty compensation (salaries and benefits) for assistant, associate, and full professors at Ph.D. degree-awarding public universities (AAUP, public class I universities). Figure 7.2 also presents an index of the price of research relative to the price index for state government purchases.

Figure 7.2 shows that the relative price of research declined marginally from 1970 to 1984 and thereafter rose slowly but steadily to return in 2000 to about where it started. Thus, the research price index seems to have been a drag on the demand for SAES research during 1984–2000. In comparison Figure 7.2 also shows the index of real faculty compensation: the index of faculty salaries divided by the implicit price index of goods and services purchased by consumers. This index shows how the purchasing power of faculty salaries changes over time relative to the price of consumer goods and services. During the 1970s, the index of real faculty compensation followed closely the cost-of-living index, and in the 1980s it rose slightly more rapidly. In 2000, it is about 20 percent higher than in 1982, which means that real faculty compensation is higher in 2000 than in 1970.

The Relative Size of SAES Research at the State Level

In each state, two public institutions primarily undertake agricultural research.[4] The USDA conducts research through the Agricultural Research Service and the states conduct agricultural research through the SAES. The relative importance of SAES research in total public research varies from state to state.[5] In the Lake States, Corn Belt, Northern Plains, Southern Plains, and Pacific regions, all states had an SAES share that was more than 80 percent of the total for the period 1951–1985 (see Table 7.3, column 1). In the Mountain, Delta States, Appalachian, and Northeast regions, the SAES share for this period was generally lower than 80 percent. In the Northeast, six of the states have an SAES share that is less than 70 percent. In the state of Maryland, where the USDA conducts a large share of its research, and we expect it has spillover benefits to other states, the SAES share was only 11 percent of the total.

Second, let's turn to the difference in relative importance to the individual agricultural experiment stations of non-CSRS/CSREES sources of revenue. The bottom line of Table 7.3 shows this average share of SAES funds across the 50 states in 1969, 1984, and 2000. The overall non-CSRS/CSREES share is 79 percent in 1969, decreased to 72 percent in 1984, but was 86 percent in 2000.

Consider the extremes in 1969 and 2000: greater than 2 standard deviations, or about 28 percentage points from the mean. In 1969, although no state had a non-CSRS share larger than 2 standard deviations from the mean, California, Florida, New York, and Nebraska had the largest shares. For New Hampshire and West Virginia, the non-CSRS shares in 1969 were smaller than 2 standard deviations from the mean.

In 2000, the Louisiana and Florida stations had non-CSRS/CSREES shares that were larger than 2 standard deviations from the mean. At the opposite spectrum are West Virginia, Vermont, and Rhode Island, where the share is less than 2 standard deviations below the mean. Thus, we see some changes between 1969 and 2000 in the states that have unusual revenue shares. Florida ranks high both years. At the other end, West Virginia had a small non-CSRS/CSREES share on both dates. Between 1969 and 2000 Wisconsin made a dramatic move from near one extreme to the other— from large to small non-CSRS government revenue share. The change for Georgia is almost equally dramatic, but it is in the opposite direction.

Table 7.3 Relative sources of revenue for public agricultural research[a] by state for selected years, 1950–2000

Regions/States	SAES Share (avg. 1951–85)	SAES Share from Other than Regular (OES-, CSRS-, or CSREES-adm.) Sources					
		All Nonregular Federal			State Govt. Appropriations Only		
		1969	1984	2000	1950	1984	2000
Lake States							
Wisconsin	86.2	81.6	60.1	87.8	50.4	49.1	34.9
Minnesota	83.8	83.4	76.8	90.2	62.9	68.7	55.9
Michigan	81.8	76.3	68.7	81.0	73.0	52.1	40.4
Corn Belt							
Illinois	85.7	80.6	67.6	86.3	63.3	46.8	50.7
Iowa	83.2	78.4	68.2	88.2	43.3	41.0	42.9
Indiana	81.6	76.6	68.1	88.9	39.4	40.4	45.6
Ohio	81.6	74.0	74.1	82.5	67.5	74.1	65.4
Missouri	84.5	81.6	58.6	81.2	23.2	36.4	33.9
Northern Plains							
Kansas	89.4	82.6	76.3	89.4	47.1	48.2	50.5
Nebraska	88.6	88.6	76.9	89.5	40.6	38.6	46.3
N. Dakota	87.4	80.8	83.8	87.3	65.3	66.3	49.1
S. Dakota	82.7	77.2	68.8	81.4	40.5	58.1	46.9
Mountain							
Colorado	78.9	73.5	36.7	85.0	40.9	22.8	25.0
Arizona	78.5	83.1	74.4	91.6	57.7	60.6	47.8
Utah	74.7	68.1	69.0	76.6	55.8	54.8	46.4
Wyoming	69.3	69.8	63.1	77.9	52.3	58.0	69.9
Idaho	76.0	78.8	65.4	83.7	75.1	52.9	49.8
Montana	78.5	78.8	77.8	80.4	52.0	48.9	34.9
N. Mexico	69.4	65.5	71.0	69.6	46.7	58.8	56.7
Nevada	71.6	65.1	64.0	88.3	12.3	53.0	48.5
Pacific							
California	93.2	92.7	78.7	91.6	88.8	68.9	51.2
Oregon	81.1	87.0	60.3	86.8	62.4	43.9	52.0
Washington	81.1	84.4	71.5	81.5	72.3	42.3	34.3
Alaska	M[b]	59.1	74.5	65.0	49.7	74.5	54.6
Hawaii	M	83.7	69.5	85.5	67.8	63.9	51.5
Southern Plains							
Texas	90.6	77.5	81.5	89.4	43.4	65.5	55.1
Oklahoma	80.7	75.6	76.3	86.2	57.9	62.8	62.1
Delta States							
Louisiana	88.8	84.1	86.8	91.2	58.8	69.4	60.0
Mississippi	77.6	72.3	79.4	83.8	45.3	54.9	46.6
Arkansas	82.6	73.3	75.4	90.5	46.8	61.0	63.6
Southeast							
Florida	87.2	92.0	84.0	92.9	76.1	72.1	62.7
Georgia	84.5	78.0	84.3	90.5	22.0	77.0	58.2
Alabama	70.0	73.8	74.9	85.6	41.5	42.9	59.9
S. Carolina	77.1	64.1	79.0	99.8	52.0	79.0	52.4
Appalachian							
N. Carolina	82.5	74.6	75.9	90.2	64.9	60.8	61.9
Virginia	77.2	74.5	72.7	90.2	60.8	56.4	50.0
Tennessee	75.1	63.2	71.1	82.8	37.2	47.5	61.3
Kentucky	76.9	69.1	68.9	82.6	32.6	65.9	73.4
W. Virginia	72.8	49.3	51.7	62.3	53.9	35.3	40.3

Northeast							
New York	88.8	88.4	66.1	85.1	77.3	39.4	29.0
New Jersey	78.2	85.7	71.2	81.4	56.8	57.2	65.4
Penn.	85.1	52.4	59.7	81.2	37.6	47.3	52.1
Maryland	11.0	72.3	72.2	73.9	49.5	67.1	29.8
Mass.	71.9	61.3	60.0	70.3	69.9	44.4	23.5
Conn.	74.8	77.3	64.8	87.6	83.0	62.9	59.0
Vermont	67.2	52.5	59.1	74.6	25.1	40.5	39.9
New Hamp.	67.5	44.1	48.3	65.5	23.5	40.8	64.5
Rhode Island	64.1	56.2	43.7	43.9	22.3	40.7	38.8
Delaware	67.4	62.1	66.7	71.4	30.1	50.1	32.3
Maine	69.6	59.5	63.9	73.9	48.4	36.8	47.9
U.S.	—	79.5	72.1	85.9	59.1	55.8	50.1

Source: U.S. Department of Agriculture 1951, 1970, 1985, 1990. CRIS 2000.

[a]Public agricultural research is defined here as SAES and USDA research (primarily ARS and ERS) conducted within a state.

[b]M indicates the amount is missing.

The three right-hand columns in Table 7.3 reproduce for convenience the state government appropriations share of SAES revenue in 1950, 1984, and 2000 (from Table 4.17). If we think of ranking states by the non-CSRS and state appropriations revenue shares, the ranks are somewhat different. The rank correlation between states ranked by non-CSRS revenue share and state government appropriations share is 0.65 in 1950 and 0.68 in 1984. Furthermore, tests of the null hypothesis that the ranks are perfectly correlated (correlation coefficient of 1.0) or of zero (no association) are both rejected at the 5 percent significance level. Thus, the rankings are positively associated.

Theories of Government Finance of Agricultural Research

Agricultural research in the public sector produces public goods—largely discoveries and inventions. This section draws upon the field of public finance for theories of public-goods provision. Although the categorizing of theories is subjective, we place the theories into two groups: (1) the agents of the government (officials or bureaucrats) behave in the "public interest;" and (2) the government's business is managed by individuals who are regularly elected, and private interest groups are effective in exerting political pressure on elected officials.

Behavior in the Public Interest

Four different perspectives for modeling public provision of agricultural research are presented. These are models of the public provision of a good that has public-good attributes. Part of the discussion focuses on the choice of appropriate jurisdictional authority for a public good produced at one location but having a definable range of impacts in other areas.

First, consider the government officials behaving as socially benevolent bureaucrats. They make decisions on the investment in agricultural research and other public goods so as to correct market failures and to provide the socially optimal quantities of all public goods. They have all the information about current and future events needed to make resource allocation socially or Pareto efficient.[6] In this model, governmental behavior is clearly exogenous to the behavior of private interest groups and decisions of farmers and others. Furthermore, no feedback from private interest groups to government policies occurs. Hayami and Kikuchi (1978) present such a model of public decisions on irrigation system investments in which they argue that government officials take account of economic conditions that affect the social rate-of-return to public investment in infrastructure.

Second, government officials are assumed to follow the will of the median voter in their state for deciding on public goods. In this model all voters of a state are assumed to be potential beneficiaries of agricultural research and other public goods. Voters are also assumed to have life spans that exceed the life of current investments in agricultural research and other public goods, to share in the cost of providing public goods, and to have a symmetric distribution for the individual quantity demanded of public goods. Furthermore, voting issues about the provision of public goods are assumed to be reducible to a single dimensional scale.

Under these conditions the median voter's preferences represent the boundary between support and opposition on the provision of public goods, and the median-voter model leads to consistent social choice on public goods (Cornes and Sandler 1996, pp. 489–492; Sandler 1995, pp. 158–161). The government officials are assumed to mechanically carry out these preferences.[7] In this model the characteristics of the median voters drive the demand for agricultural research and for other public goods, because they are the decisive marginal voters. In this sense, the quantity of public agricultural research is no longer exogenous to private variables. For example, if agricultural research is a normal good to the median voter, then an increase in the median voter's income or wealth will increase the demand for public agricultural research.

The usefulness of the median-voter model, as a basis for empirical analysis of investment decisions by state governments, is limited, largely by the need for voters to be potential beneficiaries of agricultural research. In poor countries where public agricultural research is primarily a national government activity and more than 50 percent of the population is in agriculture, this condition will be met. However, for developed countries like the United States where less than 5 percent of the population is in agriculture and exports are a sizeable share of agricultural production, the farmers (and possibly agribusinesses) are the primary direct beneficiaries. Thus at the state or national level, the frequency distribution of the desired quantity of agricultural research is likely to be very skewed, so the median voter is almost certainly going to be a nonfarmer. For the median-voter model to receive empirical support from U.S. data, state and federal government decisions on public agricultural research would need to be related to the characteristics of the median voter in each of the states and over time. This seems unlikely.

An additional problem is that the length of time over which benefits from a given expenditure are distributed is considerably longer than the term of political officials and than the remaining life span of the median voter. Under these conditions significant under-investment in agricultural research is likely to occur. The median voter and public officials have an incentive to ignore research benefits that extend into the next generation and to keep current taxes or costs low. The reason is the current median voter is a member only of the current generation. This kind of reasoning contributes to significant under-provision of agricultural research, unless a mechanism exists to internalize the benefits to the next generation into current decisions (see Cornes and Sandler 1996, Chapter 16).[8]

Third, there is a jurisdictional issue about which level of government should provide funding for inter-state and regional spillovers. If public R&D possesses an identifiable range of benefits, the public-finance principal of "fiscal equivalence" applies (Olson 1969, 1986; Cornes and Sandler 1996; Kanbur, Sandler, and Morrison 1999; Huffman and Just 1999). This is the theory of matching jurisdictional authority and the range of public-good spillovers. This theory dictates that the decision-making body's jurisdiction should coincide with the spillover region, or that region receiving the benefit (Olson 1969, 1986). If research on human health and nutrition benefits all U.S. citizens, then the institution that channels resources to this type of research would be the federal government. In contrast, if the soils of a particular state present unique challenges to farmers, then this state's government should channel resources to this research. If a pest affects the crops of states in the Midwest, then an institution that channels funds to fight this pest should include just these Midwestern states. The notion of fiscal equivalence suggests that a system of possibly overlapping jurisdictions would provide a relatively efficient mechanism to channel resources to a diverse set of agricultural research topics. This gives rise to the principle of *subsidiarity,* which implies a close match between the providing institution's jurisdictional authority and the range of benefit spillover of the public good (Peou 1998; Kanbur, Sandler, and Morrison 1999, p. 85).

Fourth, if agricultural research is an impure public good, having private-good attributes—such as benefits that are specific to one state due to climate and soils or area and benefits that spill over to some but not all states—then we can apply Sandler's theory of impure public goods (Sandler 1995, pp. 145–166).[9]

Assume that a state has revenue to spend from state tax collections, transfers of cash and in-kind from the federal government, and in-kind transfers from SAES research in other states. Public research and other public goods are assumed to have fixed prices. The state autocrat is assumed to behave as if he or she maximizes the state government sector's utility function subject to the state's budget or full income constraint.

If Nash strategy—noncooperative behavior among autocrats across state governments—is followed in deciding on the quantities of the public goods, then the demand function for agricultural research by a state government depends on the relative price of agricultural research; real full income of the state, including the value of public agricultural research from other states in the same region; and the quantity of agricultural research spill-ins or "borrowing" from other states. If a Nash equilibrium is achieved, then all states within a common spill-in or borrowing area will demand the same total quantity of public agricultural research, when allowance is made for spill-in benefits from other states. As long as agricultural research has some state-specific benefits, say due to unusual local geoclimatic conditions, no state will choose to borrow all of its agricultural research inventions from other states. (See Khanna, Huffman, and Sandler 1994 for additional details on this type of model of state government decisions on agricultural research.)

In this framework, agricultural research investment decisions are driven by rather traditional variables: relative prices of government purchased goods and real state income and by agricultural research "borrowed" or spilling in from other states. Hence, the primary hypothesis is that price and income effects explain the quantity demanded for public agricultural research and other public goods. If in addition, the spill-in research variable does not have an effect on demand for research that is separate from its effect through full income (i.e., all cash receipts plus value of spill-ins of public goods from other states), then agricultural research is a pure rather than an impure public good. Given the discussion and results that will be presented in Chapter 8, the absence of state-specific effects would be highly unusual.

The models and modeling strategy where it is assumed that government officials always behave in the public interest have come under heavy criticism. The reason is that during the late 1960s and early 1970s, some perceptive economists saw that many goods having public characteristics seemed to be provided by private institutions and that the government or government officials sometimes failed to behave in the public interest. Buchanan (1965) and Stigler (1974) elaborated on the first observation, and Martin, Zacharias, and Lange (1991) present a model of private-interest direct funding (i.e., check-off programs) of agricultural research. Tullock (1967) was one of the first to raise the second issue. He emphasized that the social costs of many public policies are much larger than the simple welfare losses associated with (correcting) externalities and (busting) monopolies and thieves. The reasons suggested by Tullock were that interest groups are formed to obtain (prevent) income or welfare transfers by public policies and that these groups frequently allocate large amounts of resources to maintaining policies or to trying to exert or counter pressures for public-policy changes. This is a theme expanded upon and refined by Olson (1965), Stigler (1971), Posner (1974), Peltzman (1976), Reid (1977), Becker (1983), and others.

Private firms, individuals, and other institutions have different interests about the quantity of public goods (e.g., public agricultural research benefits a relatively small share of a state's population, but a large share can expect to be taxed to pay for it). Interest groups form around common private interests and exert pressure to affect public expenditures on agricultural research and other public policies. Small groups that are efficient in exerting pressure can have a big impact on public policy, and one that is very different from that desired by an objective and informed median voter. Thus, the "preferences of government" officials are not stable, fixed, or unaffected by private interests. Furthermore, for many voters there is little economic incentive to become informed on public policies, so they collect little information and are vulnerable to misinformation or manipulation by pressure group activities. Hence, voter preferences, including the median voter's, may be easily manipulated (Reid 1977; Becker 1983).

Behavior of Interest Groups and Elected Government Officials

When government officials are regularly elected by some type of democratic process and voters have diverse private interests, government policy on public-goods provision is subject to interest group pressures. In these models, although elections occur at discrete points of time and create some discontinuities of public decision making, the formation, administration, and enforcement of public policies are relatively continuous and expensive processes.

We first and primarily consider the case where the only private interest of individuals seeking public office is election or re-election. Private firms, consumers, individuals, and other institutions are assumed to have different private interests concerning public goods. Because public policies on provision affect the private welfare of these agents, they are assumed to form groups that have relatively similar interests about the public-goods provision. These groups are assumed to use the resources of their members (voters in public elections, wealth, and status) to exert pressure that furthers the primary interests of their members. The target of this pressure group is other interest groups and elected officials. Thus, interest groups compete for favorable treatment in public policy-making and by elected officials. Enacted public policies are a compromise, i.e., a weighted average of the positions taken by interested pressure groups.

The weights are determined by relative (not absolute) pressure exerted. Relative pressure is not necessarily proportional to the number of votes an interest group directly controls, because a small, well-organized, and wealthy interest group can affect the votes of members of neutral interest groups by effectively using (mis)information and forming political coalitions (Olson 1965; Becker 1983). This is a form of advertising by pressure groups.

With this version of the interest-group model, state government decisions on agricultural-research provision are related to the characteristics (or power) of major interest groups. Characteristics that can be hypothesized to be favorable or unfavorable include share of the population living on farms, size distribution of farms, income distribution of nonfarm population, and share of the population required to elect a majority in the state legislature. State policy on agricultural research is also expected to be related to the size of state government revenue (with an allowance for borrowing research from other states) and the relative price of public research. Thus, this interest-group theory of agricultural-research funding suggests additional variables beyond those of the benevolent-government models. These variables include characteristics of groups likely to be favorable to and opposed, or less favorable, to expenditures on public agricultural research. See Guttman (1978, 1980); Huffman and Miranowski (1981); Hayami and Ruttan (1985, pp. 87–88); Evenson and Rose-Ackerman (1985); and Huffman and McNulty (1985); and de Gorter and Zilberman (1990) for interest-group models applied to public-research and extension decisions.[10]

One special case of the competitive-interest group model is the "capture theory" of public policy. This theory concludes that one interest group's pressure is so strong on a policy outcome that no other group's preferences matter; i.e., its position has a relative weight of one (see Stigler 1971; Posner 1974; Peltzman 1976; Reid 1977), while the weight of other groups is zero.

Turning to public decisions on agricultural research, the capture theory seems unlikely to represent public policy on funding of agricultural research—i.e., the "farm interest group" gets all the public agricultural research it wants. The theory does, however, seem to represent a view expressed by some critics of the way the public agricultural-research budget is allocated among alternative research projects. For example, agricultural-research administrators have sometimes received complaints that applied-research interests dominate or "capture" their research agenda (see Chapter 2). Also, Hightower (1973) and his Agribusiness Accountability Project claimed that public agricultural research only considered research interests of large farms and agribusiness, but interests of small farms and agribusiness and of consumers were ignored. These claims amount to a weight of unity on big-business agricultural-research interests.

An economy seems likely to function at a much lower level of efficiency if (elected) government officials are also an interest group that pursues private self-interests that go beyond their own interest in election to public office. Here votes and policy favors are purchased (sold) directly, and financial bribes for private gain are common.[11] Clearly, many individual public decisions will get made faster when officials are bribed, but public officials are redirecting public resources to their own private gain, and this will be a bad social investment in many cases. The degree of inefficiency seems likely to increase as the term of office of the official comes to a close because they will be ignoring the long-term social costs of their actions (see Knack and Keefer 1995). The coercive power of the state is also likely to be used by public officials for their own private financial gain. Under these conditions, privatization of virtually all public goods would seem likely, if it could occur, to lead to greater social efficiency and a higher rate of economic growth over time.

Empirical Evidence about State Provision of Agricultural Research

The provision of public agricultural-research discoveries by the 50-plus agricultural experiment stations has been the focus of a small set of important empirical studies. This section first examines the existing econometric evidence for support/rejection of the various public-interest and interest-group theories of U.S. state-government expenditures on agricultural research.

Empirical Support for the Theories

The empirical evidence is most supportive of the competitive pressure-group theory of state government decisions on agricultural research. However, when we acknowledge the costly nature of obtaining information for public-sector decisions and the uncertainty associated with the research-production process, other theories could also be generating the observed behavior. Some evidence in favor of and against the various models presented in the previous section is examined.

The public-interest model of benevolent government is one where public officials' decisions on investments in public goods, for example agricultural research, extension, and education, can be viewed as dictated by the analyses of very good economists. They would perform economic social-cost-benefit analyses of various options, using methods described by Alston, Norton, and Pardey (1995). When economic impacts of major types of projects are independent, the benevolent-government model suggests that public funds should be allocated so as to equalize the marginal social rate-of-return. When the economic costs or benefits of major projects are interdependent, as investments in public agricultural research, extension, and education seem to be (see Chapter 8; Huffman and Evenson 1989), then these interrelations need to be accommodated in the analyses.

Empirical evidence exists on the marginal social rate-of-return for only a few public-investment projects. We, however, may be able to reach some tentative conclusions based on this evidence. Let's consider the U.S. evidence on equality of marginal social rates-of-return on investments in agricultural research, extension, and education.

Evidence from more than forty-plus studies on U.S. agriculture spanning 1915–1999, summarized in Evenson (2001) and in Chapter 9, show a marginal real social rate-of-return to public agricultural research of 45–65 percent. One has to be impressed with the consistency of the evidence from a fairly large number of studies. For agricultural extension, the empirical evidence on the marginal social rate-of-return is less consistent, ranging from 0 to 100 percent for broad measures of extension. The number of studies that have been published is smaller than for agricultural research (see Chapter 9; Evenson 2001), but it is impossible to say with much confidence that the social rate-of-return lies within a small range. Psacharopoulos and Woodhall (1985, p. 59), Psacharopoulos (1985), Welch (1999), and Huffman (2001) present empirical evidence on the marginal social real rate-of-return to investments in U.S. secondary and higher education. For the period after 1950, the social rate-of-return has been about 6–12 percent at both levels, declining in the 1970s but rising in the 1980s and 1990s. Furthermore, it is our appraisal that a strong consensus exists among labor economists about the magnitude of these estimates. Hence, we place a high degree of confidence in these estimates for education.

Now let's return to the issue of equality of marginal rates-of-return and investments in agricultural research, extension, and education. We conclude that the evidence casts considerable doubt on equality of marginal social rates-of-return across these public goods. It seems most likely that public officials are not behaving as if they were attempting to allocate resources so as to correct externalities and to equalize marginal social rates-of-return on public investments. It is, however, possible that problems of imperfect information rather than intent have thwarted benevolent government officials from equalizing returns. Any bias due to excluded ability is approximately offset by reporting error in years of schooling completed (Krueger and Lindahl 2001).

The second public interest model is one represented by median voter behavior. No published econometric evidence exists that this is a plausible model of state government behavior in providing agricultural research for U.S. states. Several studies, however, have shown that it is a useful model for explaining expenditures of local governments on elementary and secondary education, police and fire protection, sanitary services, and public parks (see Borcherding and Deacon 1972; Sandler 1995, pp. 158–159).

Some evidence exists that state officials behave as autocrats pursuing the public interest. Khanna, Huffman, and Sandler (1994) report estimates of 48 demand functions for SAES research—1 for each of the contiguous states. The model is fitted to annual data, 1951–1985. Their model, which is estimated under the assumption of Nash noncooperative behavior across state governments, performs well in terms of yielding estimates of price and income (or revenue) elasticities that are of expected signs and plausible size. They also find empirical support for public agricultural research being an impure public good (one having both state-specific and general benefits that spill across state boundaries); i.e., the spill-in of agricultural research from other states or borrowing has effects on the demand for agricultural research that are separate from its effects through the full income variable. They do not, however, include interest-group variables in their demand equations, and hence do not test for private interest-group effects. Thus, although the Khanna, Huffman, and Sandler (1994) results are supportive of the benevolent autocrat model of state government decision making, their results do not seem to disprove interest-group behavior.

The focus now shifts to empirical support for (competitive) interest-group behavior. Marcus (1987) reports long-term evidence dealing with conflicts among competing interest groups in the United States. He sights a large amount of evidence that is consistent with a solution to public agricultural-research issues resulting in a compromise. His evidence starts with conflicts between farm groups and agricultural scientists during the 1870s and 1880s over who should advance the stock of knowledge about the science of agriculture.

Most of our attention, however, focuses on published studies using data for the post–World War II period: Guttman 1978; Huffman and Miranowski 1981; and Evenson and Rose-Ackerman 1985; and new research. These studies have focused on estimating a single function that applies to all states. Huffman and Miranowski (1981) and Guttman (1978) use as explanatory variables a state-revenue or income variable, one or more variables representing the potential effects on a given state's provision by the prospects of borrowing from other states, and additional variables that represent the relative strength of groups within states likely to have opposing views and to be influential in reaching a compromise on funding of SAES research.

In three early studies, in which models are fitted to data for years between 1960 and 1980, the interest group variables have a statistically significant and expected effect on SAES funding. For example, a larger share of large or owner-operator farms increases a state's expenditures on agricultural experiment station research (Guttman 1978; Huffman and Miranowski 1981). Court-mandated reapportionment of the state legislatures starting in 1962, which reduced the political power of the farm population, reduced the impact of farm interest-group variables and increased the impact of nonfarm interest-group variables on SAES funding. When a larger share of a state's legislators is farmers, expenditures on agricultural research increased (Evenson and Rose-Ackerman 1985). In addition, the state revenue (Guttman 1978; Huffman and Miranowski 1981) and spill-in or research-borrowing variables (all three studies) are also statistically significant variables for explaining state funding of SAES research. All three studies indirectly controlled for research price-change effects by including year-dummy variables.

No published empirical evidence exists that one interest group, for example, operators of large farms, has captured the interests of state officials and that only its interests matter in determining state expenditures on agricultural research. Variables that represent the interests of other groups also seem to matter significantly. (See Guttman 1978; Huffman and Miranowski 1981; and Evenson and Rose-Ackerman 1985; Huffman and Evenson 2004.)

A Model of Funding Shares for SAES Research

We make a fairly strong assumption that a state government's decisions on agricultural research expenditures are separable from other state expenditure decisions. We permit some benefits of agricultural research to be private (a commodity that is a private good), in the sense that they are state specific and other benefits to be public (a commodity that is a public good) and to spill over to other states. To capture a key aspect of agricultural research, the model includes voluntary and involuntary (federal, state, and private) contributions to a state government's expenditures on agricultural research. Moreover, we argue that different types and sources of contributions to a state's agricultural-research expenditures can be expected to differ in their potential for private and public-good production.

Rather than focus on each state's decision for the research-associated public and private goods, we shift the emphasis to the demand for research. A state legislature is assumed to maximize its utility from research resources subject to a budget constraint, including in-kind transfers.[12] Local scientific, agricultural, and demographic conditions will affect the translation of research input into public and private goods and, hence, affect the translation of research inputs from voluntary and involuntary contributions into utility of a state legislature.

We focus on the demand for four different types of SAES research resources: (1) federal formula funds; (2) federal grants, contracts, and cooperative agreements; (3) state government appropriations; and (4) private industry, commodity groups, and NGO contracts and grants. We assume that the preferences of the state legislature for agricultural research inputs can be approximated by a nearly ideal demand system (Deaton and Muelbauer 1980). The demand system is represented by the following research-share equations:

$$s_{it} = \alpha_i + \beta_1 \ell n\left(\frac{F_t}{P_t}\right) + \gamma_{i1} K_{1t} + \gamma_{i2} K_{2t} + \mu_{it}, \quad i = 1, 2, 3, 4, \tag{7.1}$$

where s_{it} is the i-th resource share in year t, F_t is total SAES revenue/expenditures from all sources in year t (or the budget constraint), and P_t is the research-price index in year t. K_{1t} is a vector including variables in the federal-research-funding formula, indicator of interstate public agricultural research spill-in potential due to SAES and USDA research conducted in other states, and an indicator of within-state private agricultural research spill-in potential. K_{2t} is a vector of translating variables (a state's scientific, agricultural, and political conditions).[13] The variable μ_{it} is a zero-mean random-disturbance term.

In each time period, the input shares sum to one, i.e., $s_1 + s_2 + s_3 + s_4 = 1$. For estimation purposes, one of the four share equations can be deleted, and its coefficients can then be recovered from the coefficients of the other three equations. For example, let's drop the fourth share equation, then $\alpha_4 = -\alpha_1 - \alpha_2 - \alpha_3$, $\beta_4 = -\beta_1 - \beta_2 - \beta_3$, and $\gamma_4 = -\gamma_1 - \gamma_2 - \gamma_3$. Note that equation (7.1) also imposes the condition of homogeneity of degree zero in total expenditures (F_t) and the price index (P_t), i.e. revenue shares are a function of the size of total revenue/expenditures in constant rather than current dollars.

Given equation (7.1), the elasticity-of-demand for each of the four research types can be summarized as follows:

$$\eta_{iF} = 1 + \frac{\beta_i}{s_i} \tag{7.2}$$

$$\eta_{iK} = \frac{\gamma_i}{s_i K}. \tag{7.3}$$

Equation (7.2) gives the income elasticity-of-demand for the i-th type of research activity, and equation (7.3) gives the elasticity-of-demand for the i-th type of research activity with respect to a 1 percent change in K.

The share equations (7.1) are fitted to data for a panel of states, covering the 48 contiguous states, 1970–1999, or 1,440 observations. The dependent variables are the SAES-input shares, and the regressors are the real budget constraint (i.e., total SAES expenditures or revenue divided by the Huffman and Evenson research-price index), variables associated with the federal formula, interstate and within state research spill-ins, and translating variables. Each state's lagged share of the U.S. farm population and of the rural population is included to capture federal formula determinants. Our indicator of spill-in potential of interstate public agricultural research is the stock of public agricultural research from other states within the same region. The regional subgroups are the same as those used by Khanna, Huffman, and Sandler (1994).[14] The indicator of spill-in potential of private agricultural research is a private agricultural stock variable constructed from agricultural patents awarded in each state (see Johnson and Brown 2002).

The quality of local graduate education and research is measured by two indicators. One is the Gourman ranking of a local LGU's graduate program in agricultural sciences. Dummy variables are assigned to an LGU being in the "Top-10," and "2nd-10" (relative to "3rd-10" or lower).[15] A second is from National Research Council quality ratings of doctorate program faculty in biochemistry, microbiology, and botany. One of four dummy variables is assigned if the local LGU is ranked Good-to-Strong (compared to Strong-to-Distinguished), Adequate-to-Good, Marginal-to-Adequate, and Insufficient-to-Marginal. Also, some agricultural experiment stations have a heavier emphasis on basic or pre-invention science than others, and this reputation and capacity may affect the demand for research inputs. This factor is represented by the lagged value of SAES resources allocated to basic biological-science research.

Additional variables are the share of a state's population that is farm and rural, composition of farm sales in 1982, and seven regional indicators that represent regional fixed effects which are time invariant.

Because resource shares sum to one, a shock to any one equation will be at least partially transmitted to the other three share equations. We account for this in estimation.[16] The estimated coefficients for the federal grants and contracts, federal formula funds, and state government are reported in Table 7.4, and the implied value of the coefficients for the "other" expenditure category is reported in the last column of the table. The results are surprisingly strong.[17] Hence, our model of state demand for these three types of public agricultural-research activity has explanatory power.

Turning to individual regressors, the budget constraint is a statistically significant explanatory variable in the three fitted share equations, holding public and private agricultural research spill-ins constant. Its coefficient is positive for federal grants and contracts and other SAES sources and negative in share equations for state and federal formula resources. These coefficients also have important implications for the income elasticity of demand for research inputs, which will be presented later.

The estimated coefficient for a state's share of the U.S. farm populations is positive and significant in the equation for the SAES federal formula-funding share but also for the SAES state government and federal grants and contracts shares. The impact on the SAES federal grants and contracts share is somewhat surprising; one might expect no effect.[18] These results also imply that the impact of a larger U.S. farm population share is to reduce the share from "other" SAES sources. The estimated coefficient for the state's share of the U.S. rural population is also positive in the federal formula-funding share equation.

Spill-ins of interstate public (SAES and USDA) agricultural research, or of instate private agricultural research, reduce the demand for federal formula funds. One interpretation is that these spill-ins are substitutes for federal formula funds. Also, within state private-research-spill-ins substitute for voluntary private contributions to SAES research. Within a state, private research spill-ins also increased the demand for federal grants and contracts. These private agricultural research spill-ins, which are mainly applied-research or invention related, seem to be complementary with less applied federal grant and contract research.

A Gourman ranking of top-10 or 2nd-10, relative to a lower ranking in the graduate agricultural-science programs at local LGUs, increases agricultural research funding at the state government level significantly—17.9 percent for top-10 and 9.5 percent for 2nd-10. These large positive effects are offset by negative impacts on the other three funding shares. Hence, the state government reacts as if it places significant weight on the Gourman index, even if it faces some academic criticism.

NRC ratings of a LGU's doctorate program faculty in the basic biological sciences (i.e., average of the rankings of biochemistry, microbiology, and botany) are important. Being rated below the top category, which is "Strong-to-Distinguished," reduces the federal-grants share by 6–7.5 percentage points, with little difference in the size of the reduction as a university moves down to "Marginal-to-Adequate" or "Insufficient-to-Marginal." However, being below the "Strong-to-Distinguished" cat-

Table 7.4 Econometric estimates of a nearly ideal demand system for state agricultural experiment resources, 48 states: 1970–1999 (t-values in parentheses; N = 1,440)

Regressors[a]	Revenue/Input Shares			
	Federal Grants & Contracts	Federal Formula	State Approp.	Other[b]
Intercept	−1.976	2.435	0.984	−1.443
	(11.80)	(18.84)	(3.69)	
ℓn(Total SAES Revenue, 1984 dol.)	0.067	−0.109	−0.021	0.063
	(16.08)	(34.01)	(3.24)	
U.S. Farm Population Share	0.710	0.137	0.494	−1.341
	(2.61)	(2.20)	(3.86)	
U.S. Rural Population Share	−1.165	0.075	−0.120	1.210
	(5.07)	(5.22)	(4.08)	
ℓn(Public Ag Res Spill-in Capital)	−0.003	−0.012	0.010	0.005
	(0.56)	(3.69)	(1.46)	
ℓn(Private R&D Capital)	0.229	−0.165	−0.040	−0.024
	(9.77)	(9.13)	(1.07)	
Ratings of Graduate Programs Ag Science (Gourman):				
Top 10 (=1)	−0.067	−0.002	0.179	−0.110
	(4.95)	(0.18)	(8.26)	
2nd 10 (=1)	−0.020	−0.022	0.095	−0.053
	(2.90)	(4.08)	(8.73)	
Quality Basic Biology Science Faculty (NRC):				
Good-to-Strong (=1)	−0.063	0.003	−0.030	0.090
	(6.21)	(0.37)	(1.85)	
Adequate-to-Good (=1)	−0.075	0.003	0.029	0.043
	(6.93)	(0.35)	(1.70)	
Marginal-to-Adequate (=1)	−0.064	0.021	−0.082	0.125
	(5.31)	(2.30)	(4.26)	
Insufficient-to-Marginal	−0.070	−0.007	0.000	0.077
	(5.60)	(0.74)	(0.05)	

Share SAES Research Inv. in Basic Biolog Science$_{-2}$	0.177 (5.89)	−0.136 (5.87)	−0.021 (3.24)	−0.020
State Farm Population Share	−0.352 (4.36)	0.461 2.20)	0.494 (3.86)	−0.603
State Rural Population Share	0.044 (2.38)	1.902 (10.73)	−0.123 (4.08)	1.823
Composition of Farm Sales (1982):			
Share fruits & vegetables	0.277 (5.51)	−0.283 (7.30)	−0.010 (0.12)	0.016
Share horticulture & greenhouse	0.774 (13.30)	0.028 (0.63)	−0.403 (4.36)	−0.399
Share livestock	0.253 (12.82)	0.020 (1.32)	−0.307 (9.78)	0.034
Regional Indicators:			
Northeast (=1)	−0.053 (4.54)	−0.018 (2.03)	0.036 (1.94)	0.035
Southeast (=1)	−0.088 (7.94)	0.012 (1.34)	0.155 (8.74)	−0.079
Northern Plains (=1)	−0.045 (3.54)	−0.041 (4.18)	−0.005 (0.27)	0.091
Southern Plains (=1)	−0.086 (7.80)	−0.016 (1.88)	0.138 (7.85)	−0.036
Mountain (=1)	0.020 (1.67)	−0.044 (4.81)	0.037 (1.97)	−0.013
Pacific (=1)	−0.018 (1.53)	0.033 (3.60)	−0.023 (1.23)	0.008
R^2	0.493	0.793	0.330	

[a]See Equation (7.1) in text.
[b]These coefficients are derived from the estimate coefficients in columns (1)–(3).

egory increases the SAES share from "other" SAES sources by 4–12.5 percentage points, with larger increases being for the lowest ranking. The federal formula-funding share is largely unaffected by a university's NRC faculty-quality ranking.

The capacity of an agricultural experiment station for basic biological-science research can be built through investments in this area. In our model, SAES with a large basic biological-science capacity, as reflected in the lagged value of the share of basic biological-science research, increases the demand for federal grant and contract funding. This, however, is largely offset by a reduction in federal formula and state funding.

When a state has a larger share of its population on farms, the demand for state resources and federal formula research resources increases. This implies that these resources serve farmers' interests well. The commodity mix of a state's agriculture also impacts demand for research resources. Our results show statistically significant regional effects, which are measured relative to the Central Region. They suggest, other things being equal, that the North Central Region (and Mountain Region) has a larger demand for federal grants and contracts relative to all other regions. The Southern Plains, Southeast, Northeast, and Mountain Region have a larger demand for state appropriations relative to all other regions.

Thus, consistent and relatively strong evidence exists that state governments are affected not only by the size of the resources they control, the price of agricultural research, but also by political pressure exerted by opposing private-interest groups for particular outcomes on agricultural-science policy. Hence, political power for affecting agricultural-science policy is more than a consideration of the fact that the farm share of the population is declining.

Estimates of Income and Price Elasticities

Several of the econometric studies of SAES funding have estimated revenue or income elasticities of demand for agricultural research. A few have estimated price elasticities. We summarize this evidence here. Huffman and Miranowski (1981) fitted an empirical model that contained quite a few regressors and obtained a revenue or income elasticity of 0.18. Guttman (1978) estimated a demand function for agricultural research for 1969 looking only at the allocation to poultry, dairy, and grains. He obtained state-revenue elasticities for this subset of commodity-oriented research of 0.62 and 0.75.

Khanna, Huffman, and Sandler (1994) use a very small set of regressors, only three, to explain the demand for agricultural research. Since each state was permitted a different demand function, there are a large number of different estimates of the revenue or income elasticities. To make summarization easier, regional average-income elasticities are created by weighting the income elasticity of each state by its share of regional-research expenditures. These elasticities were estimated using annual data, 1951–1985, and the elasticities are reported in the first column of Table 7.5. The weighted average-income elasticity for all states is 0.55. Except for the Northern Plains and Southern regions, the regional average-income elasticities have values in the small range of 0.49–0.57. The revenue elasticity for the Northern Plains Region is significantly smaller, and for the Southern Region significantly larger, than for the other regions.

Evenson and Rose-Ackerman (1985) do not estimate an income elasticity.[19] Although the exact magnitude of the income elasticity differs a little because of different definitions of income and of agricultural research, the evidence is overwhelmingly in favor of a positive (though significantly different from 1.0) state-income or revenue elasticity of demand for agricultural research. Thus, states or state governments that have larger real resources spend more on agricultural research, but the share of the state budget spent on agricultural research declines as (real) revenue increases.

Using parameter estimates from Table 7.4, equation 7.2 and evaluating it at the sample mean of the data set, we obtain income elasticity of demand for voluntarily contributed research. For federal-grant-research, the income elasticity is 1.58, for federal-formula-funded research is 0.4, for state-government-funded research is 0.96, and for other SAES, largely privately-funded research, is 1.35. Hence, as experiment station resources grow in real terms, the most rapid growth will be in federal grants and contracts and private sector contracts and grants.

The price of agricultural research is a key variable in Khanna, Huffman, and Sandler (1994). The weighted average-price elasticity for the United States is −1.6 (see Table 7.4). Excluding the Northeast Region, the price elasticity of demand in all regions is negative and relatively large, less than −1.0, or in other words the price is very elastic. The Northeast Region is rather unusual in that the price elasticity for New York, which is positive, dominates the regional average elasticity, although it is not significantly different from 0. Thus, the generally large price elasticity of demand implies that state officials behave as if they are very responsive to price in making decisions about the quantity of SAES research to provide.

Table 7.5 Estimates of U.S. state government income and price elasticities of demand for public agricultural research

	Regional Average Elasticity	
Region/States	Full Income[a]	Research Price[b]
Central	0.57	−1.54
(IN, IL, IA, MI, MO, MN, OH, WI)		
Northern Plains	0.36	−4.21
(KS, NE, ND, SD)		
Western	0.54	−1.66
(AZ, CA, CO, ID, MT, NV, NM, OR, UT, WA, WY)		
Southern	0.74	−2.12
(AR, LA, MS, OK, TX)		
South and Eastern Uplands	0.52	−2.01
(AL, FL, GE, KY, NC, SC,TN, VA, WV)		
Northeast	0.49	0.143[c]
(DE, CT, MA, MD, ME, NH,NJ, NY, PA, RI, VT)		
Overall U.S.	0.55	−1.64

Source: Adapted from Khanna, Huffman, and Sandler 1994.

Note: Elasticities are weighted averages of individual state estimates, except for the Southern Region, South and Eastern Uplands Region, and the Central Region (price only). The weights are a state's share of a region's total public research. See Khanna, Huffman, and Sandler (1990) for the individual elasticities.

[a]A given state's revenue is measured as full income, which includes expenditures on public agricultural research that are made in other states located in the same region.

[b]The research price is the price index for public agricultural research (Table 4.1) divided by the price index for purchases by state and local governments.

[c]New York State carries a large weight in this computation and its price elasticity is positive but not significantly different from zero.

Conclusion

The funding of public goods is conducted in a particular institutional environment, which undoubtedly affects funding decisions. Funding decisions on agricultural research and extension are made at both the federal and state-government level. The state agricultural experiment stations conduct a majority of the public agricultural research in most of the states.

The state agricultural experiment stations obtain funds for research from a wide variety of sources. Since 1980, the most notable features are a decline in federal Hatch and other formula funds; an increase of USDA administrated competitive and special grant funds; and a large increase in grants, contracts, and cooperative agreements with non-CSRS/CSREES federal agencies. Also, private sector funding from industry, commodity groups, and foundations has increased. Although state-government funding of SAES research rose until about 1990, it has declined by almost $200 million since then. It may be very difficult to recapture these state funds for SAES research.

Each state agricultural experiment station conducts research in a particular state, but some of the benefits spill over to other states. The principle of fiscal equivalence dictates that the jurisdictional authority match the geographical range of benefits. When benefits are national in scope, then the federal government should finance public agricultural research. If the benefits are confined to a particular state, then this state's government should finance public agricultural research. If benefits extend to a group of states, e.g., the Midwest, then a collective authority covering these states should finance the agricultural research. Applying this rationale to public provision of agricultural research leads to a mosaic of possibly overlapping jurisdictions differing in their range of benefits. Moreover there are models of public finance that suggest benevolent behavior of public officials and others that suggest that private interest groups influence elections of officials and public agri-cultural-research policy.

The empirical results do confirm a state-revenue or income elasticity of demand for public agri-cultural research that is positive, but less than 1.0—most likely between 0.5 and 0.7. However, the revenue elasticity of federal government-grant and contract revenue, and of private sector grants and contracts, is somewhat above 1.0. The elasticity of state government appropriations for agricultural research is approximately 1.0, and that of federal formula funds is less than 1.0. These elasticities suggest that the composition of the revenue of state agricultural experiment stations will continue to adjust as the real size of budgets rises and falls. Some evidence also suggests that the price elastici-ty of demand for agricultural research is quite large, i.e., that the real quantity demanded of agricul-tural research at the state government level is price elastic. A rise in compensation of scientists and faculty appears to reduce the quantity demanded of public agricultural research only when scien-tists' compensation increases faster than the prices of other goods and services purchased by state and local governments. An important conclusion is that progress is being made in explaining the provision of public agricultural research using economic and econometric models. These studies add an important new dimension to our understanding of public investment in agricultural research.

Notes

1. The requirement for marketing research was dropped in 1977. Regional research required that scientists at two or more experiment stations and possibly the USDA cooperate to solve research problems spanning more than one state. In 1998, regional research was renamed multi-state research and requires multidisci-plinary and multi-functionality (research and extension) dimensions.
2. At the graduate level, where high-quality science and teaching are required to attract high-quality gradu-ate students, many LGUs do not expect to service instate students. The motivation for investing in quality science is driven by service to farmers.
3. Economists have long noted that while productivity increases for farmers can have negative impacts on farm income, productivity impacts in the food processing and post-harvest industries reduce "marketing margins" (Alston et al. 2001). This benefits both consumers and farmers. Yet we see very modest invest-ment by SAES programs in research to reduce these margins.
4. Research activities conducted in public and private universities, but outside the SAES and colleges of agricultural and home economics, also have benefits to agriculture but are not easy to measure. This includes research administered by the USDA-CSRS/CSREES and conducted in colleges of veterinary medicine, in forestry schools, and in the 1890 LGUs . In other institutions, the expenditures of resources on research are not so carefully documented.
5. The NRI received federal appropriations averaging $111.5 million (of constant 2000 dollars) over 1995 to 2002, but this amount jumped to $148 million in 2003. One possibility is that the appropriations for NRI are benefiting from the phasing out of IFAFS grants.
6. This is behavior addressed by Samuelson (1954) as the primary reason for public provision of pure public goods.
7. Given a few additional strong assumptions that do not seem likely to be fulfilled in most societies and are elaborated in Cornes and Sandler (1996), the median-voter model leads to Pareto or socially optimal pub-lic-good allocation.

8. See Downs (1957), Niskanen (1971), and Denzau and Mackay (1976) for examples of some other voting models.

9. Buchanan (1965) expanded the list of so-called public goods that could be relatively efficiently provided by private institutions. The key was realizing that goods that are not used up by one person's consumption could be provided by exclusive clubs. Stigler (1974) was one of the first to show that the "free rider" problem associated with "private provision" of public goods is partially solved when public goods are "impure" public goods. He saw that many institutions (now called clubs) really provide impure public goods—where the pure public and private goods are jointly produced. These institutions can control use of and finance the public good by monitoring the use or consumption of the private good. He coined this the "cheap rider" solution. Cornes and Sandler (1984) renamed this phenomenon the "easy rider" solution. See Andreoni (1988) for some issues associated with large economies.

10. Schultz (1971) expressed an early version of an interest group theory of state government provision of agricultural research. Hayami and Ruttan's model (1985, pp. 87–88) of induced innovation in public research institutions is one where farmers are organized into politically effective associations. These associations are viewed as registering farmers' research interests with research administrators, scientists, and legislators. Their model seems, also, to be one where some weight is given to consumer (and taxpayer) interests in public-sector decisions on agricultural research.

11. Hirshleifer (1976, p. 241) states the point quite clearly about elected officials or regulators themselves constituting an interest group. Why shouldn't they be interested in their own economic welfare? "Are Chicago city aldermen guided by goals any different from or higher than the goals of Chicago factory workers or Chicago economics professors?"

12. The private good is also a state-specific public good. The model is one with impure public goods (see Cornes and Sandler 1996).

13. Although this demand system does not contain individual prices for each of the types of research inputs, it does contain a summary research-price index across all input types (P_t). The name federal formula funds can be misinterpreted to mean that the total quantity of these funds is determined by a formula. This is not true; Congress decides the total amount of formula funds. What is fixed is the rule for allocating this total to each of the states. We take this into account in the empirical specification of the model.

14. Of course, regional grouping of states always has some arbitrariness.

15. Gourman's (1985) rankings have been criticized for their subjective nature. We take the Gourman ratings at face value. If it does not contain any useful information, it will not have any explanatory power in our demand system. In contrast, if the Gourman rating has significant coefficients, this is evidence that it matters to the state legislators. Federal funding agencies, however, may place greater stock in the NRC rankings of academic programs (National Academy of Science 1997).

16. To take account of this contemporaneous correlation of disturbances in the three fitted-share equations to be fitted, we apply an estimation procedure that is equivalent to Zellner seemingly unreality least-squares estimator (Greene 2003, pp. 340–348).

17. The null hypothesis that each of the share equations individually has no explanatory power is clearly rejected. The test has 23 and 1,410 degrees of freedom and a critical value at the 1 percent level of about 2.77. The sample value of the F-statistic is 60.0 for the federal grant-share equation, 236.1 for the federal formula-funding equation, and 30.3 for the state appropriations equation. Furthermore, if one were to pool the results across all three share equations into one joint test of no explanatory power, the null hypotheses would be soundly rejected at the 1 percent level.

18. Given the more than a century of federal-funding history of SAES research and about a 25-year history of competitive-grant funding, it may not be too surprising that federal-grant programs for agricultural research are allocated somewhat on non-merit bases. If too large of a share of total federal public agricultural-research funding goes to a few states with super science programs, the other states can form a coalition to block further increases in federal agricultural-research grant programs and pass legislation for increasing formula or earmarked funding.

19. Their dependent variable is a research budget share (of total state-government expenditures). It seems that their approach effectively constrains the revenue/income elasticity to 1.0.

References

Alston, J., P. Pardey, and M.J. Taylor. 2001. "Changing Contexts for Agricultural R&D." In *Agricultural Science Policy: Changing Global Agendas,* ed. J.M. Alston, P.G. Pardey, and M.J. Taylor. Baltimore, MD: The Johns Hopkins University Press.

Alston, J., G.W. Norton, and P.G. Pardey. 1995. *Science under Scarcity: Theory and Practice for Agricultural Research Evaluation and Priority Setting.* Ithaca, NY: Cornell University Press.

AAUP (American Association of University Professors). *Academe: Bulletin of AAUP,* various issues.

Andreoni, James. 1988. "Privately Provided Public Goods in a Large Economy: The Limits of Altruism." *Journal of Public Economics* 35(1): 57–73.

Becker, Gary S. 1983. "A Theo." *Quarterly Journal of Economics.* 97(3): 371–400.

Borcherding, Thomas and Robert Deacon. 1972. "The Demand for the Services of Non-Federal Governments." *American Economic Review* 62(5): 891–901.

Buchanan, James M. 1965. "The Economic Theory of Clubs." *Economica* 32(125): 1–14.

Cooperative State Research, Education, and Extension Service. 2004. "Initiatives for Future Agricultural and Food Systems: Important Notice About IFAFS," July 1. Available at: http://www.csrees.usda.gov/about/offices/compprogs_ifa.

Cornes, Richard and Todd Sandler. 1984. "Easy Riders, Joint Production, and Public Goods." *Economic Journal* 94(375): 580–598.

Cornes, Richard and Todd Sandler. 1996. *The Theory of Externalities, Public Goods, and Club Goods. 2nd Ed.* New York, NY: Cambridge University Press.

de Gorter, Harry and David Zilberman. 1990. "On the Political Economy of Public Good Inputs in Agriculture." *American Journal of Agricultural Economics* 72: 131–137.

Deaton, A. and J. Muelbauer. "An Almost Ideal Demand System." *American Economic Review* 70: 312–326.

Denzau, A.T. and R.J. Mackay. 1976. "Benefit Shares and Majority Voting." *American Economic Review* 66(1): 69–76.

Downs, A. 1957. *An Economic Theory of Democracy.* New York, NY: Harper and Row Publishers.

Echeverria, R.G. and H. Elliott. 2001. "Competitive Funds for Agricultural Research: Are They Achieving What We Want?" In *Tomorrow's Agriculture: Incentives, Institutions, Infrastructure and Innovations,* ed. G.H. Peers, and P. Pingali. Proceedings of the Twenty-Fourth International Conference of Agricultural Economists. Hants, England: Ashgate Press.

Evenson, Robert E. 2001. "Economic Impacts of Agricultural Research and Extension." In, *Handbook of Agricultural Economics,* ed. B.L. Gardner and G.C. Rausser. Vol. 1A (Agricultural Production), New York: Elsevier Science/North Holland.

Evenson, Robert E. and Y. Kislev. 1975. *Agricultural Research and Productivity.* New Haven, CT: Yale University Press.

Evenson, Robert and Susan Rose-Ackerman. 1985. "The Political Economy of Agricultural Research and Extension: Grants, Votes, and Reapportionment." *American Journal of Agricultural Economics* 67(1): 1–14.

Greene, W.H. 2003. *Econometric Analysis.* 5th Edition. Upper Saddle River, NJ: Prentice Hall.

Gourman, J. 1985. *The Gourman Report: A Rating of Graduate and Professional Programs in American and International Universities.* Los Angeles, CA: National Education Standards.

Guttman, Joel. 1978. "Interest Groups and the Demand for Agricultural Research." *Journal of Political Economy* 86: 467–484.

Guttman, Joel. 1980. "Villages as Interest Groups: The Demand for Agricultural Extension Services in India." *Kyklos* 33: 122–141.

Hayami, Yujiro and Masao Kikuchi. 1978. "Investment Inducements to Public Infrastructure: Irrigation in the Philippines." *The Review of Economics and Statistics* 60(1): 70–77.

Hayami, Yujiro and V. W. Ruttan. 1985. *Agricultural Development: An International Perspective.* Baltimore, MD: Johns Hopkins University Press.

Hightower, James. 1973. *Hard Tomatoes, Hard Times: A Report of the Agribusiness Accountability Project on the Failure of America's Land Grant College Complex.* Cambridge, MA: Schenkman Publishing Co.

Hirshleifer, Jack. 1976. "Comment." *Journal of Law and Economics* 19(2): 241–244.

Huffman, Wallace E. 2001. "Human Capital: Education for Agriculture." In *Handbook of Agricultural Economics*, ed. B.L. Gardner and G. C. Rausser. Vol. 1A (Agricultural Production), New York: Elsevier Science/North Holland.

Huffman, W.E. and R.E. Evenson. 1993. *Science for Agriculture: A Long-Term Perspective.* Ames: Iowa State University Press.

Huffman, Wallace E. and Robert E. Evenson. 1989. "Supply and Demand Functions for Multiproduct U.S. Cash Grain Farms: Biases Caused by Research and Other Policies." *American Journal of Agricultural Economics* 71(3): 761–773.

Huffman, Wallace E. and Robert E. Evenson. 2003(revised 2004). "Determinants of the Demand for State Agricultural Experiment Station Resources: A Demand System Approach." Iowa State University, Department of Economics, Staff Paper #03028, Dec.

Huffman, W.E. and R.E. Just. 1994. "An Empirical Analysis of Funding, Structure, and Management of Agricultural Research in the United States." *Am. J. Agr. Econ.* 76: 744–759.

Huffman, W.E. and R.E. Just. 1999. "The Organization of Agricultural Research in Western Developed Countries." *Agricultural Economics* 21(Aug.): 1–18.

Huffman, W.E. and R.E. Just. 2000. "Setting Efficient Incentives for Agricultural Research: Lessons from Principal-Agent Theory." *Am. J. Agr. Econ.* 82: 828–841.

Huffman, Wallace E. and Mark McNulty. 1985. "Endogenous Local Public Extension Policy." *American Journal of Agricultural Economics* 67(4): 761–768.

Huffman, Wallace E. and J.A. Miranowski. 1981. "An Economic Analysis of Expenditures on Agricultural Experiment Station Research." *American Journal of Agricultural Economics* 63: 104–118.

Johnson, D.K.N. and A. Brown. 2002. "Patents Granted in U.S. for Agricultural SOV, by State of Inventor, 1963–1999." Wellesley, MA: Department of Economics Working Paper, Wellesley College.

Kanbur, R., R. Sandler, and K.M Morrison. 1999. *The Future of Development Assistance: Common Pools and International Public Goods.* Overseas Development Council and Johns Hopkins University Press.

Khanna, J., Wallace E. Huffman, and Todd Sandler. 1994. "Agricultural Research Expenditures in the United States: A Public Goods Perspective." *Review of Economics and Statistics* 76(May): 267–277.

Knack, S. and P. Keefer. 1995. "Institutions and Economic Performance: Cross-Country Tests Using Alternative Institutional Measures." *Economics and Politics* 7: 207–227.

Krueger, A.B. and M. Lindahl. 2001. "Education for Growth: Why and For Whom?" *Journal of Economic Literature* 39: 1101–1136.

Marcus, Alan I. 1987. "Constituents and Constituencies: An Overview of the History of Public Agricultural Research Institutions in America." In *Public Policy and Agricultural Technology: Adversity Despite Achievement,* ed. Don F. Hadwiger and William P. Browne. London: Macmillian.

Martin, Robert E., T.P. Zacharias, and M.D. Lange. 1991. "Public Inputs in Agriculture." *Southern Economic Journal* 58(1): 129–143.

Niskanen, William A. 1971. *Bureaucracy and Representative Government.* Chicago, IL: Aldine Publishing Co.

National Academy of Sciences. 1997. *Research Doctorate Programs in the United States: Continuity and Change.* Washington, D.C.: National Academy Press.

NRC (National Research Council). 1995. *Colleges of Agriculture at the Land Grant Universities: A Profile.* Committee on the Future of the Colleges of Agriculture in the Land Grant University System, Board on Agriculture. Washington, D.C.: National Academy Press.

NRC (National Research Council). 1996. *Colleges of Agriculture at the Land Grant Universities: Public Service and Public Policy.* Washington, DC: National Academy Press.

NRC (National Research Council). 2000. *National Research Initiatives: A Vital Competitive Grants Program in Food, Fiber, and Natural-Resource Research.* Washington, D.C.: National Academy Press.

NRC (National Research Council). 2003. *Frontiers in Agricultural Research: Food, Health, Environment, and Communities.* Committee on Opportunities in Agriculture. Washington, D.C.: National Academy Press.

Olson, Mancur. 1965. *The Logic of Collective Action: Public Goods and the Theory of Groups.* Cambridge, MA: Harvard University Press.

Olson, Mancur. 1969. "The Principle of Fiscal Equivalence: The Division of Responsibilities among Different Levels of Government." *American Economic Review* 59: 479–487.

Olson, Mancur. 1986. "Toward a More General Theory of Government Structure." *American Economic Review* 76: 120–125.

Peltzman, Sam. 1976. "Toward a More General Theory of Regulation." *Journal of Law and Economics* 19: 211–240.

Peou, S. 1998. "The Subsidiarity Model of Global Governance in the UN-ASEAN Context." *Global Governance* 4: 439–459.

Posner, Richard A. 1974. "Theories of Economic Regulation." *Bell Journal of Economics and Management Science* 5: 335–358.

Psacharopoulos, G. 1985. "Returns to Education: A Further International Update and Implications." *Journal of Human Resources* 20(4): 583–597.

Psacharopoulos, G. and M. Woodhall. 1985. *Education for Development: An Analysis of Investment Choices.* New York, NY: Oxford University Press.

Reid, Joseph D. 1977. "Understanding Political Events in the New Economic History." *Journal of Economic History* 37(2): 302–328.

Ruttan, Vernon W. 1982. *Agricultural Research Policy.* Minneapolis, MN: University of Minnesota Press.

Samuelson, P.A. 1954. "The Pure Theory of Public Expenditure." *Review of Economics and Statistics* 36(4): 387–389.

Sandler, Todd. 1995. *Collective Action: Theory and Application.* Ann Arbor, MI: University of Michigan Press.

Schultz, T.W. 1971. "The Allocation of Resources to Research." In *Resource Allocation in Agricultural Research,* ed. W.L. Fishel. Minneapolis, MN: University of Minnesota Press.

Stigler, George J. 1971. "The Theory of Economic Regulation." *Bell Journal of Economics and Management Science* 2(1): 3–21.

Stigler, George J. 1974. "Free Riders and Collective Action: An Appendix to Theories of Economic Regulation." *Bell Journal of Economics and Management Science* 5(2): 359–365.

Tullock, Gordon. 1967. "The Welfare Costs of Tariffs, Monopolies, and Theft." *Western Economic Journal* 5(3): 788–846.

U.S. Department of Agriculture. 1921–1941. *Annual Report, Office of Experiment Stations.* Washington, D.C.: U.S. Government Printing Office.

U.S. Department of Agriculture. 1951. *Report on the Agricultural Experiment Stations.* Washington, D.C.: U.S. Government Printing Office.

U.S. Department of Agriculture. 1971–2004. *Inventory of Agricultural Research.* Washington, D.C.: U.S. Department Agriculture, Cooperative States Research Service, Cooperative States Research, Education and Extension Service.

U.S. Department of Agriculture. Various years. *Funds for Research at State Agricultural Experiment Stations and Other State Institutions.* Washington, D.C.: U.S. Department of Agriculture, Cooperative States Research Service.

U.S. Department of Agriculture. Office of Experiment Stations. 1901. "Report on the Agricultural Experiment Stations, 1900." *Bulletin* 97. Washington, D.C.: U.S. Government Printing Office.

U.S. Department of Agriculture. Office of Experiment Stations. 1961. *Report on State Agricultural Experiment Stations.* Washington, DC: U.S. Government Printing Office.

U.S. Department of Agriculture and Association of State University and Land-Grant Colleges. 1966. *A National Program of Research for Agriculture.* Washington, D.C.: U.S. Government Printing Office.

U.S. Department of Commerce. Various years. *Statistical Abstract of the United States.* Washington, D.C.: U.S. Government Printing Office.

Welch, Finis. 1999. "In Defense of Inequality." *American Economic Review* 89(March): 1–17.

8

Research Contributions to Agricultural Productivity

Agriculture in the United States has a remarkable record during the past century of economic growth and productivity change. Real output growth has averaged 1.61 percent per annum, and MFP increases have averaged 1.62 percent per annum in the 20th century. These are outstanding (compound) rates by any yardstick. Investments in public-sector research, extension, and schooling programs have been a primary source of U.S. agricultural productivity change, i.e., in increased output per unit of input. These productivity gains are realized when new products, processes, and practices are invented, developed for farm use, and adopted by farmers. Improved infrastructure and improved markets also contribute to efficiency. Private-sector inventions made available commercially to farmers at prices that only partially reflect the real value of the inventions contribute to farm efficiency (i.e., private agri-industrial firms capture only part of the value of their inventions in higher prices of their goods).

Methods for evaluating the specific contributions of public- and private-sector research programs have been developed and applied in a number of studies. Perhaps the simplest and most direct method was developed by Griliches (1958) in his well-known study of hybrid corn. The applications of this particular method (the imputation-accounting method) require that use of the technology in question be measured. In the case of crops, the actual adoption and use of specific varieties can be measured, and this information can be used to compute social benefits.

Unfortunately, the use of many inventions, particularly minor improvements in practices, is not easily identified. Furthermore, the availability of inventions to farmers causes changes in farm organization, scale, and specialization that take place gradually over time. Thus, the development of improved crop varieties, for example, may have long-term effects on farm efficiency not directly captured in the genetic gains and related studies. Methods have been developed to deal with cases where technology use is not directly measurable. These methods are statistical methods in which variables summarizing investments in research, extension, schooling, infrastructure, and farm programs are related to aggregate farm-sector efficiency.

In this chapter, we report on studies that have applied statistical decomposition analyses. We present evidence for the 20th century. A number of early studies of the imputation-accounting type, as well as statistical studies, have been undertaken for U.S. agriculture. We summarize new econometric evidence on state level multifactor productivity (MFP) decomposition for 1970–99.

The first section of this chapter reviews long-term trends in growth and productivity of U.S. agriculture. The second reviews the methods used in productivity analyses. Statistical productivity decomposition is used in our research, and other methods are compared to it. The third section summarizes our statistical evidence covering early and late 20th century contributions of public and private research and other variables to U.S. multifactor agricultural productivity.

The Long-Term Growth of U.S. Agriculture

The performance of U.S. agriculture in terms of growth of output and partial and MFP over the past century has been remarkable.

Highlights: 1870–2000

Over the past century the average rate of growth of U.S. total farm output has been 1.61 percent (compounded) per annum (see Figure 8.1 and Appendix Table A8.1); in other words, real farm output is 4.98 times larger in 1999 than it was a century earlier. Partial livestock-productivity measures are reported in Table 8.1—columns (4), (7), (10), and (14). These measures show that pounds of beef per cow, pounds of pork per sow, pounds of pork per sow, and pounds of lamb and mutton per ewe have risen since 1950. Pounds of milk per cow has risen steadily over time and was 3.5 times larger in 2000 than 1950. Crop yields have also risen steadily since 1950. Corn yields in 2000 were 3.5 times their 1950 value. Yields of wheat, soybeans, and hay have risen less rapidly (Table 8.2).

The U.S. price of farm output relative to all prices was relatively unchanged over 1780 to 1880, except for the rise associated with the Civil War (1861–65) (Figure 8.2). From 1880 to 1925, as the U.S. population was growing rapidly and the frontier closing, the farm-output price relative to all products rose. During this period the public agricultural-research system was developing rapidly. Since 1925, the farm-output price relative to all U.S. prices has fallen thereafter, except during World War II (1940–45) and during the world grain "shortage" years of 1973–75. Since 1975 the fall in the relative farm price is quite dramatic—about 1 percent per annum.

In contrast, over the past century real farm inputs under the control of U.S farmers have hardly changed, but the composition and quality of inputs have changed. For example, 100 years ago 86

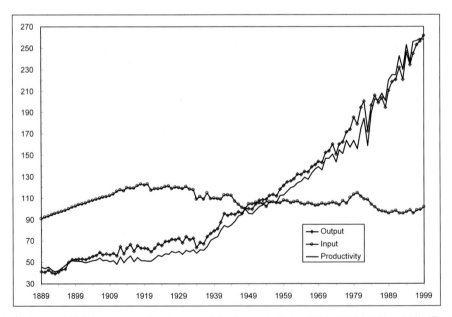

Fig. 8.1.   Indexes of U.S. farm output, input, and multifactor productivity, 1889–1999 (1948 = 100). (From Appendix Table A8.1.)

Table 8.1 Meat and milk production per female-producing unit, 1950–2000

	Beef Production			Pork Production			Sheep Production				Milk Production		
Year (1)	Beef and Veal (mill. lbs) (2)	No. Beef Cows[a] (1,000 hd) (3)	Beef per Cow[b] (lb/hd) (4)	Pork (mill. lbs) (5)	No. Sows Farrowed (Dec.–May, Jun–Nov) (1,000 hd) (6)	Pork per Sow[c] (lb/hd) (7)	Lamb & Mutton (mill. lbs) (8)	No. Ewes (≥1 yr age) (1,000 hd) (9)	Pounds per Ewe (lb/hd) (10)	Commercial Broilers (mill. lbs) (11)	Whole Milk (mill. lbs) (12)	No. Dairy Cows (1,000 hd) (13)	Milk per Cow (lb/hd) (14)
1950	10,768	16,743	569	10,714	15,106	709	597	20,059	30	1,938	116,602	21,944	5,314
1965	19,747	34,238	547	11,141	10,896	1,022	651	17,502	37	8,111	124,173	14,954	8,304
1975	24,849	45,472	546	11,779	9,928	1,186	411	10,062	41	11,096	115,326	11,140	10,352
1985	24,242	35,370	685	14,805	11,240	1,317	357	7,431	48	18,810	143,012	10,981	13,024
1995	25,541	35,190	727	17,849	11,889	1,501	285	5,404	53	34,222	155,292	9,466	16,405
2000	27,113	33,569	808	18,952	11,410	1,661	234	4,229	55	41,516	167,559	9,206	18,201

Source: U.S. Department of Agriculture, *Agricultural Statistics*, various years.
[a]Beef cows and heifers ≥ 2 years of age or beef cows and heifers with calves.
[b]Million pounds of beef divided by number of beef cows and older heifers.
[c]Million pounds of pork divided by number of sows farrowed.

Table 8.2 U.S. acreage harvested and average crop yield for corn, wheat, soybeans, and hay, 1950–2000

	Corn		Wheat		Soybeans		Hay		Total Crop-land
Year	Acres Harvested (Million)	Average Yield (Bu/Acre)	Acres Harvested (Million)	Average Yield (Bu/Acre)	Acres Harvested (Million)	Average Yield (Bu/Acre)	Acres Harvested (Million)	Average Yield T/Acre	
1950	82.8	39.0	68.6	16.6	12.6	21.3	74.0	1.39	478.3
1960	72.9	53.3	50.4	25.0	25.0	24.1	67.8	1.75	443.1
1970	62.6	77.6	48.4	31.3	42.5	27.3	62.6	2.02[a]	459.0
1980	73.0	104.6	69.7	34.0	67.5	30.0	58.9	2.22[b]	472.5
1990	73.1	124.8	49.7	37.6	56.5	33.0	61.0	2.3	462.0
2000	78.5	152.7	53.1	41.6	72.4	38.1	59.9	2.5	455.0

Source: U.S. Department of Agriculture, *Agricultural Statistics*.
Note: Crop yields are 5-year average centered on the year in the table.
[a] A 3-year average.
[b] A 1-year average.

percent of farm inputs were supplied by farmers. Only 14 percent of these inputs were purchases from nonfarmers (Kendrick 1961, p. 347). In contrast, in 1999 more than 45 percent of the farm inputs are purchased or from nonfarmers (U.S. Department of Agriculture 1991b, Table 4). Furthermore, the technology, including the genetic composition of common farm crops and animals, has changed dramatically in a century, e.g., single-cross hybrids first replaced open-pollinated corn varieties and corn varieties genetically engineered for insect resistance and herbicide tolerance are replacing the traditional single-cross hybrid varieties. U.S. agriculture has gone from horse and mule power to highly sophisticated mechanical power-tractors and self-propelled harvesters. The organization of livestock production has gone from many small herds and flocks to fewer and, on average, much larger herds and flocks. A commercial broiler industry did not exist before 1930, but the industry grew rapidly from 1950 to 1965. Furthermore, rapid growth has continued, and in 1995, the total U.S. production of broilers (mil. lbs.) exceeded by 29 percent total production of beef and veal (mil. lbs.). Furthermore, in 1995, the production of broilers by two companies (Tyson Food, Inc. and Con Agra, Inc.) accounted for roughly 23 percent of total U.S. broiler production. The skills and sophistication of successful farmers have changed. For example, the average schooling level of U.S. farmers was less than 5 years a century ago and currently is about 12 years.

In crops, the private sector has become an increasingly important source of new seed. Since 1950, virtually all of the seed corn has been from hybrids. At that time more than 20 companies provided seed, but the industry was dramatically reorganized and concentrated during the 1990s in a few companies led by Pioneer™ and DeKalb™. The potential to use genetic engineering to move genes longer "distances" increased the value of the germplasm held by private, and public, seed companies. In corn, modest commercial success has been obtained by insect resistant (Bt) and herbicide tolerant, e.g., Round-Up-Ready™ (RR) varieties. These varieties have become available since 1997 (see also Chapter 5).

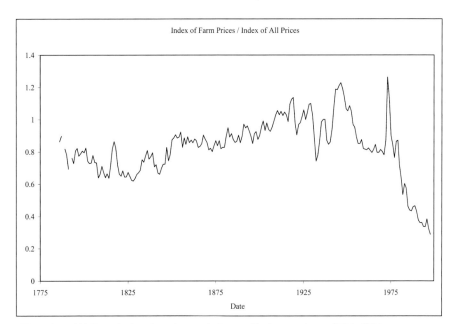

Fig. 8.2. The price of U.S. agricultural products relative to all other products, 1780–1999.

Breeding progress in soybeans was very slow through 1980, and increasingly the public sector was heavily involved in the development of new varieties. The pace of crop improvement picked up during the 1980s, and increasingly farmers purchased new seed for planting rather than saving their own seed. In 1996, herbicide tolerant (RR) varieties were first marketed. Each bag of RR soybean seed carried a technology fee (Maxwell et al. 2004) and a requirement that farmers could not save their own seed. Although there has been considerable debate about the potential of RR soybeans—especially the level of weed-seed pressure—over 1995–2003, U.S. farmers have gone from 0 to 81% of acreage planted. U.S. farmers now primarily purchase new soybean seed each year.

In cotton, herbicide tolerant (RR) and insect resistant (Bt) varieties have grown in use by farmers over 1995 to 2003 to approximately 74% of the acres. In corn, the increase has been less dramatic, 40 percent of the acreage in 2003. In the U.S., this technology has also been sold with a technology fee (Maxwell et al. 2004) and requirement of no farmer-saved seed. The increasingly sophisticated breeding techniques available to plant breeders in the private sector and the strengthening of intellectual property rights on discoveries and new developments have been important sources of new technology development to U.S. crop farms.

Advances in animal health have contributed to the organization of livestock production into large specialized units. Prevention is emphasized and sick animals are to be avoided. The discovery of the first antibiotic, penicillin, by the British scientist Fleming occurred in 1928 and has had important applications in veterinary medicine. The first organized application of artificial insemination (AI) of dairy cows occurred in New Jersey in 1938 (under direction of the New Jersey Extension Service). Adoption of this technique was relatively rapid. It had some fixed costs that gave rise to scale economies, and it held the potential for giving dairy farmers access to high-quality genes at low cost, Figure 8.3 (Johnson and Ruttan 1997). Advances in the technology of freezing and storing semen occurred in 1952 and in performance testing of dairy bulls occurred during the early 1960s.

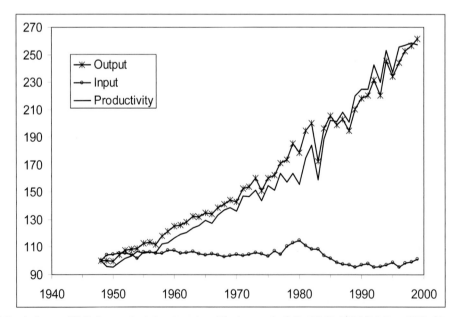

Fig. 8.3. Indexes of U.S. farm output, input, and multifactor productivity, 1948–1999 (1948 = 100). (From Appendix Table A8.1.)

Major advances have occurred in the past decade to improve reproductive technology. Advances in artificial insemination, embryo transfer, and cloning increase the incentives for private research in animal breeding. These technologies can significantly expand the market for a successful breeder of improved parent stock. They speed up the process of genetic improvement, reduce the risk of disease transmission, and expand the number of animals that can be bred from a superior parent.

Over the last decade, specialized companies have emerged to develop high-yielding breeds of broilers, layers, pigs, and dairy and beef cattle. The growth of integrated poultry, swine, and beef cattle production systems has occurred, in which production, processing, and retailing are all part of one large company.

Livestock breeding programs have become concentrated in specialized companies. In poultry, the industry has two segments—broilers and layers. Seven companies have broiler-breeding programs in the U.S. Two companies have specialized in layer breeding. These breeding programs dominate the research development and production of new poultry breeds.

The U.S. swine-breeding industry has been transformed over the past four decades from one dominated by small purebred breeders producing boars for farmers to use in rational crossbreeding programs to one dominated by larger breeding companies specialized in superior boar and gilt pure-line breeds for use in crossbreeding combinations that exploit heterosis. Ten of these companies are operating breeding programs in the U.S. Currently, larger producers tend to use the hybrid hogs while small producers continue to use rotational crossing, which is less powerful.

Poultry breeders developed a highly successful way of protecting their intellectual property investment in superior breeds by exploiting heterosis, or hybrid vigor. Hybrid vigor is the yield advantage obtained when two or more pure inbred lines are crossed in a breeding scheme. While the offspring of this cross exhibit some superior-yield performance, this yield advantage steadily declines as the offspring themselves are bred. Thus, by restricting access to the pure parent-line stock (a form of trade secret) a breeder remains the sole supplier of the hybrid. Farmers need to repeatedly purchase new stock from the breeder to maintain high yields.

For most livestock, the production of hybrid animals is not really possible in the sense of hybrid plants or even hybrid poultry. Maximum hybrid vigor is obtained when parent lines are closely inbred. However, close inbreeding is seldom practical in large mammals because reproductive fertility and health of inbred offspring is quite fragile. Cattle, for example, have very little excess reproductive capacity because a normal birth is a single calf and a small deterioration in vigor results in infertility of the inbred line. Thus, in cattle, commercial breeding is normally restricted to unrelated animals using a system of outcrossing or crossbreeding. Greater fecundity in swine means that greater reproductive capacity exists to exploit heterosis, but the health and vigor of inbred piglets is fragile.

Breeding of dairy and beef cattle concentrates on improving pure-line bulls for breeding with cows selected from farmers' herds. In beef cattle, farmers have moved away from purebred lines to crossbred lines which provide modest heterosis for yield. In dairy, the Holstein has become the dominate breed, and dairymen access superior Holstein genes through artificial insemination services. For beef and dairy, six companies dominate the AI business (Narrod and Fuglie 2000).

Quantitative Change: 1948–1999

Positive long-term trends existed in total farm output, crop output, and livestock output during this period. Growth in total farm output averaged 1.9 percent over the whole period, being somewhat faster from 1970 to 1999 than from 1948 to 1970 (Table 8.3). The decade of the 1970s was one of unusually rapid farm-output growth (2.3 percent per annum). Crop output grew slightly faster than livestock output for the whole period, and was especially rapid over the decade of the 1970s (at 2.2 percent per annum).

Farm inputs consist of capital (durable equipment and land); labor (hired and self-employed including unpaid family labor but not contract labor); and materials (farm origin, energy, chemicals, and purchased services). No long-term significant growth existed in farm inputs during this period, but the decade of the 1970s showed sizeable growth of 0.9 percent per annum followed by the 1980s with high negative growth of −1.7 percent per annum.

The trend in farm capital is positive (0.8 percent per annum) for this whole period and was quite large over 1948 to 1970 at 2.9 percent when mechanization of agriculture was rapidly moving forward. Farm capital, however, declined over the decade of the 1980s at −1.2 percent per annum and at a higher −1.5 percent over the 1990s. The long-term trend in the land input is negative (−0.5 percent per annum) with little variation over subperiods. The fluctuation in farm-capital growth over the period was due to durable equipment (Table 8.3).

Farm labor, hired labor, and self-employed (including unpaid family labor) have sizeable negative trends over the post–World War II period. Trend in farm labor was −2.4 percent per annum, but for hired labor the rate of decline was somewhat slower. From 1948 to 1970, when mechanization was proceeding rapidly, the rate of decline of farm labor was especially large at −3.3 percent per annum. This was a period when the capital-to-labor ratio was rising by 6.3 percent per annum. The decade of the 1970s was another period when the capital-to-labor ratio was rising quite rapidly (Table 8.3).

Table 8.3   Growth of U.S. farm outputs, inputs, productivity, and factor ratios (annual average percentage change), 1948–1999

Variable	Time Period					
	1948–70	1970–99	1948–99	1970–80	1980–90	1990–99
Total Output[a]	1.63	2.08	1.89	2.23	2.01	2.00
Livestock Output	1.33	1.31	1.69	0.97	1.08	1.95
Crop Output	1.20	2.32	1.78	3.13	1.84	1.63
Total Inputs	0.22	−0.11	0.03	0.90	−1.67	0.49
All Capital	2.90	−0.78	0.81	2.27	−1.23	−1.54
Durable Equipment	3.27	−1.26	0.69	2.63	−4.44	−1.93
Land	−0.59	−0.50	−0.54	−0.46	−0.62	−0.42
All Labor	−3.35	−1.66	−2.39	−2.58	−1.89	−0.41
Hired Labor	−3.23	−0.69	−1.79	0.30	−2.57	0.30
Self-employed	−3.39	−2.03	−2.62	−3.21	−1.58	−0.75
All Materials	2.37	1.06	1.63	2.49	−1.02	1.79
Farm Origin	2.07	0.90	1.41	2.03	−1.27	2.07
Energy	1.59	0.41	0.92	1.86	−1.82	1.27
Chemicals	5.29	1.26	3.00	4.39	−1.64	1.00
Purchased Services	1.26	2.17	1.78	3.77	0.53	2.21
Ratios						
Total Factor Productivity	1.41	2.19	1.86	1.32	3.69	1.51
Livestock to Crop Output	0.13	−1.01	−0.09	−2.16	−0.76	0.32
Capital to Labor Input	6.25	0.88	3.20	4.85	0.66	−1.13
Capital to Materials Input	0.53	−2.32	−0.82	−0.22	−0.21	−3.33
Materials to Labor Input	5.72	2.72	4.02	5.07	0.87	2.20

Source: U.S. Department of Agriculture, Economic Research Service, 2003.
[a]There is a third output category which is the output of inter-farm services.

The use of materials in U.S. agriculture has a strong positive trend from 1948 to 1999 (1.6 percent per annum). The period up to 1980 was one of especially rapid growth in materials use and then the trend slowed remarkably. Energy input usage has a positive trend over the whole period, being relatively rapid from 1948 to 1970 (1.6 percent per annum) with the increasing mechanization of farms. Since the energy crisis of the 1970s, the rate of growth of energy use has been much slower—0.4 percent per annum. The rate of growth of farm chemicals was 3.0 percent per annum over the whole period, and most rapid up to 1980. Chemicals usage actually declined over the 1980s (Table 8.3). Purchased services—which include machine hire, chemical applications, and contract labor—grew at 1.8 percent per annum for this period. The growth in this category was significantly more rapid from 1970 to 1999 than at the beginning of the period (Table 8.3).

Over the whole period the trend in the capital-to-materials ratio was negative, but a little growth occurred from 1948 to 1970. The materials-to-labor ratio grew at 4.0 percent over the whole period being especially rapid from 1948 to 1980. Hence, materials and capital have been substituted for labor in U.S. agriculture over the post–World War II period.

Productivity of U.S. Agriculture

Agriculture was the first sector in the U.S. for which productivity statistics were developed. As data became available on aggregate MFP statistics for other sectors, the agricultural sector ranked at or near the top.

Measurement

National annual estimates of MFP, or total factor productivity, for U.S. agriculture at the national level have existed since 1948 (Barton and Cooper 1948), and the Economic Research Service was the first federal agency to publish official estimates of total factor productivity from sector or whole economy. They started in 1960. Loomis and Barton (1961) extended the earlier Barton and Cooper estimates to cover 1910–1957. Kendrick (1961), working independently of the U.S. Department of Agriculture but using their data, pushed annual MFP estimates for agriculture back to 1889 and forward to 1957. These data, or modifications and extensions of them, have been examined using econometric models to identify the contributions to agricultural productivity of past investments in agricultural research and extension (see Evenson 1980; Cline 1975; Braha and Tweeten 1986). Regional estimates of MFP, using the 10 ERS production regions, start in 1939 and run up to the present time. Evenson (1980), Cline (1975), and others have examined these data using similar but more sophisticated models. The National Research Council (1975), Chapter 2, presents a general discussion of agricultural productivity and Antle and Capalbo (1988) an advanced discussion.

In 1980, an American Agricultural Economics Association (AAEA) task force reviewed the USDA productivity series and made several recommendations for improving the series (U.S. Department of Agriculture 1980). This lead the USDA to establish a program in the early 1990s to construct state-level estimates of MFP for the farm sector. These data series went through several revisions, and final estimates were produced in 2003. (See Ball, Butault, and Nehring (2002) for a description of the total output, total input, and MFP.) Unfortunately, the USDA decided to discontinue annual state MFP calculations after 1999. They will, however, continue to construct and publish annual national farm sector MFP numbers.

The Record

Greater output generally requires greater input. How could this large increase in real farm output of U.S. agriculture in the 20th century have been accomplished with no significant increase in farm inputs? It is because agricultural MFP has increased over the past century at an average rate of 1.62 percent per annum and 1.86 percent per annum from 1948 to 1999 (see Table 8.3 and Figure 8.3). In 1999, aggregate MFP for the agricultural sector is 2.6 times higher than in 1948.

Given that productivity need not change at all over the last 100 (or 50) years, this is a very high rate for a sector for such a long period. The average product of all inputs under the control of farmers (i.e., an output index divided by an input index) has been increasing over time. Stated another way, the rate of increase of real aggregate output has been faster than the rate of increase of real aggregate farm inputs.[1]

The record of MFP growth for the U.S. agricultural sector can also be compared with aggregate partial measures of agricultural productivity: (1) total output per unit of land; (2) total output per unit of durable equipment; (3) total output per unit of labor; and (4) crop output per unit of land. For the period 1948–1999, Figure 8.4 shows that output per unit of land and crop output per unit of land grew a little faster than MFP, but otherwise they have a similar path as MFP. Farm output per unit of durable equipment was roughly constant from 1950 to 1983 and then in 1983 starting rising more rapidly than MFP. Farm output per unit of labor rises much faster than other measures of aggregate agricultural productivity. It is 8.5 times larger in 1999 than in 1948 (Figure 8.4).

Turning to MFP at the state level, 1970–1999, the top six states for average MFP growth from 1970 to 1999 were Michigan (2.26), Oregon (2.38), Connecticut (2.35), Washington (2.32), North Dakota (2.24), and North Carolina (2.23). The bottom four states were Vermont and Wyoming (0.89), Delaware (1.08), and Nevada (1.09). (See Table 8.4.)

The contiguous 48 states can also be grouped by region (e.g., the 11 USDA regions) and average annual growth rate reported for total output computed. (See Ball et al. (1999) for a discussion of the relationship between national and state-level measures.) Growth in output was relatively high over

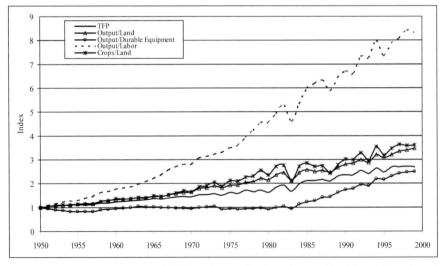

Fig. 8.4. Aggregate productivity measures: MFP, total output/land, total output/durable equipment services, total output/all labor, and crop output/land, 1950–1999 (1950 = 100). Source: Adapted from Huffman 2005, figure 4.

Table 8.4   Average annual growth rate for farm output, input, multifactor productivity, and public agricultural research stock by state, 1970–1999

Region/State	MFP Relative Level[a] 1996	Average Annual Growth Rate, 1970–1999 (%)			
		Total Output	Total Input	MFP	Public Ag Research Capital
New England					
Maine	1.026	0.08	−1.50	1.42	1.43
New Hampshire	0.865	0.13	−1.17	1.30	0.77
Vermont	1.131	0.74	−0.15	0.89	1.49
Massachusetts	0.991	0.29	−1.43	1.72	0.02
Connecticut	1.168	1.45	−0.90	2.35	0.18
Rhode Island	0.959	−0.18	−1.69	1.50	1.18
Northeast					
New York	1.070	0.50	−0.91	1.41	2.12
New Jersey	0.948	0.83	−0.60	1.43	0.96
Pennsylvania	1.032	1.69	0.17	1.52	2.24
Delaware	1.198	2.82	1.75	1.08	1.57
Maryland	1.072	1.51	0.19	1.33	2.38
Lake States					
Michigan	0.852	1.94	−0.68	2.26	3.38
Minnesota	1.053	1.94	0.00	1.94	2.49
Wisconsin	0.977	1.09	0.68	1.77	2.25
Corn Belt					
Ohio	0.846	1.33	−0.57	1.90	0.79
Indiana	1.025	1.59	−0.33	1.92	1.64
Illinois	1.057	1.29	−0.58	1.87	1.56
Iowa	1.192	1.08	−0.75	1.83	3.19
Missouri	1.002	0.78	−0.59	1.37	3.39
Northern Plains					
North Dakota	1.181	2.15	−0.09	2.24	4.06
South Dakota	1.187	1.96	−0.11	2.07	2.60
Nebraska	1.257	2.49	0.69	1.80	4.42
Kansas	1.169	2.24	0.60	1.65	3.35
Appalachia					
Virginia	0.962	1.42	−0.27	1.69	3.25
West Virginia	0.607	1.19	−0.36	1.55	2.15
Kentucky	0.984	1.56	−0.03	1.60	2.23
North Carolina	1.181	2.15	−0.09	2.23	4.50
Tennessee	0.825	1.30	−0.45	1.75	2.95
Southeast					
South Carolina	1.057	1.07	−0.81	1.88	2.13
Georgia	1.465	2.25	0.20	2.04	5.53
Florida	1.525	2.27	0.27	2.00	3.47
Alabama	1.000	1.85	−0.05	1.90	1.63
Delta States					
Mississippi	1.222	1.51	−0.39	1.90	2.69
Arkansas	1.375	2.66	0.60	2.06	3.30
Louisiana	1.188	1.12	−0.23	1.35	1.69

(*Continues*)

Table 8.4   (*Continued*)

Region/State	Relative Level[a] 1996	Average Annual Growth Rate, 1970–1999 (%)			
		Total Output	Total Input	MFP	Public Ag Research Capital
Southern Plains					
Oklahoma	0.845	1.65	0.37	1.28	1.67
Texas	0.929	1.99	0.42	1.57	2.88
Mountain States					
Montana	0.851	1.17	−0.03	1.20	2.49
Idaho	1.278	2.43	0.51	1.92	3.38
Wyoming	0.826	1.17	0.28	0.89	0.92
Colorado	1.076	1.57	0.06	1.51	3.77
New Mexico	0.964	1.98	0.43	1.55	2.49
Arizona	1.251	1.41	−0.16	1.57	4.63
Utah	0.890	1.87	0.45	1.42	2.60
Nevada	0.985	1.48	0.39	1.09	4.17
Pacific					
Washington	1.358	3.04	0.72	2.32	2.35
Oregon	0.837	2.67	0.29	2.38	2.59
California	1.445	2.64	1.18	1.46	3.02

[a]The MFP (multifactor productivity) level is relative to Alabama.

this period in the Northern Plains and Pacific regions and low in New England. Input growth was generally negative, except for the Southern Plains and Pacific regions, which showed slow growth. Total factor productivity growth was high in the Lake States, Southeast, Northern Plains, and Pacific regions but low in the Mountain region.

Later in the chapter we will show that changes in agricultural MFP are related to past investments in public and private agricultural research and extension, and in farmers' schooling. The correlation between state average growth rates of MFP and public agricultural research capital, 1970–1999, is 0.25. Some productivity increases could also come from exploiting economies of size in farm production.

Methods for Identifying Sources of Productivity

Starting in the 1950s productivity statistics showed seemingly costless increases in aggregate output. T.W. Schultz (1953), John Kendrick (1961), and E. F. Denison (1962), however, started to search for underlying sources of productivity increases at this time. Their work focused on the general economy and on agriculture, where the data were better. For agriculture, the early studies by Griliches (1963a, 1963b, 1964) are best known. There are three classes of methods that have been applied in sources of productivity analyses. They are (1) imputation-accounting methods; (2) statistical meta-function methods; and (3) statistical productivity decomposition methods. Each of these methods is examined below.

The imputation-accounting method uses evidence from experiments and professional judgments to impute the relation between productivity and likely sources, e.g., research programs. The classic hybrid-corn study by Griliches (1958) utilized farm yield data to estimate the productivity advantage of hybrid corn over the open-pollinated varieties that were previously grown. These "benefits" were then attributed to hybrid-corn research programs in the public and private sectors. The arbitrariness of these procedures, however, was one reason for the development of statistical-econometric methods. The meta-production function and productivity decomposition methods are two closely related methods. In the meta-function method, "decomposing" variables, e.g., public and private nonfarm research and extension, are included directly in a production or profit function.

The productivity decomposition method entails two steps. First, a measure of productivity is calculated. Second, productivity is statistically related to decomposing variables. This approach has some advantages. A particular functional form for the production function need not be specified. A flexible index number function, (e.g., Divisia, Fisher) is a second-order Taylor's series approximation to any underlying production function. Diewert (1976), however, has shown that particular index-number functions are exact for some functional forms. Implicitly, production coefficients on inputs controlled by farmers are not constrained to be the same across observations. However, it might be reasonable to look for common causes of productivity across observations in doing an econometric decomposition analysis.

Imputation Fundamentals

The methodology for studies concentrating on evaluating the contribution of agricultural technology using imputation methods entails the following steps:

1. Identifying the invented technology. (In most cases this is a set of inventions rather than a single invention. For example, in the hybrid-corn study, many hybrid varieties were considered.)
2. Documenting all costs associated with producing, developing, and distributing the invention(s). With hybrid corn, this included all public and private costs. These costs were incurred as long as 25 or 30 years before the realization of benefits.
3. Estimating the cost advantage for early adopters. Some studies have used experiment-station trials to make controlled "with-without" yield and cost comparisons. These comparisons, however, are generally not representative of farmers' fields, and most studies have attempted to obtain farm-level comparisons. (In the hybrid corn study both experiment-station and farm data were used.)
4. Estimating the parameters of the adoption pattern and the adoption-advantage interaction. In general, a new invention will be adopted first on economic units where the cost advantage is greatest. As an adoption diffuses in the population, the advantage typically declines (unless, as with hybrid corn, the technologies defined are undergoing continuous change).
5. Converting (3) and (4) to a benefits stream. This conversion is illustrated in Figure 8.5.

Figure 8.5 depicts two types of firms and the aggregate supply curves of each (assuming equal numbers of both types of firms) for two alternative technology scenarios. In the first scenario the technology, when adopted, is scale-neutral in that it lowers average variable cost (AVC) and marginal cost in a parallel fashion. In the second scenario the technology is scale-biased, because it shifts the minimum points of the AVC curves to the right. As depicted it does not change the slope of the marginal cost curves, although it might. Note that fixed costs are not relevant to a short-run analysis, but are important to a long-run analysis (see below). For the original technology in either scenario, the aggregate supply curve is the summation of the marginal cost curve above the minimum point of the AVC curves. Firms from the first scenario (type I) that have high AVC curves do not produce until price is above their minimum AVC; hence, they are not "on the curve" for the first units on the curve. (This has some implications for the parallel nature of the shift of the supply curve.)

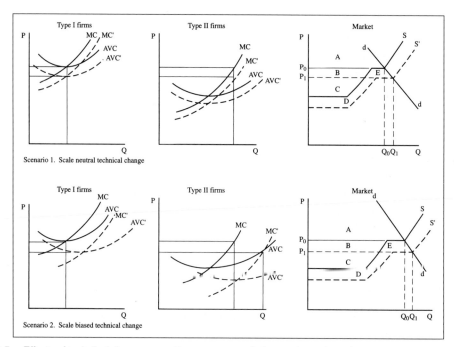

Fig. 8.5. Effects of technical change on welfare: consumers' plus producers' surplus.

If both types of firms experienced a shift in average variable cost and marginal cost curves of K percent, the supply curve would simply shift downward by K percent in a parallel fashion. If type I firms shift by K_1 percent and type II by K_2 percent, the shift will be slightly nonparallel. The segment up to the new minimum AVC for the type I firm would shift by K_2 percent; therefore a parallel shift by $S_1K_1 + S_2K_2$ percent would occur. In the second scenario, with a cost-curve shift that moves minimum average cost to the right, the shift will be somewhat more nonparallel.

For the purposes of benefit/cost analysis, the areas that lie below the market-demand curve and above the market-supply curve in Figure 8.5 are labeled.

a. The original consumers' surplus is A.
b. The original producers' surplus (payment to fixed factors) is B + C.
c. The new consumers' surplus is A + B + E.
d. The new producers' surplus is C + D.
e. The change in consumers' surplus is B + E.
f. The change in producers' surplus is D − B.
g. The change in total surplus is D + E.

(Note that the area E includes the triangle beside the demand curve.)

These studies have generally treated the change in total surplus as the benefits from the project (see Norton and Davis 1981; Alston, Norton, and Pardey 1995). The slopes of the demand and supply curves play a minor role in determining the size of these benefits of the new technology. The area D is well approximated by $K \times Q_0$, the original quantity. If demand is perfectly inelastic, D will be approximately $K \times Q_0$ and E will be small. If demand is perfectly elastic, E will disappear, but D will be larger.

In the long run, of course, farms will enter and exit the production of the commodity in response to changes in profitability. If average variable costs fall by K percent for all producers (or for "enough" producers to reach equilibrium), the new equilibrium price will fall by K percent. In this case the benefits will simply be $K \times Q_0 + E$. If not enough producers are able to obtain (use) the technology to reach this new equilibrium, the price of Q will fall by less than K percent, and those producers using the technology will collect a technology rent. The producers who are not users or adopters will have losses, unless the market price for Q does not decline.

Imputation studies then have first sought to estimate K, the shifts in supply curves. They have also estimated (or, all too often, simply assumed) the units over which K applies. Adoption rates are generally useful for determining these parameters.

6. Computing present values of benefits and costs and internal rates of return. The standard procedure of benefit/cost analysis is to compute a present value of benefits and costs over time using some discount rate. The ratio of these present values is widely used as a financial measure of return on investment. Alternatively one can solve for the "internal" rate of return, i.e., the discount rate that equates the present value of benefits to the present value of costs. For most imputation-accounting studies, the analyst must make some assumptions regarding the continuation of the benefit stream beyond the period of analysis. Will benefits continue in future periods?

The answer to this question is somewhat arbitrary in imputation-accounting studies. In the hybrid corn study, Griliches (1958) assumed that the 1957 (the ending period of the study) level of benefits would continue indefinitely but that the 1957 level of costs would also continue because those costs were required to maintain the benefits. This was obviously a conservative assumption, as subsequent yield changes for corn show. Thus, his study established a tradition for "conservative" calculations.

Unfortunately, an unusual computation in Griliches's study created severe problems for interpreting his results and for comparing them to later studies. Griliches computed the ratio:

$$\frac{\sum_{t=1900}^{1957} b_t (1+r)^t \times r + b_{57}}{\sum_{t=1900}^{1957} c_t (1+r)^t \times r + c_{57}} \tag{8.1}$$

The numerator, the cumulated benefits converted to a return r plus the 1957 level of benefits, was interpreted to be a constant financial flow to continue into future years. The denominator is the comparable cost flow. The ratio, 7, was then interpreted by many as a 700 percent rate of return on investment. This interpretation is simply wrong. The ratio is essentially a benefit/cost ratio and should be interpreted as such. It is highly sensitive to r, the discount rate used. Griliches also calculated an internal rate of return of approximately 43 percent, but the 700 percent figure is still widely quoted.[2]

Statistical Methods

Statistical methods for MFP decomposition using meta-production or meta-profits function analysis are a means for estimating rather than imputing the contribution to productivity change of developments in technology, infrastructure, and policy environments faced by farmers. Griliches (1963a, 1964) was also a pioneer in this research. Other researchers include Evenson (1968, 1980); Peterson

(1967); Cline (1975); Bredahl and Peterson (1976); Evenson and Welch (1974); Evenson and Kislev (1975); Lu, Cline, and Quance (1979); and Braha and Tweeten (1986).

Statistical methods are generally used to analyze aggregate-productivity change when no direct measure of the inventions adopted or invention-productivity link is available. These methods rely on identifying the link between investments in research, extension, and schooling (i.e., inputs) and productivity rather than between inventions (i.e., research output) and productivity using econometric methods.

Meta-Functions

Conventional production-costs profits theory relegates certain conditions to background or "held constant" status. Let multiple outputs be positive quantities $(Y_1, \ldots, Y_m)$, which are produced using several inputs $(X_1, \ldots, X_n)$, which are also positive quantities, and let the technology be described by

$$F(Y_1, \ldots, Y_m, X_1, \ldots, X_n, t) = 0. \tag{8.2}$$

Equation (8.2) is an asymmetric representation of the transformation function giving output of Y_1 as a function of the other outputs and the inputs. Assume (8.2) is a linear homogenous function. Several things are "held constant" in the background behind this expression: the technology set available to farmers, the existing infrastructure (roads, markets), and transaction costs (legal system, etc.). If (8.2) were a "meta-function," variables measuring these functions would be included in the specifications.

Equation (8.2) can be converted into a form that summarizes relationships among growth of inputs and outputs and productivity. Differentiate (8.2) totally with respect to time (t) to obtain

$$\sum_{i=1}^{m} F_i \left(\frac{\partial Y_i}{\partial t} \right) dt + \sum_{j=1}^{n} F_j \left(\frac{\partial X_j}{\partial t} \right) dt + F_t dt = 0 \tag{8.3}$$

where F_i and F_j are first derivatives of the production function F, and Ft is the derivative with respect to t. The first-order conditions for profit maximization are $P_i = \lambda F_i$ and $-R_j = \lambda F_j$; $i = 2, \ldots m$; $j = 1, \ldots n$, where P_i and R_j are prices of inputs and outputs and λ is a Lagrange multiplier.

Substituting $F_i = P_i/\lambda$ and $F_j = R_j/\lambda$ and multiplying by $\lambda/\sum P_i Y_i$ or $\lambda/\sum R_j X_j$, we obtain

$$\sum_i \frac{P_i Y_i}{\sum_i P_i Y_i} \left(\frac{\partial Y_i}{\partial t} \right) \frac{1}{Y} dt - \sum_j \frac{R_j X_j}{\sum_j R_j X_j} \left(\frac{\partial X_j}{\partial t} \right) \frac{1}{X_j} dt + \frac{\lambda Y_1}{\sum_i P_i Y_i} \left(\frac{F_t}{Y_1} \right) dt = 0 \tag{8.4}$$

$$\sum_i S_i \left(\frac{\partial Y_i}{\partial t} \right) \frac{1}{Y_i} dt - \sum_j C_j \left(\frac{\partial X_j}{\partial t} \right) \frac{1}{X_j} dt + \frac{\lambda Y_1}{\sum_i P_i Y_i} \left(\frac{F_t}{Y_1} \right) dt = 0 \tag{8.5}$$

where S_i is the revenue share and C_j is the cost share and we make use of the property that $\sum_j R_j X_j = \sum_i P_i Y_i$, i.e., that the value of all inputs under the control of farmers equals the value of marketable outputs (this is the "zero profit" condition that holds in a competitive economy). This expression holds for small changes when the "background variables" are unchanged. It relates growth in output to growth in factors or inputs. When this equation does not hold, the logic of this development tells us that the background variables have changed. This is the basis for the equation of MFP change, $\hat{T}$, as

$$\hat{T} = -\frac{\lambda Y_i}{\sum_i P_i Y_i}\left(\frac{\partial F}{\partial t}\right)\frac{1}{Y_i}\,dt$$

$$= \sum_i S_i\left(\frac{\partial Y}{\partial t}\right)\frac{1}{Y_i}\,dt - \sum_j C_j\left(\frac{\partial X_j}{\partial t}\right)\frac{1}{X_j}\,dt = \hat{y} - \hat{x} \tag{8.6}$$

where $\hat{y}$ and $\hat{x}$ are weighted rates of growth of the outputs and inputs, respectively.

The meta-function approach entails the inclusion of decomposing or meta-type variables, e.g., research and extension, directly into equation (8.2), thus converting it from a conventional function to a meta-function. This meta-function is then estimated directly. See Huffman and Evenson (1989) for an application of the meta-function approach to explaining productivity changes of U.S. cash-grain farms.

The chief advantage of the meta-function approach is that one need not impose the profit-maximizing conditions required to move from (8.3) to (8.4) and (8.5). Meta-function parameter estimates for equation (8.2) including the meta-type variables do not impose profit-maximizing restrictions on parameter estimates.[2]

Productivity Functions

Using this method, productivity decomposition proceeds in two stages. In stage 1, $\hat{T}$ is computed for the relevant data points (e.g., for each state and year), based on equation (8.6). In stage 2, the $\hat{T}$ measures are statistically decomposed by regression methods relating them to variables measuring the missing background variables, i.e., productivity is assumed to be a function of the background variables. These are variables for public agricultural research, private agricultural research, agricultural extension, farmers' schooling variables, farm structure and farm programs.

The productivity decomposition procedure has several advantages that usually more than offset the disadvantages associated with the profit-maximizing weights. These are

1. There is no "right-hand" output or input explanatory variable. This deals with the likely endogeneity of these variables.
2. Functional forms for the production function need not be specified. "Divisia" indexes for (8.2) are second-order approximations to any underlying production technology (see Jorgenson, Gollop, and Fraumeni 1987).
3. Implicitly, production coefficients are not imposed to be constant over observations. One state, for example, can have different production technology than another state. This allows for a broad range of "pooling" of observations, and this is usually required to obtain significant variation in the meta-type variables.

Productivity Accounting Methods

Further intuition into both the imputation and statistical methods can be obtained from a brief review of productivity accounting. The basic idea of productivity accounting is that by "chipping away" at the residual MFP-growth component with enough corrections and imputations one can reach an almost complete accounting of the components of MFP growth. The pioneers in this general approach are Schultz (1953), Griliches (1960, 1963b), and Denison (1962). Jorgenson and Griliches (1967) contributed a major study of this type and engaged in a debate with Denison over procedures (Denison 1969, 1972; Jorgenson and Griliches 1972). Jorgenson, Gollop, and Fraumeni (1987) have recently completed major analyses of productivity growth in 45 U.S. industries, including agriculture.

The foundations for the accounting approach can be developed in the following simple way. Suppose that the true relationship between output and inputs is

$$Y = \delta F\ (LQ_\ell,\ MQ_m,\ HQ_h,\ Z) \tag{8.7}$$

where δ is a scale economies parameter; $F(\ )$ is the functional relationship; and L, M, and H are service years of labor, machines, and land. Q_ℓ, Q_m, and Q_h are quality indexes per service year for labor (ℓ), machines (m), and land (h). The product of the quantity and quality component of each input translates into "real" quality-constant units, over time (or across observations). Z is a vector of variables that characterizes technology and infrastructure contributions not channeled through scale or factor quality.

Now suppose that we do not observe δ, Q_ℓ, Q_m, or Q_h and simply measure the relationship between output and inputs as

$$\bar{Y} = \bar{F}\ (L, M, H) \tag{8.8}$$

The observed MFP growth rate from (8.8) will be

$$\overline{MFP} = \hat{Y} - S_\ell \hat{L} - S_m \hat{M} - S_h \hat{H} \tag{8.9}$$

where S_l, S_m, and S_h are factor cost shares, and Y, L, M, and H are rates of change of output and inputs.

The true MFP growth rate is, however,

$$\overline{\overline{MFP}} = \hat{Y} - S_\ell\left(\hat{L} + \hat{Q}_\ell\right) - S_m\left(\hat{M} + \hat{Q}_m\right) - S_h\left(\hat{H} + \hat{Q}_h\right) - Z\alpha - \hat{S}\delta \tag{8.10}$$

where α is the elasticity of product with respect to the Z variable and $\hat{S}$ is the rate of change in farm size.

Suppose further that the shares S_l, S_m, and S_h are measured with error (S_l^*, S_m^*, S_h^* are the true shares); then the difference between measured MFP (8.9) and the correctly measured MFP (8.10) is

$$\begin{aligned}
\overline{\overline{MFP}} - \overline{MFP} &= \left(S_l - S_l^*\right)\left(\hat{L} + \hat{Q}_\ell\right) + \left(S_m - S_m^*\right)\left(\hat{M} + \hat{Q}_m\right) \\
&+ \left(S_h - S_h^*\right)\left(H + Q_h\right) + S_\ell Q_\ell + S_m Q_m + S_h \hat{Q}_h + Z\alpha + \hat{S}\delta
\end{aligned} \tag{8.11}$$

The first three terms in equation (8.11) are based on errors in measuring the factor shares or marginal products, and the second three on the failure to correct for factor quality change. The technology-infrastructure term unassociated with factor quality and the scale term are also included.

Griliches (1973) and others who have utilized a framework-like equation (8.11) have noted that the simple specification of this model does not, by itself, mean much. To be useful, more information and work are required to obtain better measures of each of the separate components (i.e., of shares and quality indexes). The definitions themselves are a tautology unless this is done.

Thus, stringent data demand, and subjectivity of analyses, have limited the applications of imputation-accounting methods. The basic purpose of statistical decomposition methods, however, is not to identify the specified components of (8.11) but rather to associate MFP with the investments in research and other things that have caused it to change by using econometric or statistical methods.

Statistical Decomposition of MFP

The empirical evidence on the contributions of public and private research, extension, and farmers' education to the growth of U.S. MFP is presented and discussed.

The Period 1870–1950

For the purpose of analysis this period is split into two subperiods: one for the pre-modern growth period 1870–1925 and the second for the early modern growth period 1926–1950. The data for the first period on aggregate U.S. (gross output) MFP are from Kendrick (1961) and build directly on his annual data starting in 1889 and benchmark estimates for 1879 and 1869. Evenson (1980) derived annual estimates of national-level MFP to fill in the annual series starting in 1870. The productivity analysis for 1926–1950 uses data on MFP constructed from USDA data for four geographic regions: North Central, South, Pacific, and Mountain (see Loomis and Barton 1961).

For the period 1870–1950, Evenson (1980) uses (1) an agricultural invention index; (2) an estimate of the stock of real public-agricultural research; and (3) a land-quality index to explain aggregate U.S. agricultural MFP. The invention index is the cumulative weighted number of agricultural patents lagged 10 years. The public-agricultural research stock is the weighted summation of real public-agricultural research expenditures during the previous 18 years, starting in 1850.

His regression results showed that agricultural productivity during 1870–1925 was strongly and positively related to the level of the agricultural-invention index and the real stock of public-agricultural research (Evenson 1980, Table 13). Agricultural productivity, however, was not significantly affected by changes in land quality or soil exhaustion that occurred with rapid expansion and intensive use of some soils.

Thus, these results showed that the agricultural research and development process during this pre-modern growth period increased farm productivity. The activities of inventors—largely farmers and blacksmiths who did not have any training in science but were facing practical production problems as the frontier moved westward (see Chapters 1 and 5)—raised agricultural productivity. Also, basic scientific discoveries by Liebig, Koch, and Pasteur in Germany and France were rapidly advancing the stock of knowledge about the science of agriculture during the last half of the 19th century, and their scientific discoveries seem to have been effectively transferred to the United States by enterprising graduate students like Samuel W. Johnson (Yale), Harry Russell (Wisconsin), and others (see Chapter 3).

Until U.S. agricultural science and technology caught up to German and French advances, relatively rapid U.S. agricultural-productivity advances occurred. The evidence suggests that the U.S. caught up with Europe about the turn of the century. The great depression of the 1890s, primarily 1892–1894 (see Friedman and Schwartz 1963, p. 678) also seems to have done some economic pruning that contributed to a rapid rise in U.S. agricultural productivity during 1894–1899 (see Figure 8.1). Starting about 1899, however, further advances in U.S. agricultural productivity became difficult to achieve. The SAES system had to develop into a successful science system before it could develop a steady stream of scientific discoveries that would fuel further increases in agricultural productivity (see Chapter 2).

For the period 1926–1950, the explanatory variables are (1) the invention index; (2) the real stock of public applied or technology-oriented agricultural research; (3) the real stock of public agriculturally related scientific research; and (4) a weather index. Furthermore, the applied agricultural-research stock and the related science-research stock were multiplied together to create an interaction variable. These five variables and three time-period dummy variables were used to explain MFP.

The econometric model was fitted to data pooled together for the four regions, 1926–1950, and showed positive and statistically strong effects of the invention index and the public applied agricultural research stock on agricultural productivity (Evenson 1980, p. 199). The coefficient of the applied agricultural research and related science-research interaction variable was positive, but not statistically strong. Productivity, also, was shown to be related to the weather index.

Furthermore, Figure 8.1 showed an unusually rapid rise in MFP for agriculture during 1934–1948, and the regression results confirmed that the rate of increase was large relative to predicted productivity. Although the economic hardships in agriculture and elsewhere were severe and widespread during the Great Depression of 1929–1933 (see Friedman and Schwartz 1963), the economic pruning and reorganizing that occurred during and immediately following those years seems to have set the stage for an unusually high rate of MFP growth in U.S. agriculture during 1934–1948.

The Period 1948–1982

In the first edition, we described the derivation of a two-sector, crop and livestock, state aggregate MFP series, 1948–1982, for 42 U.S. states (excludes New England, Alaska, and Hawaii). This series explained state productivity indexes, we used two public agricultural-research stock variables, one for applied public-agricultural research and one for pre-technology science research. Both are constructed using timing weights and spatial spill-in weights. Public extension, farmers' school, and private agricultural-research variables are other key factors.

First, we expected the signs of coefficients for applied and pre-technology research-interaction effects to be positive or complementary. In the case of public crop research, these expectations are fulfilled. However, for public livestock research, the coefficient is negative, which suggests that applied livestock and pre-technology science research are substitutes. There are different interpretations that one might place on these findings. One is that for livestock research, pre-invention science and applied research are weakly linked. Recall that we showed in Chapters 2 and 3 that SAES research in animal science did not advance rapidly until the 1940s. Rapid advances in SAES crop-science research started at least a decade earlier (see Table 3.3), so the livestock sciences may be immature relative to crop sciences. We view the substitute nature of livestock-related pre-invention and applied science as unsatisfactory for enhancing livestock productivity over the long term.

Second, the results for the public applied research and extension interaction are mixed. Public applied livestock research and livestock extension are complementary as expected, but public applied crop research and crop extension are substitutes. The explanation might lie in differences in geoclimatic specificity of crop and livestock technology. Because crop technology is more geoclimatically specific, public applied crop research and extension seem more likely to be substitutes.

Third, all the schooling and extension interaction effects are negative. These are results that are consistent with a number of other studies (Hayami and Ruttan 1985; Birkhaeuser, Evenson, and Feder 1991; Evenson 1992). Farmers' schooling and public extension are substitutes for efficiently processing information about new technology that affects productivity.

Some seemingly negative results for the effects of public applied livestock research and livestock extension on agricultural productivity were uncovered for the post-1950 period. Although we unexpectedly found that public livestock-related pre-technology science research and public applied livestock research were substitutes rather than complements in affecting productivity, public livestock-related pre-technology science research and private livestock R&D have been performing much better in affecting agricultural productivity. Major changes in the livestock sector have occurred since about 1950. Small-scale units produced a major share of the livestock output in 1950, but they account for very little by the end of the 1980s. Given that farmers' average schooling level increased by about 50 percent during 1950–1982 and the bulk of livestock production has become geographically concentrated and physically concentrated in large to very large producing units, our results may be predictable.

The Period 1970–1999

Here we present new econometric evidence of the impacts of public agricultural research, private agricultural R&D, and public extension on state agricultural MFP for the period 1970–1999. In particular, it emphasizes a test of the hypothesis that the composition of public agricultural-research funding (i.e., share from federal competitive grants and contracts and from federal-formula and state-government appropriations) has no effect on state agricultural productivity. The alternative hypothesis is that the composition has a non-linear impact on productivity. We also report simulations of a new agricultural science policy—one that shifts federal-formula funds to competitive-grant funding, or the reverse.

The Model

To obtain the model of state aggregate MFP assume a state aggregate production function with disembodied technical change where Q is an aggregate of all types of farm outputs from farms within a state aggregated into one output index; A(RPUB, RPRI, EXT) is the associated technology parameter; and F( ) is a well-behaved production function (Chambers 1988, p. 181). K is state aggregate quality-adjusted physical capital input, L is state aggregate quality-adjusted labor input, and M is state aggregate quality-adjusted materials input. The technology parameter A( ) is hypothesized to be a function of state public agricultural-research capital (RPUB), private agricultural-research capital (RPRI), and public agricultural-extension capital (EXT). The state aggregate production function is then:

$$Q = A(RPUB, RPRI, EXT) \, F(L, K, M). \tag{8.12}$$

Now we define state multifactor productivity (MFP) as:

$$MFP = Q/F(L, K, M) = A(RPUB, RPRI, EXT). \tag{8.13}$$

Taking natural logarithms of both sides of equation (8.13) and adding a random disturbance term u, we have

$$\ell n \, MFP = \ell n \, A(RPUB, RPRI, EXT) + u. \tag{8.14}$$

Our goal is to test for a positive impact of public agricultural research stock on MFP and for a significant impact of public agricultural research composition, e.g., shares due to major funding sources (see Huffman and Just 1994), on MFP. To accomplish this, the funding shares are interacted with the public agricultural-research stock. Hence, the embellished version of the state agricultural MFP equation is:

$$\ell n\, MFP = \beta_1 + \beta_2\, \ell n\, RPUB + \beta_3\, [\ell n\, RPUB]\, SFF$$
$$+ \beta_4\, [\ell n\, RPUB]\, (SFF)^2 + \beta_5\, [\ell n\, RPUB] GR + \beta_6\, [\ell n\, RPUB]\, (GR)^2$$
$$+ \beta_7\, \ell n\, EXT + \beta_8\, [\ell n\, RPUB]\, \ell n\, EXT + \beta_9\, \ell n\, RPUBSPILL$$
$$+ \beta_{10}\, \ell n\, RPRI + u$$

(8.15)

where SFF is a state's share of SAES funding from federal-formula and state-government appropriations (i.e., programmatic funding); GR is a state's share of SAES funding from federal grants, contracts, and cooperative agreements (i.e., federal grants and contracts); RPUBSPILL is a state's public agricultural-research capital spill-in. The elasticity of state agricultural MFP with respect to RPUB, RPUBSPILL, EXT, and RPRI is:

$$\frac{\partial \ell nMFP}{\partial \ell nRPUB} = \beta_2 + \beta_3\, SFF + \beta_4\, (SFF)^2 + \beta_5\, GR + \beta_6 (GR)^2 + \beta_8\, \ell n\, EXT,$$

(8.16)

$$\frac{\partial \ell nMFP}{\partial \ell nRPUBSPILL} = \beta_9,$$

(8.17)

$$\frac{\partial \ell nMFP}{\partial \ell nEXT} = \beta_7 + \beta_8\, \ell n\, RPUB,$$

(8.18)

$$\frac{\partial \ell nMFP}{\partial \ell nRPRI} = \beta_{10}.$$

(8.19)

The elasticity of state agricultural productivity (MFP) with respect to a change in the state's own public agricultural research capital is given in equation (8.16), and clearly this elasticity takes on different values as the composition of state agricultural experiment station funding changes, i.e., SFF and GR, and the amount of local extension activity (EXT).[3] The elasticity of a state's agricultural MFP with respect to the public agricultural research capital spill-in is displayed in equation (8.17), and it is a constant. The elasticity of state agricultural MFP with respect to EXT is given by equation (8.18), and it clearly varies as ℓn RPUB. In particular, if state public agricultural research and extension are "complements," β_8 is expected to be positive; or if they are "substitutes," it will be negative. The elasticity of state agricultural MFP with respect to RPRI is given in equation (8.19), and it is a constant.[4]

The unique feature of equation (8.15) is that the productivity of RPUB depends on and is proportional to the composition of SAES funding sources—SFF and GR:

$$\frac{\partial \ell nTFP}{\partial SFF} = (\beta_3 + 2\beta_4\, SFF)\, \ell n\, RPUB,$$

(8.20)

$$\frac{\partial \ell nTFP}{\partial GR} = (\beta_5 + 2\beta_6\, GR)\, \ell n\, RPUB.$$

(8.21)

Equations (8.20) and (8.21) show how composition of public agricultural-research funding affects state agricultural MFP. The proportional change of state agricultural MFP due to a 1 percentage-point change in SFF is given in equation (8.20). Likewise, the proportional change of state agricultural MFP due to a 1 percentage-point change in GR is given by equation (8.21). The inclusion of squared terms in these equations [(SFF)^2, (GR)^2] permits us to examine potential nonlinear impacts of funding composition on the productivity of public agricultural research at the state level.

We test the null hypothesis that SAES funding composition has no impact on state agricultural MFP; i.e., discoveries from all types of funds—federal formula and state government appropriations, federal grants and contracts, and "other" funding—are equally productive for causing technical change leading to growth in state agricultural MFP. This is the joint null hypotheses: $\beta_3 = \beta_4 = \beta_5 = \beta_6 = 0$. If this hypothesis is accepted, then the state agricultural MFP equation, equation (8.15), will be of a traditional form. If, however, this hypothesis is rejected, a public agricultural research policy that changes both the size of a state's public agricultural research capital and its composition, as reflected in SSF and GR, will affect state agricultural MFP.

The total impact on ℓn MFP of a marginal change of ℓn RPUB, SFF, and GR on MFP is:

$$\text{d } \ell\text{nMFP} = [\delta\ell\text{nMFP}/\delta\ell\text{nRPUB}] \text{ d } \ell\text{nRPUB} + [\delta\ell\text{nMFP}/\delta\text{SFF}] \text{ d SFF}$$
$$+ [\delta\ell\text{nMFP}/\delta\ell\text{n GR}] \text{ d GR} \tag{8.22}$$

However, if changes are larger than marginal ones, taking differences between beginning and ending values of ℓn MFP gives results that are more reliable. An alternative approach is to use a difference equation approach. First, evaluate equation (8.15) at the sample mean values for each state to establish a baseline. Second, define new values of the public R&D policy variables as ℓn RPUB$'$ = ℓn RPUB$^\circ$ + Δ ℓn RPUB, where an "o" superscript is used to designate the starting value of a variable or baseline and a "$'$" is used to designate the new value of a variable. Likewise, let SFF$'$ = SFF$^\circ$ + Δ SFF, and GR$'$ = GR$^\circ$ + ΔGR. Third, compute the difference between new and baseline estimates as:

$$\Delta\ell\text{nMFP} = \ell\text{nMFP}' - \ell\text{nMFP}^\circ = \beta_2 \ell\text{nRPUB}' + \beta_3(\ell\text{nRPUB}')\text{SFF}'$$
$$+ \beta_4 (\ell\text{nRPUB}')(\text{SFF}')^2 + \beta_5(\ell\text{nRPUB}')\text{GR}' + \beta_6(\ell\text{nRPUB}')(\text{GR}')^2$$
$$+ \beta_8 (\ell\text{nRPUB}') \ell\text{nEXTF}^\circ - \beta_2\ell\text{nRPUB}^\circ - \beta_3(\ell\text{nRPUB}^\circ)(\text{SFF}^\circ)$$
$$- \beta_4 (\ell\text{nRPUB}^\circ)(\text{SFF}^\circ) - \beta_5(\ell\text{nRPUB}^\circ)(\text{GR}^\circ)^2 - \beta_8(\ell\text{nPRUB}^\circ)\ell\text{nEXT}^\circ \tag{8.23}$$

With the use of public funds allocated to agricultural research having alternative uses, it is interesting to ask what the social rate of return on these investments is. For example, if one million dollars of additional public funds is invested today in an average state, it will have benefits distributed over the next 34 years in this state and other states in the same area, which are recipients of spill-in effects. By setting the net present value of the benefits equal to the cost, we can solve for the internal rate-of-return. When benefits are in constant prices, we obtain a real rate-of-return on the public investment. The internal rate-of-return (r) computation is:

$$\left[\frac{\partial\ell\text{nMFP}}{\partial\ell\text{nRPUB}}Q/T+(n-1)\frac{\partial\ell\text{nMFP}}{\partial\ell\text{nRPUBSPILL}}Q/S\right]\sum_0^m w_i\left[1/(1+r)^t\right]=1 \tag{8.24}$$

where Q is the sample mean value for state agricultural output, T is the sample mean for a state's own public agricultural research capital, (n − 1) is the number of states into which agricultural research spill-in effects flow, S is the sample mean of the public agricultural research spill-in capital, w_is are timing weights used to create the stock of public agricultural research, and r is the internal rate-of-return, including impacts of R&D spillover.

The Data Set and Method of Estimation

The data set is a state panel on aggregate agriculture, 1970–1999, for 48 contiguous states, or 1,440 observations. We use new annual state MFP data obtained from the USDA (see Ball et al. 2002). See

Table 8.4 for average annual rates of MFP growth by state for the 48 contiguous states over the period 1970–1999. The MFP growth rate was highest in Oregon (2.38) and Washington (2.32) and lowest in Vermont and Wyoming (0.89). Not all public agricultural research expenditures can be expected to impact agricultural productivity, e.g., research on "post-harvest activities," and studies of households, families, and communities. Huffman et al. have constructed public agricultural research data having an agricultural productivity focus by state for 1927–1997. These data have been converted to constant dollar values using the Huffman and research-price index presented in Table 4.1. By using real research expenditures rather than nominal expenditures, the trend is greatly reduced.

The right-most column of Table 8.4 shows the average rate of increase of the public agricultural-research stock variable by state, 1970–1999. The highest average annual rates of growth are for Georgia at 5.53 percent and North Carolina at 4.5 percent, and the lowest rate of growth is for Massachusetts and Connecticut at slightly positive rates. The Spearman rank correlation of the average MFP growth and public stock of public agricultural research over 1970–1999 for these 48 states is 0.30, which is significantly different from zero at the 5 percent level (provided a normal distribution can be justified). This result suggests a modest degree of association between the growth of MFP and public agricultural-research stock.

The science of constructing research stock variables from constant dollar-value expenditures remains in its infancy (Griliches 1979, 1998). Although a few researchers have included many lags of public agricultural-research expenditures without much structure, e.g., Alston, Craig, and Pardey (1998), this generally asks too much of the data. Hence, by imposing prior beliefs about the shape of timing weights, we reduce the demands on the data to identify parameters. Griliches (1998) concludes that R&D most likely has a short gestation period, then blossoms, and is eventually obsolete. We approximate these patterns with a gestation period of two years during which the impacts are negligible; the next seven years during which impacts are assumed to be positive and are represented by increasing weights; followed by six years of maturity during which weights are high and constant; and then 20 years during which weights are declining and then fade out to zero. See Evenson (2001) and Alston and Pardey (2001) for a discussion of timing weights. This weighting pattern is known as trapezoid-shaped time weights (see Figure 8.6), and they are used to translate the real public agricultural research expenditures into a real public agricultural research stock (RPUB).[5] Table 8.4 also presents data on the average annual rate of increase of the stock of public agricultural research by state, 1970–1999.

Regional grouping of states in which spill-in effects might occur are arbitrary; we choose to define spillovers using the geo-climate sub-region map (see Figure 8.7).[6]

To construct state private agricultural R&D capital, we apply timing weights that are similarly shaped but shorter in length. The total length is 19 years, which is consistent with U.S. patent length. The number of agricultural patents issued (see Johnson and Brown 2002) is used to approximate private agricultural R&D in each state. A measure of public agricultural-extension capital is constructed from staff days of agricultural and natural-resource extension activity (Ahearn, Lee, and Bottom 2002). We assume that one-half of the impact of extension occurs in the current year, and the balance is allocated with declining weights over the next four years. See Table 8.5 for definition of symbols and summary definitions.

Interaction terms between a state's public agricultural-research stock and SAES funding shares were created, i.e., the SFF and GR were multiplied by ℓn RPUB. However, given that the public agricultural-research stock was derived using 34 years of data, we lagged SFF and GR by 12 years

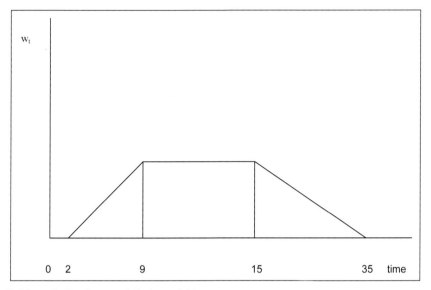

Fig. 8.6. Public agricultural research timing weights.

Legend:
1. Northeast Dairy Region
2. Middle Atlantic Coastal Plain
3. Florida and Coastal Flatwoods
4. Southern Uplands

5. East-Central Uplands
6. Midland Feed Region
7. Mississippi Delta
8. Northern Lake States

9. Northern Great Plains
10. Winter Wheat and Grazing Region
11. Coastal Prairies
12. Southern Plains

13. Grazing-Irrigated Region
14. Pacific Northwest Wheat Region
15. North Pacific Valleys
16. Dry Western Mild-Winter Region

Fig. 8.7. U.S. agricultural geo-climatic regions and sub-regions (Adapted from U.S. Department of Agriculture 1957 and Evenson 1996).

Table 8.5   Variable names and definitions, econometric productivity analysis

Name	Symbol	Mean (Sd.)	Description
Multifactor Productivity	MFP	−0.205[a] (0.254)	Multifactor productivity for the agricultural sector (Ball et al. 2002)
Public Agricultural Research	RPUB	16.129[a] (0.870)	The public agricultural research stock for an originating state. The research capital is a summation of past public-sector investments in agricultural research with a productivity enhancing emphasis (Huffman, McCunn, and Xu) in 1984 dollars (Huffman and Evenson 1993). Stock obtained by summing past research expenditures with a 2 through 34-year lag and trapezoidal shaped timing weights
Private Agricultural Research	RPRI	6.076[a] (0.248)	A state's stock of private patents of agricultural technology. The number of patents for each year (Johnson and Brown 2002) obtained by weighting the number of private patents in crops (excluding fruits and vegetables and horticultural and greenhouse products) and crop services, fruits and vegetables, horticulture and greenhouse products, and livestock and livestock services by a state's 1982 sales share in crops (excludes fruits, vegetables, horticultural and greenhouse products), fruits and vegetables, horticulture and greenhouse products and livestock and livestock products, respectively. The annual patent total is summed with a 2 thru 18 year lag using trapezoidal timing weights.
Public Ag Extension Capital	EXT	1.292[a] (0.976)	A state's stock of public extension is created by summing public full-time equivalent staff in agriculture and natural resource extension applying a weight of 0.50 to the current year and then 0.025, 0.125, 0.0625, and 0.031 for the following four years.
Budget Share from Federal Grants and Contracts	GR_{t-12}	0.096 (0.76)	The share of the SAES budget from National Research Initiative grants and contracts; other CSRS funds; USDA contracts, grants and cooperative agreements; and non-USDA federal grants and contracts (USDA), lagged 12 years.
Budget Share from Federal Formula Funds	$SFF1_{t-12}$	0.230 (0.112)	The share of the SAES budget from Hatch, Regional Research, McIntire-Stennis, Evans-Allen, and Animal Health (USDA), lagged 12 years.
Budget Share from State Government Appropriations	$SFF2_{t-12}$	0.521 (0.123)	The share of the SAES budget from state government appropriations (USDA), lagged 12 years
Budget Share from Federal Formula and State Appropriations	SFF_{t-12}	0.751 (0.132)	$SFF1_{t-12} + SFF2_{t-12}$
Budget Share from Other Funds	OR_{t-12}	0.165 (0.132)	The share of the SAES budget from private industry, commodity groups, NGO's and SAES sales (USDA), lagged 12 years
Share for Basic Biological Science Research	$ShBBS_{t-8}$	0.259 (0.087)	Share of SAES resources allocated to basic biological science research in year $t-8$ (See Table 4.7)

Public Ag Research Capital Spill-in	RPUBSPILL	17.763* (0.567)	The public agricultural research spill-in stock for a state is constructed from state agricultural sub-region data (see Figure 8.7)
Regional Indicators	Northeast		Dummy variable taking a 1 if state is CT, DE, ME, MD, MA, NH, NJ, NY, PA, RI, or VT
	Southeast		Dummy variable taking a 1 if state is AL, FL, GA, KY, NC, SC, TN, VA, or WV
	Central		Dummy variable taking a 1 if state is IN, IL, IA, MI, MO, MN, OH, or WI
	North Plains		Dummy variable taking a 1 if state is KS, NE, ND, or SD
	South Plains		Dummy variable taking a 1 if state is AR, LA, MS, OK, or TX
	Mountains		Dummy variable taking a 1 if state is AZ, CO, ID, MT, NV, NM, UT, or WY
	Pacific		Dummy variable taking a 1 if state is CA, OR, or WA

[a]Numbers reported in natural logarithms.

to place them roughly at the weighted midpoint of the total lag length. We also created an interaction term between ℓn RPUB and ℓn EXT.

Equation (8.15) is to be fitted to the annual data, 1970–1999, pooled over the 48 contiguous states, giving a total number of 1,440 observations. Although heteroscedasticity across states, i.e., different variances for the disturbance in Equation (8.15), and autocorrelation over time might seem to be important, we take the following approach. In recent years it has become popular and acceptable to estimate models by ordinary least squares (OLS) but to correct the standard errors and t-values for a general form of autocorrelation and heteroscedasticity (see Davidson and MacKinnon 1993, p. 548–556; White 1980; MacKinnon and White 1985; Wooldridge 1989, 2003, p. 410). Even though we know that OLS will be inefficient, there are advantages to this approach. First, the explanatory variables may not be strictly exogenous. If they are not, the feasible-generalized least squares (FGLS) estimator is not even consistent, let alone efficient. Second, in most applications of FGLS, the errors are assumed to follow a first-order autoregressive process [AR(1)] and quasi-first differences applied before estimation. Since rho (ρ), the first-order autocorrelation coefficient, and variance of the disturbances are unknown, the best-case scenario with FGLS is a consistent estimator, which requires that the sample size go to infinity. In panel-data over time, we will, however, be in the small-sample situation. In this case, FGLS has unknown statistical properties and can hardly be claimed to be better than OLS. Hence, it is frequently better and more acceptable to compute standard errors for the OLS estimates that are robust to general forms of serial correlation and heteroscedasticity (Wooldridge 2003, p. 410). This is the route that we take. Both traditional and White t-values are reported.

The Results

Table 8.6 displays ordinary least-squares estimates of the parameters of the total factor productivity model. Regression 1 reports the estimate of parameters of equation (8.15), which are the main results. All of the coefficients are significantly different from zero at the 1 percent level.[7] The R^2 is 0.52, and a joint test of no explanatory power of the equation gives a sample F-statistic of 140. This test has 15 and 1,424 degrees of freedom, and the tabled F-value is about 2.0 at the 1-percent significance level. Hence, the state aggregate MFP model has significant explanatory power.[8]

Table 8.6   Least-squares estimate of total multifactor productivity equation, 48 states: 1970–1999[a] [N = 1,440]

Regressors[b]	Regression 1			Regression 2	
	Coefficient	t-Value		Coeff	t-Value
		White[c]	Traditional		
Intercept	−8.701	17.60	18.38	−8.366	17.10
ℓn(Public Ag Res Capital)$_t$	0.290	10.39	12.15	0.279	11.50
ℓn(Private Ag Res Capital)$_t$	0.113	4.03	4.40	0.102	3.96
ℓn(Public Extension Capital)$_t$	1.364	6.44	7.07	1.306	6.74
ℓn(Public Ag Res Capital)$_t \times$ SFF$_{t-12}$	0.123	5.46	4.60	0.123	4.61
ℓn(Public Ag Res Capital)$_t \times$ (SFF$_{t-12}$)^2	−0.087	5.61	4.85	−0.088	4.96
ℓn(Public Ag Res Capital)$_t \times$ GR$_{t-12}$	−0.073	6.58	6.11	−0.088	6.62
ℓn Public Ag Res Capital)$_t \times$ (GR$_{t-12}$)^2	0.154	5.36	4.55	0.174	5.02
ℓn(Public Ag Res Capital)$_t \times \ell$n (Public Extension Stock)$_t$	−0.075	5.59	6.11	−0.071	5.78
ℓn(Public Ag Res Capital Spill-in)$_t$	0.123	10.02	10.33	0.117	9.63
ℓn(Private Ag Res Capital)$_t \times$ ShBBS$_{t-8}$				0.030	2.63
Regional Indicators					
Northeast (= 1)	0.176	5.79	5.84	0.179	5.93
Southeast (= 1)	0.066	3.14	2.89	0.072	3.24
Northern Plains (= 1)	0.342	12.55	11.48	0.335	11.12
Southern Plains (= 1)	0.079	3.42	3.41	0.087	3.73
Mountain (= 1)	0.226	8.83	8.70	0.234	8.96
Pacific (= 1)	0.112	3.46	4.20	0.126	4.64
R^2	0.524			0.526	

[a]The Central Region is the excluded region.
[b]The dependent variable is ℓn(MFP)$_t$.
[c]White t-values are computed using White (1980) and Wooldridge (1989) for heteroscedasticity- and autocorrelation-robust standard errors (Wooldridge 2003, p. 57).

Using sample mean values of the data, the elasticity of MFP with respect to RPUB, RPUB-SPILL, EXT, and RPRI is 0.231, 0.123, 1.267 [= 1.364 − 0.075(1.292)], and 0.113, respectively. These elasticities are all positive. Public agricultural-research capital and extension capital interact negatively, i.e., the estimate of β_8 is -0.075. Public agricultural research and extension are substitutes, which is similar to what Huffman and Evenson (1993) found for the livestock sub-sector. Hence, public agricultural research and extension seem to have become stronger substitutes over time. The coefficients of the variables describing the composition of SAES agricultural research funding, β_3, β_4, β_5, and β_6 are each significantly different from zero at the 1 percent level.[9] Hence, the productivity of the state public agricultural research capital is affected significantly by the composition of SAES funding, i.e., all types of funding are not equally effective with respect to impacts on state agricultural productivity.

To gain insight, we graph $\partial \ell$n MFP/∂SFF against SFF. If the marginal effect is zero, then a change in the share of state agricultural experiment-station funding coming from federal formula and state appropriations (block grants) would not affect the productivity of public agricultural research. If $\partial \ell$n MFP/∂SFF is a constant and positive (negative), then any incremental changes in SFF and $\partial \ell$n MFP will move in the same (opposite) direction. If the marginal effect of SFF on MFP is initially low and then increases over some range of SFF, eventually topping out, and turning downward again, then $\partial \ell$n MFP/∂SFF will have an inverted "U" shape pattern as SFF increases. This type of pattern has an easily computable value of SFF that gives maximum marginal effect of

SFF on $\partial \ell$n MFP. Similar statements can be made about how $\partial \ell$n MFP/∂GR changes as GR increases. The patterns of marginal effects of SFF and GR on $\partial \ell$n MFP are displayed in Figure 8.8.

The empirical estimates of the marginal impact of SFF on ℓn MFP has in fact an inverted "U" shape that peaks at 0.702 (see Figure 8.8, Panel A) and the impact of GR on ℓn MFP has a "U" shape (see Figure 8.8, Panel B) that has a minimum at 0.237. In contrast, the sample mean value of SFF_{t-12} is actually 0.75 and of GR_{t-12} is 0.096 (Figure 8.8, panel B). Hence, at the sample mean,

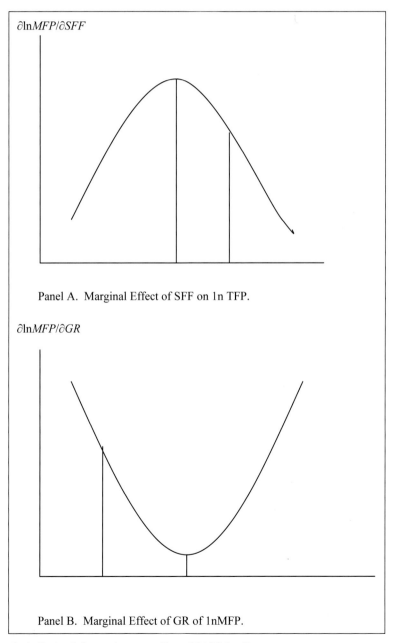

Panel A. Marginal Effect of SFF on ℓn TFP.

Panel B. Marginal Effect of GR of ℓnMFP.

Fig. 8.8. Marginal effects of change in composition of SAES funding.

the evaluation of equation (8.20) gives a marginal effect of changing SFF on ℓn MFP of -0.124 [= $-0.073 + 2(0.154)0.751$] 16.129 and of GR on ℓn MFP of -0.701 [=$-0.073 + 2$ (0.154)0.096] 16.129. Hence, an incremental reallocation of funds from SFF, say 5 percentage points, to GR, i.e., a decline in the share of programmatic funding offset by an equal increase in federal grants and contracts, will lower state agricultural MFP (See Chapter 7 for a model of funding shares).

The private sector has very weak incentives for research in the basic biological sciences (see Table 4.7 for fields of science), but discoveries from this area of science are important to public and applied research because they restore the inventive potential of applied research and establish the scientific foundations for some successful innovations. In regression 2, Table 8.6, we report a test of null hypothesis that the share of SAES research in the basic biological sciences lagged 8 years ($ShBBG_{t-8}$) interacts positively with the stock of private agricultural research. The estimated coefficient is 0.03, and it is significantly different from zero at the 1-percent level. This result suggests that private agricultural research and SAES basic biological-science research are complements for impacting state MFP. Furthermore, the introduction of this new regressor in Regression 2 has little impact on size or statistical significance of the other estimated coefficients as reported in Regression 1.

Using equation (8.24), sample mean values of variables from Table 8.5 and parameter estimates from Table 8.6, Regression 1, we obtain an estimate of the real rate-of-return to an increment in public funds allocated to public agricultural research of 76 percent. This real rate-of-return is adjusted for inflation, and exceeds by a large margin the real return on government treasury bills of about 3 over the past 30 years. Additional discussion of the payoff to public investments is taken up in Chapter 9.

Simulation of a New Science Policy

Simulation of a new policy is possible using equation (8.23). The current blend of federal formula and state appropriations, as opposed to federal competitive grants, contracts, and cooperative agreements, provides SAES directors with considerable flexibility in using the resources and providing direction for research programs that meet local and regional needs. Directors have the advantage of building reputations with state clientele and their scientists, which tends to increase the efficiency of the public agricultural-research organization. Generally, state legislatures expect their LGUs to spend state appropriations on finding solutions to local problems. Failure of state agricultural experiment-station directors to deliver on discoveries needed locally will most likely result in a future weakening of state legislative support, which has occurred in some states, e.g., Wisconsin and Colorado.

Some officials have suggested reducing federal-formula funds for experiment-station research. One option is to offset the reduction of federal-formula funds with increases in competitive-grant programs, although Congress has been reluctant to pursue this scenario. We will show that such a shift would lower agricultural productivity in general and benefit only a few states, while reducing funds and agricultural productivity for all the other states.

One possible scenario would be to reduce total federal SAES funding from federal-formula funding (SFF1) by 10 percentage points, and, hence, reduce SFF ($\Delta SFF = \Delta$). These funds then could be re-allocated to the USDA's competitive-grant programs, such as the National Research Initiatives, to increase GR. We assume that these funds actually go to the state agricultural-experiment stations.[10] Two things are of significant interest, the long-run impact on SAES funding and the impact on state agricultural MFP.

To implement this policy at the state level, we assume that each state will have their baseline federal-formula funds re-scaled by 13/23 and their federal grants and contracts funding will increase by a factor of 2.04 times the baseline value.[11] Following this policy, 26 states would have an increase in their public agricultural-research stock, and six states (California, Indiana, Michigan, New York, Oregon, and Wisconsin) would have more than a 10 percent increase. (See Table 8.7.) Twenty-two states would have a decline; and in six states (Kentucky, Massachusetts, New Hampshire, South Carolina, Vermont, and West Virginia), the decline would be by more than 10 percent. Using equation (8.23) and Regression 1, Table 8.6, we compute the implied change in ℓn MFP for each state.[12] This change is not proportional to the change in RPUB because SFF and GR are also changing, and we have shown that they impact ℓn MFP, too. Forty-five states would experience a decline in ℓn MFP from this policy change; the largest—approximately an 8-percent decline—would occur in Alabama, Nebraska, and West Virginia.[13] Only three states would experience an increase in state agricultural productivity—California, Oregon, and Wisconsin. These latter states have a history of significant reliance on federal grants and contracts for SAES funding (see Chapter 7).

When public agricultural research is funded by federal competitive grants and contracts, the research agenda is set by the funding agency in Washington, D.C., and decisions are based on proposals rather than completed projects. In addition, the federal competitive-grants programs do not pay for research-proposal writing, so the risk of federal research-grant programs is borne by the competing scientists or their institutions and the somewhat distorted incentive structure increases transactional costs, while lowering the scientists' productivity. Furthermore, federal funding agencies tend to fund less than 100 percent of funded research-project costs, so other funds, most notably state-appropriated or federal-formula funds, are used to subsidize research sponsored from outside the state. These are reasons why, from a social perspective, federally funded competitive grants do not look nearly as economically attractive as they do to the federal funding agencies who generally take a "private benefits" perspective.

Social scientists have periodically noted that public agricultural research, cooperative extension, farmers' education, private agricultural research, infrastructure, and government all contribute to productivity change. Over the past two decades, a number of studies have examined the effect of public investments in agricultural research and development, and all have demonstrated a positive and significant impact on agricultural productivity. This is thought to be, in part, because the state agricultural-experiment stations have a long-term focus on addressing local problems. As a result, the positive reputation earned through these long-term relationships creates strong incentives for discovery (Huffman and Just 1999, 2000) and incentives that are different for one-time or inconsistent funding (or contracting) from a federal competitive-grant program.

Conclusion

We have provided evidence covering more than a century showing that economic growth in U.S. agriculture has been rapid. Much of this growth, however, has been without an increase in the quantity of inputs under the control of farmers. It has, however, required investments in public agricultural research and private R&D. The record of MFP growth for the century has been large by any standard, and for the post–World War II period, it has been large relative to other sectors of the U.S. economy. For the period 1970–1999, considerable variation exists in agricultural productivity across U.S. states, but a relatively small set of variables has successfully decomposed or explained

Table 8.7   Baseline and simulation results: Re-allocation of 10 percentage points of federal formula funds to federal grants and contracts[a]

State	Mean Values, 1970–1999					Simulated Outcome			
	$SFF1_{t-12}$	$SFF2_{t-12}$	GR_{t-12}	OR_{t-12}	REV1P[b]	$\Delta \ell nRPUB3$	ΔGR_{t-12}	ΔSFF_{t-12}	$\Delta \ell nMFP$
AL	0.2327	0.4141	0.0564	0.2967	19777.07	−0.0435	0.0636	−0.0769	−0.0841
AR	0.2087	0.5313	0.0333	0.2267	19657.41	−0.0581	0.0381	−0.0532	−0.0508
AZ	0.1203	0.6202	0.1157	0.1437	20078.46	0.0654	0.1038	−0.0944	−0.0358
CA	0.0520	0.6967	0.1662	0.0850	95084.71	0.1402	0.1286	−0.1209	0.0104
CO	0.2290	0.4831	0.1753	0.1126	18937.18	0.0487	0.0751	−0.0657	−0.0080
CT	0.2022	0.6167	0.1266	0.0545	7591.56	0.0394	0.1119	−0.1091	−0.0135
DE	0.3461	0.4132	0.0570	0.1836	4255.42	−0.0962	0.0695	−0.0883	−0.0660
FL	0.0625	0.7594	0.0600	0.1180	52301.81	0.0344	0.0571	−0.0531	−0.0284
GA	0.1913	0.5937	0.0427	0.1721	31351.12	−0.0397	0.0480	−0.0564	−0.0453
IA	0.1766	0.4234	0.1444	0.2557	31220.50	0.0709	0.1297	−0.1126	−0.0615
ID	0.2169	0.5562	0.0520	0.1749	10357.87	−0.0443	0.0495	−0.0568	−0.0467
IL	0.2049	0.4889	0.1137	0.1925	22467.72	0.0285	0.1107	−0.1059	−0.0602
IN	0.1562	0.3719	0.1980	0.2740	29789.08	0.1295	0.1568	−0.1236	−0.0318
KS	0.1285	0.5142	0.0991	0.2583	25025.89	0.0456	0.0919	0.0804	−0.0617
KY	0.3454	0.6314	0.0023	0.0210	15319.23	−0.1603	0.0031	−0.0068	−0.0432
LA	0.1290	0.7666	0.0350	0.0694	25123.29	−0.0198	0.0379	−0.0392	−0.0241
MA	0.3629	0.5227	0.0464	0.0680	6012.92	−0.1184	0.0529	−0.0573	−0.0419
MD	0.2382	0.6250	0.0638	0.0729	10431.63	−0.0406	0.0626	−0.0657	−0.0333
ME	0.3264	0.3967	0.0660	0.2110	7038.70	−0.0786	0.0730	−0.0884	−0.0725
MI	0.1653	0.4958	0.1731	0.1658	30027.38	0.1026	0.1443	−0.1286	−0.0207
MN	0.1750	0.6304	0.1052	0.0894	32194.02	0.0328	0.1017	−0.0986	−0.0266
MO	0.2186	0.4676	0.1036	0.2102	20672.59	0.0124	0.1040	−0.1011	−0.0672
MS	0.2587	0.4480	0.0657	0.2275	22716.40	−0.0454	0.0753	−0.0851	−0.0790
MT	0.1819	0.4495	0.0900	0.2786	10514.76	0.0142	0.0899	−0.0862	−0.0764
NC	0.1804	0.5510	0.1350	0.1335	41020.45	0.0601	0.1234	−0.1157	−0.0335
ND	0.1776	0.6347	0.0528	0.1349	15453.58	−0.0226	0.0575	−0.0599	−0.0421
NE	0.1081	0.3799	0.0947	0.4174	29478.07	0.0500	0.0880	−0.0681	−0.0877
NH	0.5510	0.3618	0.0062	0.0811	2877.41	−0.2669	0.0084	−0.0333	−0.0331
NJ	0.1551	0.6264	0.0836	0.1349	14620.41	0.0189	0.0824	−0.0809	−0.0403
NM	0.2752	0.5318	0.0960	0.0970	7335.32	−0.0212	0.0998	−0.1022	−0.0412
NV	0.2740	0.4841	0.1055	0.1364	5072.64	−0.0102	0.1087	−0.1111	−0.0476
NY	0.1121	0.5443	0.1685	0.1750	44864.97	0.1185	0.1342	−0.1121	−0.0249
OH	0.2300	0.7187	0.0230	0.0283	23238.50	−0.0793	0.0275	−0.0294	−0.0338
OK	0.2204	0.5730	0.1081	0.0985	15940.21	0.0162	0.1069	−0.1058	−0.0337
OR	0.1207	0.4549	0.2232	0.2012	23783.39	0.1640	0.1576	−0.1291	0.0053
PA	0.2894	0.5384	0.0836	0.0886	20678.89	−0.0398	0.0934	−0.0970	−0.0465
RI	0.3765	0.3625	0.1791	0.0819	2678.67	0.0194	0.1700	−0.1694	−0.0099
SC	0.3242	0.6072	0.0033	0.0653	14304.31	−0.1495	0.0043	−0.0178	−0.0234
SD	0.2439	0.5530	0.0359	0.1671	8441.07	−0.0714	0.0420	−0.0542	−0.0492
TN	0.2811	0.3434	0.1395	0.2360	16770.50	0.0168	0.1181	−0.1181	−0.0730
TX	0.1616	0.5076	0.0895	0.2413	50730.37	0.0225	0.0889	−0.0836	−0.0693
UT	0.2343	0.4819	0.1743	0.1095	9576.64	0.0762	0.1538	−0.1464	−0.0076
VA	0.2059	0.4897	0.1423	0.1622	24834.60	0.0569	0.1318	−0.1228	−0.0418
VT	0.4728	0.4322	0.0299	0.0651	3172.35	−0.1953	0.0374	−0.0460	−0.0545
WA	0.1698	0.5274	0.0993	0.2035	21602.13	0.0279	0.0943	−0.0880	−0.0596
WI	0.1512	0.4933	0.2490	0.1065	36581.87	0.1765	0.1755	−0.1582	0.0594
WV	0.4821	0.3523	0.0512	0.1144	5300.50	−0.1700	0.0724	−0.0937	−0.0790
WY	0.2993	0.5706	0.0655	0.0645	4731.49	−0.0648	0.0752	−0.0793	−0.0415

[a]See text for full discussion of the context of the change.
[b]Total value of SAES funds for all uses in thousands of 1984 dollars.

these differences. Some unusual differences in contributions of public agricultural research to productivity have been uncovered. Also, private sector research, public extension, and farmers' schooling play an important role in the productivity story.

We have presented new econometric evidence of the determinants of state MFP. Both the stock of public agricultural research—within state and spill-in from other states—and the composition of SAES funding affect MFP significantly over 1970–1999. The results showed that complex interactions effects occur between a state's public agricultural research-capital stock and SAES funding composition (shares of federal-formula and state appropriations, or programmatic funding, and of federal grants and contracts). These results show that a marginal percentage-point transfer of federal funds from formula to competitive-grant programs, while holding total funding constant, would, on average, reduce state agricultural productivity. A more complex simulation would cause non-marginal adjustments in ℓn MFP across the states: e.g., a 10-percentage point reduction in federal-formula funds (a re-scaling by 13/23) and a transfer to federal competitive-grants programs (a re-scaling by 2.04). The results show that only 3 states would experience an increase in ℓn MFP while the other 45 would face a decrease. Hence, it is not hard to imagine that most agricultural experiment-station directors would be opposed to this re-allocation.

From the perspective of socially efficient resource allocation, the public sector should only undertake research that the private sector will not undertake, primarily because of absence of profitability. In other words the public sector should concentrate its efforts on general and pre-invention sciences and leave much of the applied research and technology development to the private sector. This research would include more basic research activity in which scientific breakthroughs are going to be highly unpredictable, are largely public goods, and in which narrowly targeting research efforts may lead to fruitless activities. We also provide evidence that a larger share of SAES research in the basic biological sciences is complementary with intrastate private agricultural research. This is evidence of the important role of SAES basic biological science in indirectly supporting invention for profit in the private sector.

For agricultural-experiment stations to maintain or increase state support, they must maintain rapport with their within-state clientele groups. This implies that state agricultural-experiment stations will perform more applied research that has locally visible products than what is socially efficient. Selling pre-invention scientific discoveries to local clientele groups seems more difficult than selling applied discoveries. However, firms that are engaged in agricultural invention should be a supportive constituency. Experiment-station administrators as a group, however, can improve their packaging and selling of pre-invention science research to clientele groups and state legislators. It does require placing research priorities in a long-term, rather than an immediate, perspective.

Notes

1. Jorgenson and Stiroh (2000) provide post–World War II evidence that the MFP of the U.S. agricultural sector ranks well compared to 32 other sectors of the U.S. economy.
2. The treatment of private costs can be handled in different ways. The hybrid-corn study computed a return to combined public and private research spending. This is not the return realized by the private firms in the hybrid seed-corn industry. They realized a lower rate of return because the higher prices they could charge for superior seed did not capture the full value of the improved technology.
3. In particular, Huffman and Evenson (1993) found that public agricultural research and extension stock interacted positively in the crop sub-sector and negatively in the livestock sub-sector.

4. Significant interaction effects did not exist between public and private agricultural research stocks.

5. Our public agricultural-research stock variable is a much better proxy for useful technical change than a time trend. And because it is constructed from real public agricultural-research expenditures it is not strongly trended over the study period. Similar weights were used by Khanna, Huffman, and Sandler (1994), McCunn and Huffman (2000), and in Huffman and Evenson (1993).

6. We have followed a common convention in constructing interstate public agricultural-research spillover variables using all productivity-oriented public agricultural-research expenditures (see Huffman and Evenson 1993; Khanna, Huffman, and Sandler 1994; and McCunn and Huffman 2000). Of course other options are possible, but federal grant funds are frequently insufficient to complete a project, which means that federal-formula and state-appropriated funds may be diverted to these objectives.

7. If heteroscedasticity and/or autocorrelations were serious in our data, we would expect the "robust t-values" to be much smaller in absolute value than for the traditional t-values. In contrast, for our data these two t-values are, on average, approximately the same—about 7.13.

8. Using the residuals from this equation, the estimate of ρ, the first-order autocorrelation coefficient, averaged across all 48 states is only 0.11, which is near 0 and far from 1, the nonstationary value.

9. Also, the joint test of no-fund composition effects, i.e., $\beta_3 = \beta_4 = \beta_5 = \beta_6 = 0$, is soundly rejected. The sample F-statistic for this joint test is 16.8, and the critical value of the F-statistic with 4 and 1424 degrees of freedom at the 1-percent significance level is about 3.4.

10. Given that the National Research Initiate (NRI) Program is a national competitive program, some of the funded projects are for individuals who are not at a LGU and, hence, not associated with a state agricultural experiment station. In only two cases is a state agricultural experiment station not directly affiliated with a LGU.

11. The percentage change in RPUB and the size of the total real SAES budget in 1984 dollars is assumed to be the same.

12. We treat this scenario as a non-marginal change and apply the difference equation (8.23).

13. We have ignored the impact of the policy change on RPUBSPILL, because it is difficult to approximate how it would change. In addition to public agricultural research impacting state agricultural productivity, it may have other largely independent effects on basic scientific discoveries, effects that are socially valuable but not related to agricultural productivity (Committee on Opportunities in Agriculture 2003). Hence, our simulation results may not capture all of the social benefits of a reallocation of federal funds between formula and grants and contracts.

References

Ahearn, M., J. Lee, and J. Bottom. 2002. "Regional Trends in Extension Resources." Paper presented at Southern Agricultural Economics Association Meeting, Orlando, FL, February 2002.

Alston, J.M., B. Craig, and P.G. Pardey. 1998. "Dynamics in the Creation and Depreciation of Knowledge, and the Returns to Research." International Food Policy Research Institute EPTD Discussion Paper No. 35.

Alston, J., G.W. Norton, and P.G. Pardey. 1995. *Science under Scarcity: Theory and Practice for Agricultural Research Evaluation and Priority Setting.* Ithaca, NY: Cornell University Press.

Alston, J.M. and P.G. Pardey. 2001. "Attribution and Other Problems in Assessing the Returns to Agricultural R&D." *Agricultural Economics* 25: 141–52.

Antle, John M. and Susan M. Capalbo. 1988. "An Introduction to Recent Developments in Production Theory and Productivity Measurement." In *Agricultural Productivity: Measurement and Explanation,* ed. S.M. Capalbo and J.M. Antle. Washington, DC: Resources for the Future.

Ball, V. Eldon, J.P. Butault, and R. Nehring. 2002. "U.S. Agriculture, 1960–96: A Multilateral Comparison of Total Factor Productivity." In *Agricultural Productivity: Measurement and Sources of Growth,* ed. V.E. Ball and G.W. Norton. Boston, MA: Kluwer Academic Publication.

Ball, V. Eldon, F.M. Gollop, A. Kelly-Hake, and G.P. Swinand. 1999. "Patterns of State Productivity Growth in the U.S. Farm Sector: Linking State and Aggregate Models." *American Journal of Agricultural Economics* 81: 164–179.

Barton, Glen T. and Martin R. Cooper. 1948. "Relation of Agricultural Production to Inputs." *Review of Economics and Statistics* 30: 117–126.

Birkhaeuser, Dean, R.E. Evenson, and G. Feder. 1991. "The Economic Impact of Agricultural Extension: A Review." *Economic Development and Cultural Change* 39: 607–650.

Braha, Habta and L. Tweeten. 1986. "Evaluating Past and Prospective Future Payoffs from Public Investments to Increase Agricultural Productivity." Technical Bulletin T-163. Stillwater, OK: Agricultural Experiment Station, Oklahoma State University.

Bredahl, M. and W. Peterson. 1976. "The Productivity and Allocation of Research: U.S. Agricultural Experiment Stations." *American Journal of Agricultural Economics* 58(4): 684–692.

Chambers, R.G. 1988. *Applied Production Analysis: A Dual Approach.* New York: Cambridge University Press.

Cline, Philip. 1975. "Sources of Productivity Change in U.S. Agriculture." Ph.D. dissertation. Stillwater, OK: Oklahoma State University.

Committee on Opportunities in Agriculture, NRC. 2003. *Frontiers in Agricultural Research: Food, Health, Environment and Communities.* Washington, D.C.: National Academic Press.

Davidson, R., and J.G. MacKinnon. 1993. *Estimation and Inference in Econometrics.* New York: Oxford University Press.

Denison, E.F. 1962. *The Sources of Economic Growth in the United States and the Alternatives before Us.* New York: Committee for Economic Development.

Denison, E.F. 1969. "Some Major Issues in Productivity Analysis: An Examination of Estimates by Jorgenson and Griliches." *Survey of Current Business* 49(5) Part 2: 1–2.

Denison, E.F. 1972. "Final Comments." *Survey of Current Business* 52(5) Part 2: 95–110.

Diewert, W.E. 1976. "Exact and Superlative Index Numbers." *Journal of Econometrics* 4: 115–145.

U.S. Department of Agriculture, Economic Research Service. 2003. "Data: Agricultural Productivity in the United States." Available at: http://www.ers.usda.gov/data/agproductivity.

Evenson, Robert E. 1968. "The Contribution of Agricultural Research and Extension to Agricultural Production." Ph.D. dissertation. Chicago, IL: University of Chicago.

Evenson, Robert E. 1980. "A Century of Agricultural Research and Productivity Change Research, Invention, Extension and Productivity Change in U.S. Agriculture: An Historical Decomposition Analysis." In *Research and Extension Productivity in Agriculture,* ed. A.A. Araji. Moscow, ID: Department of Agricultural Economics, University of Idaho.

Evenson, Robert E. 1992. "Research and Extension in Agricultural Development." San Francisco, CA: ICS Press for the International Center for Economic Growth.

Evenson, R.E. 1996. "Two Blades of Grass: Research for U.S. Agriculture." In *Papers in Honor of D. Gale Johnson, The Economics of Agriculture,* Vol. 2. Chicago: University of Chicago Press.

Evenson, R.E. 2001. "Economic Impacts of Agricultural Research and Extension." In *Handbook of Agricultural Economics,* Vol.1A, ed. B.L. Gardner and G.C. Rausser. New York: Elsevier.

Evenson, Robert E. and Y. Kislev. 1975. *Agricultural Research and Productivity.* New Haven, CT: Yale University Press.

Evenson, Robert E. and Finis Welch. 1974. "Research, Farm Scale, and Agricultural Production." New Haven, CT: Economic Growth Center, Yale University.

Friedman, Milton and Anna J. Schwartz. 1963. *A Monetary History of the United States, 1867–1960.* Princeton, NJ: Princeton University Press, for the National Bureau of Economic Research.

Griliches, Zvi. 1958. "Research Costs and Social Returns: Hybrid Corn and Related Innovations." *Journal of Political Economy* 66(5): 419–431.

Griliches, Z. 1960. "Measuring Inputs in Agriculture: A Critique." *Journal of Farm Economics* 42: 1411–1427.

Griliches, Zvi. 1963a. "Estimates of the Aggregate Agricultural Production Function from Cross Sectional Data." *Journal of Farm Economics* 45: 419–428.

Griliches, Zvi. 1963b. "The Sources of Measured Productivity Growth: United States Agriculture, 1940–1960." *Journal of Political Economy* 71: 331–346.

Griliches, Zvi. 1964. "Research Expenditures, Education and the Aggregate Agricultural Production Function." *American Economics Review* 54: 961–974.

Griliches, Zvi. 1973. "Research Expenditures and Growth Accounting." In *Science and Technology in Economic Growth,* ed. B. R. Williams. London: Macmillan Press.

Griliches, Z. 1979. "Issues in Assessing the Contributions of Research and Development to Productivity Growth." *Bell J. Economics*. 10(1979): 92–116.

Griliches, Z. 1998. *R&D and Productivity: The Econometric Evidence*. Chicago, IL: The University of Chicago Press.

Hayami, Y. and V.W. Ruttan. 1985. *Agricultural Development*. Baltimore, MD: Johns Hopkins University Press.

Huffman, Wallace E. and R.E. Evenson. 1989. "Supply and Demand Functions for Multiproduct U.S. Cash Grain Farms: Biases Caused by Research and Other Policies." *American Journal of Agricultural Economics* 71: 761–773.

Huffman, W.E. and R.E. Evenson. 1993. *Science for Agriculture: A Long-Term Perspective*. Ames: Iowa State University Press.

Huffman, W.E. and R.E. Evenson. 2004. "New Econometric Evidence on Agricultural Total Factor Productivity Determinants: Impact of Funding Sources." Iowa State University, Department of Economics, Working Paper #03029.

Huffman, W.E. and R.E. Just. 2000. "Setting Efficient Incentives for Agricultural Research: Lessons from Principal-Agent Theory." *American Journal of Agricultural Economics* 82: 828–841.

Huffman, W.E. and R.E. Just. 1994. "An Empirical Analysis of Funding, Structure, and Management of Agricultural Research in the United States." *American Journal of Agricultural Economics* 76. 744–759.

Huffman, W.E., A. McCunn, and J. Xu. Forthcoming. "Public Agricultural Research with an Agricultural Productivity Emphasis: Data for 48 States, 1927–1995." Iowa State University, Department of Economics, Staff Paper.

Johnson, D.K.N. and A. Brown. 2002. "Patents Granted in U.S. for Agricultural SOV, by State of Inventor, 1963–1999." Wellesley, MA: Department of Economics Working Paper, Wellesley College.

Johnson, N., L. and V.W. Ruttan. 1997. "The Diffusion of Livestock Breeding Technology in the U.S.: Observations on the Relationship between Technical Change and Industry Structure," *Journal of Agribusiness* 15: 19–36.

Jorgenson, Dale, F. Gollop, and B. Fraumeni. 1987. *Productivity and U.S. Economic Growth*. Cambridge, MA: Harvard University Press.

Jorgenson, Dale and Zvi Griliches. 1967. "The Explanation of Productivity Change." *Review of Economic Studies* 34: 249–283.

Jorgenson, Dale W. and Z. Griliches. 1972. "Issues in Growth Accounting: A Reply to Edward F. Denison." *Survey of Current Business* 52(5) Part 2: 65–94.

Jorgenson, Dale W. and K.J. Stiroh. 2000. "U.S. Economic Growth at the Industry Level." *American Economic Review* 90: 161–167.

Kendrick, John W. 1961. "Productivity Trends in the United States." National Bureau of Economic Research, New York. Princeton, NJ: Princeton University Press.

Khanna, J., W.E. Huffman, and T. Sandler. 1994. "Agricultural Research Expenditures in the United States: A Public Goods Perspective." *Review of Economics and Statistics*, 76(May): 267–277.

Loomis, R.A. and G.T. Barton. 1961. *Productivity of Agriculture*. U.S. Department of Agriculture-Economic Research Service.

Lu, Yao-chi, Philip Cline, and Leroy Quance. 1979. *Prospects for Productivity Growth in U.S. Agriculture*. Washington, DC: U.S. Department of Agriculture, Economics, Statistics, and Cooperatives Service, Agricultural Economics Report 435.

MacKinnon, J.G. and H. White. 1985. "Some Heteroscedasticity Consistent Covariance Matrix Estimators with Improved Finite Sample Properties." *Journal of Econometrics* 29: 305–325.

Maxwell, B.J., W.W. Wilson and B.L. Dahl. 2004. "Marketing Mechanisms in GM Grains and Oilseeds." *Agribusiness and Applied Economics Report, No. 547*. North Dakota State University, Department of Agribusiness and Applied Economics.

McCunn, A. I. and W.E. Huffman. 2000. "Convergence in U.S. Productivity Growth for Agriculture: Implications of Interstate Research Spillovers for Funding Agricultural Research." *American Journal of Agricultural Economics* 82: 370–388.

Narrod, C.A. and K. O. Fuglie. 2000. "Private Investment in Livestock Breeding with Implications for Public Research Policy." Washington, DC: Office of Risk Assessment and Cost-Benefit Analysis, USDA.

National Research Council. 1975. *Agricultural Production Efficiency.* Washington, DC: National Academy of Sciences.

Norton, G.W. and J.S. Davis. 1981. "Evaluating Returns to Agricultural Research: A Review." *American Journal of Agricultural Economics* 63: 685–699.

Peterson, W.L. 1967. "Returns to Poultry Research in the United States." *Journal of Farm Economics* 49: 656–669.

Schultz, T.W. 1953. *The Economic Organization of Agriculture.* New York, NY: McGraw-Hill Book Company.

Schultz, T.W. 1961. "Education and Economic Growth." In *Social Forces Influencing American Education*, ed. N.B. Henry. National Society for the Study of Education. Chicago, IL: The University of Chicago Press.

U.S. Department of Agriculture. *Farm Labor.* U.S. Department of Agriculture-Economic Research Service, various issues.

U.S. Department of Agriculture. 1926–1991. *Agricultural Statistics.* Washington, DC: U.S. Government Printing Office.

U.S. Department of Agriculture. 1957. *Yearbook of Agriculture, 1957: Soils.* Washington, DC: U.S. Government Printing Office.

U.S. Department of Agriculture. 1980. *Measurement of U.S. Agricultural Productivity: A Review of Current Statistics and Proposals for Change.* TB-1614. Economics, Statistics, and Cooperatives Service.

U.S. Department of Agriculture. 1991a. *Economic Indicators of the Farm Sector: Production and Efficiency Statistics, 1989.* U.S. Department of Agriculture-Economic Research Service ECIFS8.5.

U.S. Department of Agriculture. 1991b. *Economic Indicators of the Farm Sector: State Financial Summary, 1990.* Rockville, MD: Economic Research Service-National Agricultural Statistics Service.

White, H. 1980. "A Heteroskedasticity-Consistent Covariance Matrix Estimator and a Direct Test of Heteroskedasticity." *Econometrica* 48: 817–834.

Wooldridge, J.M. 2003. *Introductory Econometrics: A Modern Approach.* Mason, OH: South-Western.

Wooldridge, J.M. 1989. "A Computationally Simple Heteroskedasticity and Serial Correlation-Robust Standard Error for the Linear Regression Model." *Economic Letters* 31: 239–243.

Appendix Table A8.1 Data on multifactor productivity for U.S. agriculture, 1889–1999

Year	Q	X	MFP	Year	Q	X	MFP	Year	Q	X	MFP
1889	41.42	90.78	45.65	1926	69.66	121.07	57.56	1963	132.55	106.99	123.89
1890	40.77	92.21	44.19	1927	71.54	118.44	60.40	1964	131.93	105.05	125.58
1891	42.50	93.29	45.59	1928	71.03	120.11	59.13	1965	135.03	104.20	129.59
1892	40.40	94.49	42.56	1929	72.41	119.76	60.46	1966	134.18	105.20	127.55
1893	39.39	95.45	41.29	1930	68.43	118.80	57.62	1967	138.98	104.39	133.14
1894	40.84	96.41	42.38	1931	74.08	120.71	61.37	1968	141.32	103.03	137.17
1895	43.45	97.60	44.50	1932	71.03	118.32	60.04	1969	144.08	103.67	138.98
1896	43.75	98.80	47.28	1933	72.56	117.48	61.73	1970	142.93	104.84	136.33
1897	50.18	100.00	50.18	1934	63.79	108.74	58.64	1971	152.68	103.76	147.15
1898	52.50	101.43	51.75	1935	68.93	111.13	62.03	1972	154.00	104.79	146.97
1899	52.50	102.63	51.15	1936	67.05	108.98	61.55	1973	160.14	105.90	151.22
1900	53.08	103.83	51.09	1937	74.15	114.49	64.75	1974	162.19	105.10	143.80
1901	52.86	104.55	50.54	1938	77.26	109.22	70.74	1975	160.21	103.55	154.72
1902	52.50	105.62	49.70	1939	79.51	109.70	72.49	1976	162.19	107.15	151.36
1903	54.16	106.94	50.66	1940	81.24	109.58	74.12	1977	171.36	104.73	163.62
1904	55.68	107.66	51.69	1941	87.18	108.98	79.99	1978	173.84	110.57	157.23
1905	56.41	108.74	51.87	1942	95.44	112.93	84.52	1979	185.37	113.23	163.70
1906	59.52	109.82	54.17	1943	93.63	112.81	83.01	1980	178.68	114.79	155.66
1907	56.91	111.01	51.51	1944	94.93	112.09	84.70	1981	194.55	111.11	175.10
1908	58.15	111.37	52.18	1945	94.42	106.70	88.33	1982	200.33	108.67	184.35
1909	57.06	112.33	50.78	1946	97.03	104.19	93.11	1983	172.48	108.57	158.87
1910	58.65	113.53	51.63	1947	95.51	100.95	94.62	1984	196.60	104.02	189.00
1911	55.83	116.04	48.12	1948	100.00	100.00	100.00	1985	205.77	101.63	202.47
1912	64.73	117.48	55.08	1949	100.13	104.46	95.86	1986	198.86	98.63	201.63
1913	57.71	116.88	49.39	1950	99.63	104.54	95.30	1987	203.17	97.68	207.99
1914	63.36	119.40	53.08	1951	104.35	105.72	98.70	1988	194.77	96.94	200.92
1915	66.62	119.28	55.86	1952	107.66	105.58	101.97	1989	210.20	95.57	219.94
1916	60.39	119.04	50.72	1953	108.61	104.87	103.56	1990	218.45	97.14	224.89
1917	65.97	121.43	54.29	1954	109.04	101.98	106.92	1991	220.51	98.05	224.88
1918	63.14	122.87	51.39	1955	112.51	106.49	105.65	1992	231.74	95.45	242.80
1919	63.21	122.15	51.75	1956	113.43	106.46	106.54	1993	220.45	95.82	230.07
1920	63.00	123.11	51.15	1957	111.82	105.55	105.93	1994	245.77	97.10	253.10
1921	59.74	117.00	51.03	1958	118.36	105.51	112.93	1995	234.32	98.79	237.18
1922	63.43	118.68	53.45	1959	121.71	107.72	112.99	1996	244.37	95.50	255.88
1923	67.05	118.56	56.53	1960	125.16	107.55	116.37	1997	252.86	98.45	256.84
1924	65.89	118.92	55.38	1961	126.20	105.59	119.52	1998	256.40	99.14	258.62
1925	69.66	120.47	57.80	1962	128.16	106.17	120.71	1999	261.49	101.52	257.57

Source: Adapted from Kendrick 1961, pp. 365–366, U.S. Department of Agriculture 1991a, Appendix (revised), and U.S. Department of Agriculture, Economic Research Service, 2003.

Note: Q is an index of real U.S. farm outputs; X is an index of inputs under the control of farmers; and MFP(= Q/X) is U.S. multifactor agricultural productivity.

9

The SAES-USDA System: Challenges for the 21st Century

In the first edition of this book our final chapter was devoted largely to responses to four critical reviews of the SAES-USDA research system: The Pound Report (Committee on Research Advisory to the U.S. Department of Agriculture 1972); The Office of Technology Assessment (OTA) report (1981); the Winrock Report (Rockefeller Foundation 1982); and National Research Council reports (1989, 1992). We defended the SAES-USDA system using rate-of-return study evidence showing variable but high returns to taxpayers' dollars and argued that the SAES-USDA system was responsive to its critics. Our strategy in this final chapter will be different. The SAES-USDA system is not under serious challenge by critics today, but it is under challenge by events and by the changing conditions within which it must function.

Clearly, the major challenge to the SAES-USDA system and to the LGU system today emerged from changes in the biological sciences. Advances in molecular biology have transformed not only the way biological science is done, but the way related invention and innovation are made. The SAES-LGU-USDA system should have been prepared to lead the technological dimension of the "biotechnology revolution." Most LGUs, as noted in Chapter 2, developed "pre-invention science" capacity. This should have positioned the LGUs to take advantage of the developments in molecular biology and to engage in the activities required to bring developments in science into the invention realm.[1]

But in practice, the basic science departments themselves, i.e., the molecular biology science departments in research universities, both non-LGU and LGU, developed the pre-invention science contributions that made invention possible. Some LGUs responded to these developments quickly; others did not. Private firms responded even more quickly to the invention-innovation opportunities afforded them.

The net result of this "scrambling" to keep up with rapid developments in the non-LGU and private sectors is that the SAES-USDA system was left in catch-up mode. At present, observers of the biotech field clearly recognize that first generation agricultural biotech products have been produced by private industry. All of the first-generation products are cost-reducing products for farmers. There is broad agreement that second-generation agricultural biotech productions with consumer-quality enhancement (low saturated fat, low pesticide residual etc.) will reduce consumer resistance to GM foods. And there is further agreement that the production of such products by public-sector programs would do much to alleviate political concerns regarding GM foods. But, to date, the SAES-LGU-USDA system has not produced such products.[2]

The second major challenge for the SAES-USDA system is the remarkably high rate of "structural" change in U.S. agriculture. At the beginning of the 20th century, U.S. agriculture was still organized around "horse and buggy" technology. Most farm units were single units, many still of the 160-acre Homestead Act size. Rural communities and school systems were still viable. By the end of the 20th century this had changed drastically.

Farms today are several times the size they were in the early part of the 20th century. Many farmers have incorporated several farms into their operation. Modern farm equipment now allows for the traditional 160-acre unit of the Homestead Act to be planted in less than a day (Huffman and Evenson 2001). Modern transportation has disadvantaged the small rural towns and advantaged larger communities. Small towns have lost their schools and most businesses in much of the country.

Perhaps the most drastic changes have occurred in the food-marketing sector. The first wave of supermarkets effectively eliminated the "mom and pop" grocery stores in the 1950s and 1960s. A second wave of "big box" supermarket expansion (led by WalMart™, the world's largest retailer) is now creating larger supermarkets. And while "niche" market products (organic foods) have grown in importance as incomes have risen, these products are also handled by supermarkets. And as noted in Chapter 5, the agricultural biotech industry has settled into an equilibrium with just six or seven multinational firms.

Livestock production has effectively been "industrialized" over the past several decades. The broiler and egg-producing systems were beginning to be industrialized in the 1950s. By the 1960s and 1970s, confined animal feeding units (CAFUs) were the norm. Hog-production CAFUs followed in the 1970s and 1980s. Dairy cattle units were similarly transformed. Beef cattle feed lots today are much larger than in the 1960s, and most are located in favorable winter climate regions. Most livestock production today is no longer associated with grain production. Contract production now dominates livestock production. In effect, livestock production has been industrialized.[3]

In this final chapter we address six broad questions associated with the future of the SAES-USDA system. The first is the question of system response to the agricultural biotechnology revolution. The second is the question of system response to the changing structure of U.S. agriculture. The third is the increasingly complex question of private support for public research programs. The fourth is the system productivity record. The fifth is the issue of federal-formula funds versus competitive-grant funding of SAES research. The sixth is the commodity allocation record.

SAES-USDA System Response to the Biotechnology Revolution

As noted in Chapter 5, the SAES-USDA system has been (or should be) challenged to respond to developments in basic science and the new post–Bayh-Dole Act era. These have lead to new developments in IPRs and the growth in private-sector R&D.

Developments in Science

Consider the response to developments in the basic sciences. As noted above, the SAES-USDA system cannot be regarded as the initiator and leader of the invention component of the agricultural biotech revolution. The actual closing of the gap between developments in science and invention potential occurred rapidly, and the normal role of the pre-invention sciences was effectively bypassed. Private enterprises moved rapidly to exploit the new invention and innovation potential offered by new discoveries in the basic sciences.

The more basic sciences are generally differentiated from the pre-invention sciences by an interest in producing knowledge for its own sake and a tendency to either ignore or downplay the invention or commercial potential of their work (see Chapter 2). This logic would imply that the basic scientist might have developed discoveries that were then converted to invention potential by the pre-invention agricultural scientists. But that is not what actually happened in the recombinant DNA (rDNA) revolution. The rDNA revolution was produced by the basic sciences. As the biological sciences began to explore molecular questions, many biological sciences departments split into specialized molecular biology departments and evolutionary and ecological science departments. The molecular biology departments not only developed the science of molecular biology, but they developed much of the technology as well. The evolutionary sciences developed very slowly, and the ecological sciences are still in their infancy.[4]

The agricultural pre-invention sciences, while quick to pick up on the technological aspects of rDNA transformation, were nonetheless caught "asleep at the switch" to some extent. Many private firms, notably Monsanto, DuPont, and other firms in the agricultural chemicals industry, were quick to see the potential in agricultural biotechnology and invested heavily in product development (see Chapter 5). As developments in IPRs afforded IPR protection to transgenic plant varieties and gene constructs as well as to transgenic techniques, the private sector entered the plant-breeding sector in a major way. The net result of these changes is that most SAES-LGU units are now minor players in almost all fields of invention: mechanical, electrical, chemical, and now genetic. In spite of the obvious value of a public SAES-USDA biotech "product," no product has been produced by the system. And it is now more than 20 years after the private sector decided to make major investments in biotech product development.

Will SAES-LGU units lose support if they produce fewer and fewer inventions that can be identified with their research programs? Through the 1980s, most farmers identified specific varieties of wheat, soybeans, rice, and other crops with SAES-USDA research, but now, as with mechanical, electrical and chemical inventions, most SAES programs risk being crowded out of the genetic invention field.

In head-to-head invention competition with private-sector firms, SAES-USDA programs do not fare well. Almost all new models of farm machinery are developed by a few private companies. SAES-USDA programs are not set up to innovate (i.e., to develop products from inventions). The Bayh-Dole Act has generated SAES-USDA inventions, but it is not clear that SAES-USDA units will be very successful in licensing to innovators.

So, with genetic invention now on a par with other forms of invention, the issue for SAES-USDA units is whether they are now positioned to make private-sector R&D more productive. The productivity study reported in Chapter 8 clearly indicated that high investments in pre-invention research were called for (see page 274).

The SAES-USDA system did, however, respond quite rapidly in terms of incorporating scientific advances into graduate coursework in LGUs and incorporating research programs into SAES units. In Chapter 5, the active participation of SAES-USDA programs in biotechnology field trials is documented. The invention record in biotechnology by SAES-USDA research units has also been good.[5]

The failure of the SAES-USDA system to produce agricultural biotechnology products may be due to the fact that SAES-USDA units are not set up to commercialize (innovate) inventions. But this problem should be addressed by the Bayh-Dole provisions. An SAES unit can offer an exclusive license to a firm willing to invest in commercializing an invention. Although some inventions have been licensed, it is a little surprising that product development has been so slow.

In summary then, the SAES-USDA system's response to the biotechnology revolution has good performance in the field of scientific discoveries, middling performance in invention, and poor performance in innovation. But this assessment should be placed in historical context. It may be unreasonable to expect the SAES-USDA system to compete with private multinational firms such as Monsanto and DuPont. These firms have far higher R&D budgets than do SAES-USDA units and a comparative advantage in developing and marketing new products.[6]

The ultimate question for SAES-USDA programs is whether they can find ways to complement private-sector R&D programs without being taken over by them (Busch et al. 2004). The comparative advantage of SAES-USDA programs vis-à-vis private-sector innovation is in the following areas:

a.   Testing products for public safety, etc.
b.   Economic evaluation of products
c.   Assessments of environmental and resource quality
d.   Pre-invention science, i.e., closing the gap between scientific developments and invention potential (especially for second generation products)

Can the SAES-USDA programs find similar roles in the biotechnology fields?

IPRs

Two major developments in IPRs have taken place since 1980. The first was the expansion of the scope of patent rights to cover living multicellular plants and animals (see Chapter 5). This expansion of scope was the consequence of court decisions, hence of "case law." No legislative change was involved. The second development, involving legislative change, was the Bayh-Dole Act (see Chapter 5). The Bayh-Dole Act ended the requirement that universities share licensing revenue coming from federally funded research projects with the federal government. The Bayh-Dole Act also ended the requirement that licensing of patent (or other) rights be non-exclusive.

Both changes have had a profound impact on the SAES-USDA system. The expansion of patent scope to cover living multicellular plants and animals has effectively placed "genetic invention" on a par with mechanical, chemical, and electrical inventions. The invention-innovation process in mechanical, chemical, and electrical invention has long been dominated by (for profit) private firms. The SAES-USDA system has effectively operated in two areas vis-à-vis the private industry. The first is to engage in "pilot invention" to close the "gap" between applied science and invention. The second is to engage in scientific discovery in the pre-invention sciences and to restore the inventive potential for private innovation. These are appropriate roles for public-sector programs.

The Bayh-Dole Act clearly changed incentives regarding inventions in the SAES-USDA system and in university research generally (Thursby et al. 2001; Foltz et al. 2003; Just and Huffman 2005). Many university research programs are now reporting modest patent licensing (and, in some cases, breeders' rights) income. In the post–Bayh-Dole Act era, many universities have set up technology-transfer offices to facilitate the licensing of university patents. These have worked reasonably well in the mechanical, chemical, and electrical invention fields. The operative question for the SAES-USDA system is whether this can be duplicated for genetic inventions based on biotechnology sciences.

Weakening Ties to the State

In the new science and intellectual-property environment, the private sector has rapidly expanded its R&D for agriculture. The private sector has taken over new product development in areas that the SAES once dominated, e.g., corn inbred lines, soybean varietal development. Although universities are obtaining some revenue from licensing new technologies—frequently from out-of-state firms—they have withdrawn from other technology development areas (see Chapter 5). Inventive activity with a goal of obtaining licensing income and research discovery in the pre-invention and basic sciences is filling this void. This new direction for SAES research has changed the distribution of benefits, extending them largely beyond a state's boundaries.

For publicly provided agricultural research, the principle of fiscal equivalence dictates that the jurisdictional authority of public agricultural research match the geographical range of benefits. When these benefits are to licensees, state governments are unlikely to see the need to provide funds. When benefits extend beyond a single state to a region, then a collective authority covering this region should finance agricultural research. When benefits are national in scope the federal government should finance public agricultural research. Hence, we have provided two reasons why state government support for agricultural research may weaken. We might, however, find states in particular regions or with particular problems pooling resources to fund agricultural research on multi-state problems. Unless a sophisticated mosaic of new overlapping jurisdictions is established, our conjecture is that it would be unusual for state support of SAES research to grow during the 21st century.

SAES-USDA System Response to Structural Changes in U.S. Agriculture

The structure of agriculture has been changing due to a variety of technical and economic forces (e.g., see Huffman 1977; Hoban et al. 1998; Boehlje 1999; Huffman and Evenson 2001; NRC 2002; Huffman and Orazem 2005). The salient features of structural change are:

a.   Growing size of farms. The small commercial farm has virtually disappeared from U.S. agriculture. Programs to save these farms, or to facilitate new entrants into farming, have been ineffective. With biotechnology products, scale economies in crop production are significant.

b.   Industrialization of livestock production. Livestock production is no longer tied to crop production. Livestock production units are specialized, large, and generally increasing in size (Martinez 1999). Most livestock production is now undertaken under contracting arrangements with processors. These large enterprises raise new environmental problems (Hoban et al. 1998).

c.   Increased concentration in the marketing-processing sectors. This includes plant and animal breeding where the supply of genetic material is increasingly concentrated.

d.   The decline of rural communities. The "horse and buggy" towns of the 1950s and even the 1960s are in decline unless they provide non-farm employment opportunities.

e.   Decline of political influence. The proportion of the U.S. labor force engaged in agricultural production is now well below 2 percent. The number of bona fide full-time farmers has declined (USDA 2000). Farm political influence, however, remains remarkably strong because of the increasingly homogeneous interests of operators of specialized farms and improved information technology for coordinating the membership of farm organizations. These factors have largely offset the declining numbers of farm producers (Olson 1965). But American agriculture no longer can claim that small commercial farms (owned by "widows and orphans") are a significant part of rural America.

f.   Growing resentment regarding farm subsidies. With the passage of the 2002 Farm Bill, resentment regarding agricultural subsidies has increased. Agricultural subsidies do not meet standards of fairness or equity. Nonfarm taxpayers are being asked to make income transfers to farmers who have higher incomes and far larger wealth than they do.

Has the SAES-USDA system responded in a constructive way to changes in the structure of U.S. agriculture? When one looks realistically and collectively at these factors, it is quite remarkable that SAES programs have been maintained in many states. The New England states, for example, produce little agricultural product, and some of the Great Plains states also have low product values.

All states have an interest in maintaining a state LGU because these universities are seen as important to states in providing higher-education training services and perhaps more importantly, in maintaining membership in the global community of scholars. The Colleges of Agriculture (or Natural Resources) may be seen as low-cost components of the modern LGU.

We are probably not far from the time when some colleges of agriculture will be consolidated into regional colleges (e.g., in New England), and this will probably also mean SAES consolidation. SAES programs do have minimum threshold sizes and the management of truly modern biotechnology research programs probably increases the threshold size for an effective program as expensive and specialized research equipment is required to furnish biotech laboratories. Private-sector firms, for example, do not maintain field experiment stations in all states, and SAES programs may be increasingly under pressure to achieve scale economies.

Although the social science programs (economics and sociology) have addressed the changing structure of farm production and of rural communities one senses that the magnitude of the problems hasn't been fully appreciated. The university social science programs addressing the concentration of agricultural marketing also appear to have an element of understatement as to the magnitude of the problem.

A final point on the marketing and food technology question is that the SAES units have generally had relatively low investments in research programs for the post-harvest areas: e.g., food safety, food technology (the search for more effective food-processing technology). Food-industry firms are closer to the markets and have a comparative advantage in developing new food products. However, with the prospects for second-generation GM products in the relatively near future, the need for food-safety evaluation and processing technology development argues for a substantial increase in SAES research in these fields.

All in all, the SAES-USDA system has managed to preserve its budgetary position remarkably well in view of the decline in the number of fulltime equivalent (FTE) farmers in the United States (see Chapter 7). Employment in McDonald's restaurants is approaching the number of FTEs in U.S. agriculture. Colleges of agriculture have broadened their focus somewhat, taking up environmental and natural-resource concerns, and foreign graduate students have replaced U.S. graduate students in graduate programs.

Yet, we are "uneasy" about future changes. Rural communities have been on the decline in most of the U.S. for decades. We also sense that SAES-USDA research programs have been laggards in responding, not only to the biotechnology revolution but to structural change in agriculture and communities as well. The ultimate question is the length of the lag in response and the feasibility that alternative research programs will address the researchable questions entailed.

On the latter point, we note that no other public-sector program in state governments or in the federal government other than the SAES-USDA system is positioned to address the research problems associated with the structural changes in agricultural and rural communities. The private sector is now positioned to handle much of the invention and innovation for U.S. agriculture, but it is not positioned to undertake pre-invention science, research on resource and environmental quality, and public-policy research that will continue to be vital to agricultural sector growth in the 21st century.

Issues of Private Support for Public Research

The SAES-USDA system has had a history and tradition of strong public service (Chapter 1). Before the Bayh-Dole Act of 1980, SAES-USDA units made no attempt to "charge" for the technology they developed. New publicly developed technologies were treated largely as a public good and made widely available. Furthermore, support for the public agricultural-extension service was justified on the grounds that new technologies, particularly "best practice" technologies, required public information services to achieve rapid adoption (Huffman 1974, 2001). This changed with the Bayh-Dole Act. The fundamental argument of the Bayh-Dole Act was that unless an "exclusive license" was granted to an innovating firm, the innovating firm would not make the investments required to commercialize the invention. Since licensing revenues could be used to reduce the public support from state taxpayers, this system was not seen as a violation of the public role of the LGUs (Just and Huffman 2005).

Public universities have also had support from private individuals in the form of gifts and endowments, and these two are generally seen as consistent with LGU's public role unless the donor attempts to exercise influence on the nature of the activities supported. The proportion of industry-funded research in the SAES system has risen in recent decades and is now quite substantial at 15 percent (see Chapter 7). Some of this industry support is relatively innocuous and consistent with public responsibilities. In some cases, industry support is a form of "payment for services". In other cases, the research in question has public value, as for example, when the research bridges the gap between scientific development and invention potential. Many industries are willing to support this type of research.

But some industry support may be seen as being in conflict with public interest or responsible behavior of a public institution. This would be the case where public research programs were designed to benefit specific private firms. A form of industrial support that some may see as a model for future support for agricultural biotechnology research is exemplified by the Novartis support for research at the University of California at Berkeley. The agreement specifies that Novartis Agricultural Development Institute, a private firm, contributes research program support (in this case, 25 million dollars over a period of 5 years) for a range of agricultural biotechnology projects of the Department of Plant and Microbial Biology in the College of Agriculture and Natural Resources. In return for this support, Novartis received "first negotiating rights" to license inventions emerging from the research supported by their funds and inventions from all publicly funded research during the contract period of November, 23, 1998 to November 23, 2003 (Busch et al. 2004). This is a case of selling option rights in future discoveries to obtain support for the research associated with these future discoveries.[7]

The Novartis-UC Berkeley program has recently been the subject of an external evaluation. The evaluation raises a number of concerns, including that this form of private-sector contract is inconsistent with the public role and responsibilities of a land grant public university (see Busch et al. 2004).

As SAES programs grapple with their support base, they will inevitably face conflicts of interest between private support and public responsibility. Certain categories of private support will be relatively easy to accept as being without serious conflict. Others, such as the Novartis-UC Berkeley arrangement, will be more troubling.

Returns to Research: U.S. Studies

More than 60 studies of the agricultural extension and research system in the U.S. have now been conducted and published in some form (Evenson 2001). All of these studies measure the impact of extension programs, applied research programs, pre-invention science programs, and private-sector R&D spill-in on the productivity of farmers. Three methods are utilized (see Chapter 8). The first method is a version of "project evaluation" methods (sometimes referred to as imputation-accounting methods). In cases where the adoption of the research product (e.g., the adoption of hybrid corn varieties) can be measured, estimation of the productivity impacts can be linked to adoption data. This enables the analyst to construct a stream of benefits that can be related to a stream of costs. Given annual benefit and cost estimates, a benefit/cost ratio using an "external rate of discount" can be computed. An "internal rate-of-return" can also be computed as the rate at which the present value of benefits equals the present value of cost.

The second and third methods are required when one cannot measure the adoption of the products of the research program. These methods require statistical estimates of the relationship between production or productivity and a research-stock variable constructed from investment data. The construction of the research-stock variable requires specification of the geographic scope or range of research-program benefits and of the "timing-weight pattern" of the relationship between the timing of investment and later production impacts.

Typically the geographic issue is realized by identifying an SAES program with a specific state, although a number of studies have used geo-climate regions as shown in Figure 8.6. The 16 geo-climate regions allow the investigator to test for geographic spillovers from one region to another. In the first edition of this book we concluded that crop research spilled freely within subregions and that livestock research spilled freely within regions. (In Tables 9.1–9.5 we use the notation [G] to indicate estimates of geographic spill-in.)

When creating research-stock variables, most studies have imposed arbitrary timing weights. Some, however, have used interactive techniques to actually estimate the shape of timing weights (these studies are indicated by (T) in Tables 9.1–9.5). The typical timing-weight pattern imposed indicates that extension programs have an impact in the current production year with some lag effects in the following year. For research programs, however, a lag of two to three years between research expenditure and the start of any productivity impact is usually imposed. Then a period of

Table 9.1   Returns to agricultural extension (statistical studies)

Studies	Production structure	Period		MIRR[a]
A. Agricultural Extension				
Huffman (1974)	MFP	1959–74		16
Huffman (1976)	MFP	1964		110
Evenson (1979)	MFP	1971		100
Huffman (1981)	MFP	1979		110
Evenson (1994)	MFP (T)	1950–72	Crops	101
	MFP (T)	1950–72	Livestock	89
	MFP (T)	1950–72	Aggregate	82
Norton and Paczkowski (1993)	MFP	1993		37
B. Combined Extension—Research				IRR
White and Havlicek (1982)	MFP	1943–77		7–36
Lu et al. (1979)	MFP	1939–72		25
Evenson (1979)	MFP	1948–71		110
	Mean MIRR			66
	Median MIRR			82

[a]MIRR is the marginal annual internal rate of return to society.

Table 9.2 Returns to SAES-USDA research: Project evaluation methods

Study	Period	Research field	IRR[a]
Griliches (1958)	1940–55	Hybrid corn	35–40
Griliches (1958)	1940–57	Hybrid sorghum	20
Peterson (1967)	1915–60	Poultry	21–25
Schmitz & Seckler (1970)	1958–69	Tomato Harvester	37–46
Peterson & Fitzharris (1977)	1937–42	Aggregate	50
	1947–52	Aggregate	51
	1957–62	Aggregate	49
	1957–72	Aggregate	34
Bengston (1984)	1975	Forestry products	19–22
Bares & Loveless (1985)	unknown	Forestry products	9–40
Brunner & Strauss (1986)	unknown	Forestry products	73
Chang (1986)	unknown	Forestry products	35
Haygreen et al. (1986)	1972–81	Forestry products	14–36
Newman (1986)		Forestry products	0–7
Westgate (1986)	1969–2000	Forestry products	37–111
	Mean IRR		35
	Median IRR		35

[a]IRR is the annual internal rate of return to society.

rising cumulative impact on productivity of seven to ten years is realized (Griliches 1998). This may or may not be followed by a period of declining impact.

Actually the shape of the timing weights imposed on the research stock is important because they influence the calculation of the "internal rate-of-return" to investment in extension or research programs. A given research (extension) expenditure in time t generates a benefit stream over future periods. The present value of this benefit stream (and this depends on time weights) is set equal to the expenditure—the discount rate at which expenditures in time t are equal to the present value of benefits in the "internal rate-of-return" calculation (e.g., see Equation 8.24).

Tables 9.1–9.5 report summaries of rate-of-return studies. Table 9.1 summarizes six studies of agricultural-extension services and three studies of a combination of agricultural extension and research.

All are statistical studies. Only one estimated the shape of timing weights. Table 9.2 summarizes 12 studies utilizing the project evaluation method; six of these studies are for forest products. And Table 9.3 summarizes 35 studies of SAES-USDA research programs. This is the major body of evidence for SAES-USDA system productivity. Almost all of these studies are actually of the SAES system. This evidence supports an unchanging rate-of-return over time.

The median IRR is over 40 percent in Table 9.3. The range is quite variable, but these estimates indicate that the taxpayer is investing in a productive research system. As we have noted in Chapter 7, the SAES-USDA research system is supported by farmers, not by consumers. But consumers captured virtually all of the benefits from SAES-USDA investments (see Figure 8.5). Farmers have had to make major economic adjustments as new technology has been delivered to them, particularly by private firms.

Table 9.4 summarizes four studies of private-sector R&D spill-in. Some of this R&D is conducted by foreign firms, but most of it is conducted by U.S. firms. Private-sector firms can capture only a part of the benefits of this research in higher prices or technology fees. Even with strong IRRs this

Table 9.3   Returns to SAES-USDA research: Statistical methods

Study	Production structure	Period	MIRR[a]
Griliches (1964)	MFP	1949–59	25–40
Latimer (1964)	MFP	1949–59	n.s.
Peterson (1967)	MFP	1915–60	21–25
Evenson (1968)	MFP (T)	1949–59	47
Cline (1975)	MFP	1939–48	41–50
Bredahl & Peterson (1976)	MFP	1937–42	56
	MFP	1947–57	51
	MFP	1957–62	49
	MFP	1967–72	34
Lu et al (1979)	MFP	1938–72	24–31
Evenson (1979)	APF (T) (G)[b]	1868–1926	65
	APF (T) (G)	1927–50	95
	APF (T) (G)	(South) 1948–71	130
	APF (T) (G)	(North) 1948–71	93
	APF (T) (G)	(West) 1948–71	95
Knutson & Tweeten (1979)	MFP	1949–72	28–47
Lu et al (1979)	MFP	1939–72	23–30
White et al (1978)	MFP	1929–77	28–37
Davis (1979)	MFP	1949–59	66–100
Davis & Peterson (1981)	MFP	1949	100
	MFP	1954	79
	MFP	1959	66
	MFP	1964–74	37
Norton (1981)	MFP	(Grains) 1969–74	31–44
	MFP	(Poultry) 1969–74	30–56
	MFP	(Dairy) 1969–74	27–33
	MFP	(Livestock) 1969–74	56–66
Otto & Havlicek (1981)	MFP	(Corn) 1967–79	152–212
	MFP	(Wheat) 1967–79	79–148
	MFP	(Soybeans) 1967–79	188
Sundquist et al (1981)	APF	(Corn) 1977	115
	APF	(Wheat) 1977	97
	APF	(Soybeans) 1977	118
Welch & Evenson (1989)	MFP	1969	55
White & Havlicek (1982)	MFP	1943–77	7–36
Smith et al (1983)	MFP	(Dairy) 1978	25
	MFP	(Poultry) 1978	61
	MFP	(Beef, Swine) 1978	22
Braha & Tweeten (1986)	MFP	1959–82	47
Fox (1986)	MFP	(Livestock) 1944–83	150
	MFP	(Crops) 1944–83	180
Seldon (1987)	MFP	(Forestry product) 1950–80	163
Seldon & Newman (1987)	MFP	(Forestry product) 1950–86	236
Evenson (1989)	MFP (T) (G)	(Aggregate) 1950–82	43
	MFP (T) (G)	(Crops) 1950–82	45
	MFP (T) (G)	(Livestock) 1950–82	11
Evenson (1991)	APF	(Crops) 1950–85	41
	APF	(Livestock) 1950–85	11
Huffman & Evenson (1993)	APF (T) (G)	(Crops) 1950–85	47
	APF (T) (G)	(Livestock) 1950–85	45
Alston, et al (1998)	MFP (T)	na	17–31
Chavas & Cox (1992)	MFP	na	28
Gopinath & Roe (1996)	APF		37
Makki et al (1996)	MFP	1930–90	27

Makki & Tweeten (1993)	MFP	1930–90	93
Oehmke (1996)	MFP	Pre 1930	nc
	MFP	1931–90	12
Yee (1992)	MFP	1931–85	49–58
Norton et al (1992)	APF	(Aggregate) 1987	30
	APF	(Cash grains) 1987	31
	APF	(Vegetables) 1987	19
	APF	(Fruits) 1987	33
	APF	(Other field crops) 1987	34
	APF	(Dairy) 1987	95
	APF	(Poultry) 1987	46
	APF	(Other Livestock) 1987	55
Huffman & Evenson (2005)	MFP	1970–1999	56
		Mean MIRR	53
		Median MIRR	45

[a]MIRR is the marginal annualized internal rate-of-return to society.
[b]Aggregate production function (APF).

Table 9.4   Studies of research spill-in from the private sector

Study	Production structure	Period	MIRR to private sector[a]
Huffman and Evenson (1993)	MFP	1950–85	41
Evenson (1991)	MFP	(Crops) 1950–85	45–71
Evenson (1991)	MFP	(Livestock) 1950–85	81–89
Gopinath and Roe (1996)	M	(Food processing) 1991	7
Gopinath and Roe (1996)	M	(Farm machinery) 1991	2
Gopinath and Roe (1996)	M	(Total) 1991	46
Huffman and Evenson (2005)	MFP	1970–1999	40
		Mean MIRR	45
		Median MIRR	45

[a]MIRR is the marginal annualized internal rate-of-return.

part is roughly one-third. Farmers capture most of the remaining benefits. Table 9.4 suggests that these benefits to farmers are large (In fact, farm productivity increases by approximately as much from private-sector R&D spill-ins as from SAES-USDA research).

Finally Table 9.5 summarizes three studies where an effort was made to identify the production impacts of pre-invention science. Pre-invention science has a longer time-lag with impacts distributed over more years than applied research, and these studies estimated this time-lag (the time to peak impact for applied research was roughly 9 years; for pre-invention science it was 15 years). These studies attest to the fact that pre-invention science is an important component of SAES-USDA research programs.[8]

Federal-Formula versus Competitive-Grant Funds

An on-going debate continues about the appropriate size of the federal government's competitive-grant and formula funded programs for agricultural research (e.g., see Office of Technology Assessment 1991; Huffman and Just 2000; Echeverria and Elliott 2001, NRC 2003). First, the large

Table 9.5 Studies of returns to pre-invention science

Study	Production structure	Period	MIRR[a]
Evenson (1979)	APF (T)	1927–50	110
	APF (T)	1946–71	45
Huffman & Evenson (1993)	MFP (T) (G)	(Crops) 1950–85	57
	MFP (T) (G)	(Livestock) 1950–85	83
	MFP (T) (G)	(Aggregate) 1950–85	64
Evenson (1991)	MFP (T) (G)	(Crops) 1950–85	40–59
	MFP (T) (G)	(Livestock) 1950–85	54–83
		Mean MIRR	65
		Median MIRR	59

[a]MIRR is the marginal annualized rate-of-return to society.

impacts of public agricultural research on agricultural productivity and high rates of return on investment occurred largely before the USDA had a significant competitive-grants program (see Tables 1.6, 9.2, and 9.3). These funds were used to support agricultural experiment-station research in LGUs where there is a well-defined policy for promotion, tenure, and performance of scientists. Hence, based on empirical evidence covering almost a century, one cannot conclude that federal-formula funds were unproductive.

Second, some officials have suggested reducing federal-formula funds for agricultural experiment-station research. The implication being that scientists could more extensively compete for federal competitive research grants. When public agricultural research is funded by federal competitive grants and contracts, the research agenda is set by the funding agency in Washington, D.C. This agenda will be different than the agenda set by the local SAES director for pre-invention and applied research. Recall that in Chapter 8 we simulated the likely impact of a federal agricultural-science policy in which an increase in USDA competitive-grant funds offsets a reduction in federal-formula funds. We showed that such a reallocation would lower agricultural productivity in general and benefit only a few states while reducing funds and agricultural productivity in all other states. Most notably, federal competitive-grant funding can be expected to have a different geographical distribution of benefits—less benefit locally and larger interstate spillover effects.

Third, when public agricultural research is funded by federal competitive grants and contracts, decisions are based on proposals rather than on completed projects. Huffman and Just (2000) show that this is an inefficient contact. In addition, the federal competitive-grants programs do not pay for research-proposal writing, so the risk of federal research-grant programs is borne by the competing scientists or their institutions and the somewhat distorted incentive structure increases transactional costs and can lower scientists' research productivity (Office of Technology Assessment 1991, Chubin and Hackett 1990). Furthermore, federal-funding agencies tend to fund less than 100 percent of the cost of most research projects, so other funds, particularly state-appropriated or federal-formula funds, are used to subsidize research sponsored from outside the state. For these reasons federally funded competitive grants do not look nearly as economically attractive to SAES directors as they do to the federal funding agencies.

Over the past two decades a number of studies have examined the effect of public investments in agricultural research and development, and all have demonstrated a positive and substantial impact on agricultural productivity. This occurs, in part, because the state agricultural-experiment stations have a long-term focus on addressing local problems and a long-term relationship with their scientists. As a result the positive reputation earned through these long-term relationships has strengthened incentives for discovery (Huffman and Just 1999, 2000).

And finally, given that each research-funding mechanism has its own advantages and disadvantages but also serves as a measuring stick for other mechanisms, society faces less risk to the future discovery process by having a diversified portfolio of projects and scientists working on agricultural research. U.S. agricultural-science policy should continue to support such diversity.

Comparisons: Commodity Research Allocation vs. Farm Revenue Shares

In Chapter 5, we developed the logic of the "congruence rule" for the allocation of applied research to commodities.[9] The congruence rule states that "unless there is reason to expect researcher productivity to differ between commodity programs, research resources should be allocated in proportion to the economic usage of the commodity." This rule should then result in a close matching of the share of research developed to a commodity program and to the share of cash receipt in farm income. (See the discussion in Chapter 5.)

We have attempted to compare research program shares in both the public SAES-USDA system and the private sector with cash-receipt shares in Table 9.6. The following assumptions were built into the calculations:

1. The crop research shares are based on the National Plant Breeding Study (see Chapter 5). Public-sector systems were multiplied by 1.2 to reflect the pre-invention research component.
2. Livestock-research shares for the private sector were based on the ratio of crop inventions to livestock inventions reported in Table 5.5. Roughly half as many livestock-production inventions were obtained as for crops. For sheep and wool, we assume little private R&D occurs.
3. For forestry research, we also assumed little private-sector R&D. Most private-sector forestry R&D is for forestry-product development (plywood, chipboard, etc.).
4. Cash receipt shares are taken from the most recent census of agricultural data. Forestry includes incomes from wood sales.

Consider crop-research and cash-receipt shares. We first note that the public-sector research shares do appear to be responding to private-sector shares.[10] A comparison between research shares and cash-receipts shares shows the largest positive difference (i.e. where the cash-receipt share exceeds the research share) to be in soybeans (5.8 percent) and fruits and nuts (5 percent). The largest negative discrepancy is for vegetables, where the research share exceeds the cash-receipts share by 4.7 percent. Roughly 50 percent of all agricultural research is for crop research while crop cash receipts account for 53 percent of farm cash receipt.

For livestock the research share is roughly 40 percent, while the cash-receipt share is 46 percent. Beef cattle and poultry have higher farm cash-receipt shares than research share, while the sheep and wool research share greatly exceeds its farm cash-receipt share (even though we have not attributed much private-sector research to sheep and wool). The largest discrepancy in Table 9.6 is for forestry research, where the research share is high (9 percent of public agricultural research expenditure) but forest farm income shares are very low.[11]

Some of these discrepancies can be attributed to scale economics in research. This might explain the soybean discrepancy. Research-system diseconomies of scale would explain the fruit-nuts discrepancy, and the vegetable discrepancy could be explained by highly-productive vegetable-research programs reinforced by IPRs.

For livestock research, beef cattle research could reflect scale economics. The most glaring discrepancy is for sheep and wool, where the research share exceeds the cash-receipts share by a large margin. This would appear to be inefficient.

Finally, the high share of public-sector forestry research is in great contrast to income from the forestry sector. And this discrepancy is almost surely inefficient, unless one can argue that non-pecuniary services from forests are high. It is true that many have argued for high non-pecuniary benefits for this sector. It is also the case that as the farm-production clientele of the SAES-USDA system has declined, research on natural resources has risen. Colleges of Agriculture are now frequently Colleges of Agriculture and Natural Resources. Only a few colleges have moved in the direction of the food industries (an exception is at the University of Florida), even though the value added in the food industries, including restaurants, has been rising and is now a significant multiple of farm income. But this has not happened for forest-based products.

Table 9.6. Public and private commodity research shares and cash receipt shares

Commodities	Research shares 1990s (%)			Farm cash receipt shares (1995)
	Public	Private	Total	%
Crops				
Cereals				
Corn	1.1	11.7	12.8	10.9
Wheat	2.0	1.2	3.2	4.6
Rice	0.6	0.5	1.1	0.9
Sorghum	0.4	0.9	1.3	0.9
Other	1.0	1.1	2.2	2.8
Legumes				
Soybeans	1.5	2.2	3.7	9.5
Other	0.6	0.6	1.3	1.1
Oilseed Crops	0.7	1.4	2.1	0.9
Fiber Crops	0.8	2.2	3	3.4
Forage Crops	1.9	1.1	3	2.4
Root Crops	1.5	0.6	2.1	1.2
Sugar Crops	0.5	0.6	1.1	1.0
Vegetables	1.7	5.8	7.5	2.8
Fruits-Nuts	2.4	0.7	3.1	8.1
Ornamentals	0.6	1.4	2.0	2.1
Misc. Crops	0.7	0.3	1.0	0.6
Total Crops	18.0	32.3	50.0	53.2
Livestock				
Beef Cattle	6.2	5.5	11.7	17.7
Dairy Cattle	5.1	4.5	9.6	10.4
Swine	3.6	3.4	7.0	6.4
Poultry	3.6	3.2	6.8	11.0
Sheep (Wool)	4.7	1.1	5.8	0.6
Total Livestock	23.2	17.7	40.0	46.2
Forestry	9.0	1.0	10.0	0.6

Conclusion

In this final chapter, we have addressed five questions. The first and most important question was whether the SAES-USDA system was effectively responding to the scientific revolution in the biological sciences. We do see a high degree of responsiveness on this score. Most applied science departments in LGUs now have courses incorporating recent developments in molecular biology and related sciences.

But we are disappointed about the pace of product development in the SAES-USDA system. The development of second-generation biotech products by SAES-USDA programs with attractiveness for health conscious consumers would do much to alleviate consumer anxiety and concerns over GM foods. But the SAES-USDA system has not developed such products and appears to be a number of years from doing so. Perhaps this concern is unfounded in light of the complexities of the Bayh-Dole Act and the invention-innovation linkage. The difficulty that public-sector programs face in competition with private-sector firms is well known.

The second question addressed was whether the SAES-USDA system was realistically responding to the changing structure of agriculture in the United States. U.S. agriculture has changed drastically in the past 40–50 years. Crop farms have expanded in size and become more specialized. Modern farm equipment allows for crop farming on an ever larger scale. The availability of GM crop products enables even greater scale economies.

Livestock production has changed even more than is the case for crop production. Virtually all livestock production is now industrialized. Confined animal feeding units (CAFUs) are in place today and most livestock production is now contracted by vertically integrated firms. Industry concentration has increased.

The distribution of farms by size has become bimodal with a growing share of small hobby-type farms near major metropolitan areas that generate very little cash income and large commercial farms. In rural areas that are some distance away from major metropolitan areas, small towns are disappearing. The only viable rural communities left are communities with significant size and communities with off-farm employment, commuting, or recreational opportunities.

The SAES-USDA system has responded slowly to these changes. Our sense is that the system is roughly "20 years behind the times." The colleges of agriculture in the LGUs have had to respond to declining student numbers and those that have been successful have added business programs at the undergraduate level and admitted foreign students at the graduate level. They have been creative in attracting students—particularly foreign students. Agricultural-extension programs have been slow to respond to the changing market for information.

The third question addressed private financing of SAES-USDA research (particularly SAES research). We discussed the rising share of industry-funded research and noted some elements of conflict with the responsibilities of a public-service LGU. On the whole, we concluded that most LGUs were incorporating industrial support responsibly.

Our fourth question was whether the SAES-USDA system delivered productivity gains to farmers. We reviewed more than 60 studies addressing this question. Our findings were that while returns to investment were quite variable, mean and median returns estimates were high. Investment in productivity-increasing activities in the SAES-USDA system and in private-sector firms has yielded high returns, and these returns are not declining. The U.S. agricultural sector has outperformed the nonfarm sectors in the U.S. economy in multifactor productivity growth for at least seven decades. The result has been lower real prices of farm products—roughly decreasing at a rate of

1 percent per year—a trend which benefits consumers of these products. The dynamics of these productivity gains have meant that the structural changes noted above have taken place. Farm incomes have not always been improved by new technology. But there is little doubt that the market adjustment to U.S. agricultural productivity changes has had broad benefits. These benefits are "pro-poor" because the poor spend a high proportion of their income on food.

The fifth issue addressed was federal-formula funding versus competitive grant funding of agricultural experiment station research. For more than a century, the federal government has provided formula or block-grant research funds for state agricultural experiment stations. Decisions on how to use these funds have been made at the agricultural experiment station level. These funds have been allocated to salaries of scientists and research assistants and for certain types of infrastructure needed for agricultural research. The record shows that the social rate-of-return on federal-formula funds has been quite high. CSRS/CSREES–administered competitive grant programs have been used only recently, basically since 1980. Because the research agenda for these funds has been set at the national level, they tend to fund a different type of agricultural research than federal-formula funds. Competitive grant programs are especially useful for identifying new scientific talent when new areas of research are to be explored. They also contribute to a diversified portfolio of funding mechanisms for agricultural research, which reduces the discovery risk associated with science for agriculture. The future science for agriculture will be strengthened by the federal government providing diversified funding mechanisms for public agricultural research.

The sixth question addressed the commodity alternatives of research resources in the SAES-USDA system. We found that most crop-research allocations could be "rationalized." One of the livestock-resource allocations, sheep and wool, could not. This is a source of resource misallocation.

We also found that the forestry research share on natural resources research more generally was considerably higher than the contribution of that sector to income. This might be rationalized if sufficient non-pecuniary benefits could be identified; otherwise it is too large.

We end this volume then with the judgment that the SAES-USDA system remains viable and productive. It has enough responsiveness to remain so for some time. This does not mean that we will not see new regional forms, particularly in states where SAES units and agriculture are small.

Notes

1. The pre-invention sciences contribute to the interests of inventors by addressing problems inhibiting invention prospects.
2. Critics of GM foods are often critical of multinational firms in the GM-crops industry. Public-sector development of GM-crops products would demonstrate that the public sector uses modern science.
3. This industrialization process has been quite rapid. Hog production was industrialized in a few years.
4. Ravetz (1971) discusses the immature sciences and their susceptibility to "capture" by interest groups.
5. See Chapters 5 and 6 for biotechnology invention.
6. The seven major multinational firms in the ag biotech industries spend more than $3 billion on R&D.
7. This selling of option rights differs from the selling of patent rights in that it commits public research programs to obligations when the research product has not been produced.
8. These IRR estimates are discussed in an international comparative context in Evenson (2001).
9. The logic of the congruence rule is that if the same technical parameters affecting the probability of making an invention hold for two different commodity programs, this implies that research investments should be allocated in proportion to the economic value of the commodities.

10. A regression of public-research shares on private-research shares produced the following regression results, which show that a large share of the variance in research shares can be explained by cash receipt shares:

research share = 1.5315 + 0.596 cash receipts share, (R^2 = .63)
                 (2.20)   (5.96)

11. Public perceptions are often at odds with reality. The income generated by forest products, firewood, and lumber to woodlot owners and farmers is actually very low.

References

Alston, Julian M., Michele C. Marra, Philip G. Pardey, and T.J. Wyatt. 1998. "Research Returns Redux: A META-Analysis of the Returns to Agricultural R&D." EPTD No. 38. Environment and Production Technology Division. Washington, D.C.: International Food Policy Research Institute.

Bare, B.B. and R. Loveless. 1985. "A Case History of the Regional Forest Nutrition Project: Investments, Results, and Applications." Final Report for USDA Forest Service, Project Number PNW82-248, North Central Forest Experiment Station. Seattle, WA: University of Washington.

Bengston, D.N. 1984. "Economic Impacts of Structural Particleboard Research." *Forest Science* 30(3): 685–97.

Boehlje, M. 1999. "Structural Changes in the Agricultural Industries: How Do We Measure, Analyze, and Understand Them?" *American Journal of Agricultural Economics* 81: 1028–1041.

Braha, H. and L. Tweeten. 1986. "Evaluating Past and Prospective Future Payoffs from Public Investments to Increase Agricultural Productivity." *Technical Bulletin T-165*. Agricultural Experiment Station. Stillwater, OK: Oklahoma State University.

Bredahl, M. and W. Peterson. 1976. "The Productivity and Allocation of Research: U.S. Agricultural Experiment Stations." *American Journal of Agricultural Economics* 58: 684–692.

Brunner, A.D. and J.K. Strauss. 1986. "The Social Returns to Public R&D in the U.S. Wood Preserving Industry (1950–1980)." Research Triangle Park, NC: Southeastern Center for Forest Economics Research.

Busch, L, R. Allison, A. Rudy, B.T Shaw, T. Ten Eyck, D. Coppin, J. Konefal, C. Oliver with J. Fair-Weather. 2004. *External Review of the Collaborative Research Agreement between Novartis Agricultural Development Institute and the Regents of the University of California*. East Lansing, MI: Institute for Food and Agricultural Standards, Michigan State University.

Chang, S.U. 1986. "The Economics of Optimal Stand Growth and Yield Information Gathering." Report submitted to the USDA Forest Service. St. Paul, MN: North Central Forest Experiment Station.

Chavas, J-P. and T.L. Cox. 1992. "A Nonparametric Analysis of the Influence of Research on Agricultural Productivity." *American Journal of Agricultural Economics* 74: 583–591.

Cline, P. L. 1975. "Sources of Productivity Change in United States Agriculture." Ph.D. dissertation. Stillwater, OK: Oklahoma State University.

Chubin, D.E. and E.J. Hackett. 1990. *Peerless Science: Peer Review and U.S. Science Policy*. Albany, NY: State University of New York Press.

Committee on Research Advisory to the U.S. Department of Agriculture. 1972. "Report Submitted to the National Academy of Science/National Research Committee." (Pound Report). Unpublished.

Davis, J.S. 1979. "Stability of the Research Production Coefficient for U.S. Agriculture." Ph.D. dissertation. St. Paul, MN: University of Minnesota.

Davis, J.S. and W. Peterson. (1981). "The Declining Productivity of Agricultural Research." In *Evaluation of Agricultural Research*. ed. G.W. Norton, W.L. Fishel, A.A. Paulsen and W.B. Sundquist. Miscellaneous Publication 8-1981. St. Paul, MN: Minnesota Agricultural Experiment Station, University of Minnesota.

Echeverria, R.G. and H. Elliott. 2001. "Competitive Funds for Agricultural Research: Are They Achieving What We Want?" In *Tomorrow's Agriculture: Incentives, Institutions, Infrastructure and Innovations*, ed. G.H. Peers, and P. Pingali, Proceedings of the Twenty-Fourth International Conference of Agricultural Economists. Hants, England: Ashgate Press.

Evenson, R.E. 1968. "The Contribution of Agricultural Research and Extension to Agricultural Production," Ph.D. Dissertation (University of Chicago).

Evenson, R.E. 1979. "Agricultural Research, Extension and Productivity Change in U.S. Agriculture: A Historical Decomposition Analysis." Agricultural Research and Extension Evaluation Symposium, May 21–23, 1979. Moscow, Idaho.

Evenson, R.E. 1989. "Productivity Decomposition in Brazilian Agriculture." Unpublished manuscript, Economic Growth Center, Yale University, New Haven, CT.

Evenson, R.E. 1991. "Research and Extension in Agricultural Development." *Forum Valuazione* 2 (November).

Evenson, R.E. 1994. "Analyzing the transfer of agricultural technology." In *Agricultural Technology: Policy Issues for the International Community,* ed. J.R. Anderson. Wallingford, UK: CAB International.

Evenson, R.E. 2001. "Economic Impacts of Agricultural Research and Extension." In *Handbook of Agricultural Economics, Vol 1A.* ed. B.L. Gardner and G. Rausser. New York, NY: North-Holland. 2001.

Foltz, J.D., K. Kim, and B. Barham. 2003. "A Dynamic Analysis of University Agricultural Biotechnology Patent Production." *American Journal of Agricultural Economics* 85:187–197.

Fox, G. 1986. "Underinvestment, Myopia and Commodity Bias: A Test of Three Propositions of Inefficiency in the U.S. Agricultural Research System." Department of Agricultural Economics and Business. Guelph, Ontario: University of Guelph.

Gopinath, M. and T.L. Roe. 1996. "R&D Spillovers: Evidence from U.S. Food Processing, Farm Machinery and Agriculture," Bulletin No. 96-2. Economic Development Center, Department of Applied Economics, University of Minnesota. St. Paul, Minnesota.

Griliches, Z. 1958. "Research Costs and Social Returns: Hybrid Corn and Related Innovations." *Journal of Political Economy* 66: 419–431.

Griliches, Z. 1964. "Research Expenditures, Education and the Aggregate Agricultural Production Function." *American Economic Review* 54: 961–974.

Griliches, Z. 1998. *R&D and Productivity: The Econometric Evidence.* Chicago, IL: The University of Chicago Press.

Haygreen, J., H. Gregerson, I. Holland, and R. Stone. 1986. "The Economic Impact of Timber Utilization Research." *Forest Products Journal* 36(2): 12–20.

Hoban, T., J. Molnar, M. McMillian, and J. Parrish. 1998. "Industrialization of Agriculture: Case Study Lessons for the Natural Resources Conservation Service." U.S. Dept. Agriculture Technical Report.

Huffman, W.E. 1974. "Decision Making: The Role of Education." *American Journal of Agricultural Economics* 56: 85–97.

Huffman, W.E. 1976. "The Productive Value of Human Time in U.S. Agriculture." *American Journal of Agricultural Economncis* 58: 672–683.

Huffman, W.E. 1977. "Interaction between Farm and Nonfarm Labor Markets." *American Journal of Agricultural Economics* 59: 1054–1061.

Huffman, W.E. 1981. "Black-White Human Capital Differences: Impact on Agricultural Productivity in the U.S. South." *American Economic Review* 71: 94–107.

Huffman, W.E. 2001. "Human Capital: Education and Agriculture." In *Handbook of Agricultural Economics, Vol. IA,* ed. Bruce L. Gardner and Gordon C. Rausser. Amsterdam, Netherlands: Elsevier Science/North-Holland.

Huffman, W.E. 2005. "Trends, Adjustments, Demographics, and Income of Agricultural Workers." *Review of Agricultural Economics* 27 (forthcoming).

Huffman, W.E., and R.E. Evenson. 1993. *Science for Agriculture: A Longterm Perspective.* Ames, IA: Iowa State University Press.

Huffman, W.E. and R.E. Evenson. 2001. "Structural Adjustment and Productivity Change in U.S. Agriculture, 1950–82." *Agricultural Economics* 24: 127–147.

Huffman, W.E. and R.E. Evenson. 2005. "New Econometric Evidence on Agricultural Total Factor Productivity Determinants: Impacts of Funding Composition." Iowa State University, Department of Economics, Working Paper #03029.

Huffman, W.E. and R.E. Just. 1999. "The Organization of Agricultural Research in Western Developed Countries." *Agricultural Economics* 21(Aug): 1–18.

Huffman, W.E. and R.E. Just. 2000. "Setting Efficient Incentives for Agricultural Research: Lessons from Principal-Agent Theory." *American Journal of Agricultural Economics* 82: 828–841.

Huffman, W.E. and P.F. Orazem. 2005. "Agriculture and Human Capital in Economic Growth: Farmers, Schooling and Nutrition." Iowa State University, Department of Economics, Working Paper #04016. Chapter for *Handbook of Agricultural Economics, Vol 3, Agricultural Development*, ed. Robert Evenson, T. Paul Schultz, and P. Pingali. Elsevier Science. (forthcoming).

Just, R.E. and W.E. Huffman. 2004. "The Role of Patents, Royalties, and Public-Private Partnering in University Funding." In *Essays in Honor of S.R. Johnson*, ed. J.P. Chavas and M. Holt. The Berkeley Electronic Press.

Knutson, M. and L.G. Tweeten. 1979. "Toward an Optimal Rate of Growth in Agricultural Production Research and Extension." *American Journal of Agricultural Economics* 61: 70–76.

Latimer, R. 1964. "Some Economic Aspects of Agricultural Research and Extension in the U.S." Ph.D. dissertation. W. Lafayette, IN: Purdue University

Lu, Y.C., P. Cline, and L. Quance. 1979. "Prospects for Productivity Growth in U.S. Agriculture." *Agricultural Economics Report, No. 435*. Washington, DC: USDA-ESCS.

Makki, S.S. and L.G. Tweeten. 1993. "Impact of Research, Extension, Commodity Programs, and Prices on Agricultural Productivity." Paper presented at the 1993 meetings of the American Agricultural Economics Association, Orlando, Florida.

Makki, S. S., L. G. Tweeten, and C. S. Thraen. 1996. "Returns to Agricultural Research: Are We Assessing Right?" Contributed Paper Proceedings from the Conference on Global Agricultural Science Policy for the Twenty-First Century, August 26–28, Melbourne, Australia. pp. 89–114.

Martinez, S. 1999. "Vertical Coordination in the Pork and Broiler Industries." *Agricultural Economics Report No. 777*. Washington, D.C.: Economic Research Service, U.S. Department of Agriculture.

Newman, D.H. 1986. "An Econometric Analysis of Aggregate Gains from Technical Change in Southern Softwood Forestry." Ph.D. Dissertation. Durham, NC: Duke University.

Norton, G.W. 1981. "The Productivity and Allocation of Research: U.S. Agricultural Experiment Stations, Revisited." In *Evaluation of Agricultural Research*, ed. G.W. Norton, W.L. Fishel, A.A. Paulsen, and W.B. Sundquist. Miscellaneous Publication 8-1981. Minnesota Agricultural Experiment Station. St. Paul, MN: University of Minnesota.

Norton, G.W., J. Ortiz, and P.G. Paredey. 1992. "The Impact of Foreign Assistance on Agricultural Growth." *Economic Development and Cultural Change* 40(4): 775–786.

Norton, G.W. and R. Paczkowski. 1993. "Reaping the Return on Agricultural Research and Education in Virginia." Information Series 93-3, College of Agriculture and Life Sciences, Virginia Polytechnic Institute and State University, Blacksburg, VA.

NRC (National Research Council, Board on Agriculture). 1989. *Investing in Research: A Proposal to Strengthen the Agricultural, Food, and Environmental System*. Washington, DC: National Academy Press.

NRC (National Research Council). 1992. *Plant Biology Research and Training for the 21st Century*. Commission on Life Sciences, Washington, DC: National Academy Press.

NRC (National Research Council). 2002. *Publicly Funded Agricultural Research and the Changing Structure of U.S. Agriculture*. Committee to Review the Role of Publicly Funded Agricultural Research on the Structure of U.S. Agriculture. Washington, D.C.: National Academy Press.

NRC (National Research Council). 2003. *Frontiers in Agricultural Research: Food, Health, Environment, and Communities*. Committee on Opportunities in Agriculture. Washington, D.C.: National Academy Press.

Oehmke, J.F. 1996. "The Maturation of the U.S. Agricultural Research System and its Impacts on Productivity." Staff Paper #96-85, Department of Agricultural Economics. East Lansing, Michigan: Michigan State University.

Office of Technology Assessment. 1981. *An Assessment of U.S. Food and Agricultural Research System*. Washington, D.C.: U.S. Government Printing Office.

Office of Technology Assessment. 1991. *Federally Funded Research: Decisions for a Decade*. OTA-SETA-490. Washington, D.C: U.S. Government Printing Office.

Olson, Mancur. 1965. *The Logic of Collective Action: Public Good and the Theory of Groups*. Cambridge, MA: Harvard University Press.

Otto, D. and J. Havlicek, Jr. 1981. As cited in Evenson, R.E. 1980. *Human Capital and Agricultural Productivity Change*. Draft. New Haven, CT: Economic Growth Center, Yale University.

Peterson, W.L. 1967. "Returns to Poultry Research in the United States." *Journal of Farm Economics* 49: 656–669.

Peterson, W.L., and J.C. Fitzharris. 1977. "The Organization and Productivity of the Federal State Research System in the United States." In *Resource Allocation and Productivity in National and International Agricultural Research*, ed. T. M. Arndt, D. G. Dalrymple, and V. W. Ruttan. Minneapolis: University of Minnesota Press.

Ravetz, J.R. 1971. *Scientific Knowledge and Its Social Problems*. Oxford, England: Clarendon Press.

Rockefeller Foundation. 1982. *Science for Agriculture*. (Winrock Report). New York, NY: Rockefeller Foundation.

Schmitz, A., and D. Seckler. 1970. "Mechanized Agriculture and Social Welfare: The Case of the Tomato Harvester." *American Journal of Agricultural Economics* 52: 569–577.

Seldon, B.J., 1987. "A Nonresidual Estimation of Welfare Gains from Research: The Case of Public R&D in a Forest Product Industry." *Southern Economic Journal* 54: 64–80.

Seldon, B.J. and D.H. Newman. 1987. "Marginal Productivity of Public Research in the Softwood Plywood Industry: A Dual Approach." *Forest Science* 33: 872–888.

Smith, B., G.W. Norton, and J. Havilcek, Jr. 1983. "Impacts of Public Research Expenditures on Agricultural Value-added in the U.S. and the Northeast." *Journal of the Northeastern Agricultural Economics Council* 12: 109–114.

Sundquist, W.B., C. Cheng, and G.W. Norton. 1981. "Measuring Returns to Research Expenditures for Corn, Wheat, and Soybeans." In *Evaluation of Agricultural Research*, ed. G.W. Norton, W.L. Fishel, A.A. Paulsen and W.B. Sundquist. Miscellaneous Publication 8-1981. Minnesota Agricultural Experiment Station. St. Paul, MN: University of Minnesota.

Thursby, J.G., R. Jensen, and M.C. Thursby. 2001. "Objectives, Characteristics and Outcomes of University Licensing: A Survey of Major U.S. Universities." *Journal of Technology Transfer* 26:59–72.

U.S. Department of Agriculture. 2000. *1997 Census of Agriculture*. NASS. <http://www.nass.usda.gov/census/census97/highlights/usasum/us.html>

Welch, F. and R.E. Evenson. 1989. "The Impact and Pervasiveness of Crop and Livestock Improvement Research in U.S. Agriculture." Economic Growth Center. New Haven, CT: Yale University.

Westgate, R.A. 1986. "Benefits and Costs of Containerized Forest Tree Seedling Research in the United States." In *Evaluation and Planning of Forestry Research*, ed. D.P. Burns. General Technical Report NE-GTR111. Brommall, PAII (USDA Forest Service, Northeastern Forest Experiment Station).

White, J.F. and J. Havlicek. 1982. "Optimal Expenditures for Agricultural Research and Extension: Implications on Underfunding." *American Journal of Agricultural Economics* 64(1): 47–54.

White, J.F., J. Havlicek, Jr. and D. Otto. 1978. "Fifty Years of Technical Change in American Agriculture." International Conference of Agricultural Economists, in Banff, Alberta, Canada, September 3–12, 1979.

Yee, J. 1992. "Assessing Rate of Return to Public and Private Agricultural Research." *Journal of Agricultural Economics Research* 44: 35–41.

Glossary/Acronyms

AAEA—American Agricultural Economics Association

AAUP—American Association of University Professors

AMS—Agricultural Marketing Service

APHIS—Agricultural Plant Health Inspection Service

ARS—Agricultural Research Service

ARS/USDA—Agricultural Research Service of the United States Department of Agriculture

AVC—average variable cost

BAE—Bureau of Agricultural Economics

BASF—A private chemical company

BR—breeders' rights

BST—bovine somatatrophin

CAFU—confined animal feeding unit

CBD—Convention on Biodiversity

CD—cultivar development

CGIAR—Consultative Group on International Agricultural Research

CIAT—International Center for Tropical Agriculture

CIMMYT—International Maize and Wheat Improvement Center

CRIS—Current Research Information System

CSREES—Cooperative States Research, Education and Extension Service

CSRS—Cooperative States Research Service

DNA—deoxyribonucleic acid

DUS—distinct, uniform and stable

ERS—Economic Research Service

FAO—Food and Agricultural Organization

FTE—full-time equivalent

GAO—General Accounting Office

GE—germplam enhancement

GM—genetically modified

GMOs—genetically modified organisms

IARC—International Agricultural Research Center

ICPNV—International Convention for the Protection of New Varieties of Plants

II—invention innovation

IITA— International Institute for Tropical Agriculture

IOM—industry-of-manufacture

IPC—international patent class

IPRs—intellectual property rights

IRR—internal rate of return

IRRI—International Rice Research Institute

ISI—Institute for Scientific Information

ISNAR—International Service for National Agricultural Research

LGUs—land grant universities

MFP—multifactor productivity

MNC—multinational corporation

MVs—modern varieties

NARS—National Agricultural Research System

NGOs—Non-Governmental Organizations

NICs—newly industrialized countries

NRI—National Research Initiative (Competitive Grants Program)

OECD—Organization for Economic Cooperation and Development

OES—Office of Experiment Stations

OTA—Office of Technology Assessment

PBR—plant breeders' right

PVP—Plant Variety Protection

PVPA—Plant Variety Protection Act

PVPC—Plant Variety Protection Certificate

R&D—research and development

rDNA—recombinant deoxyribonucleic acid

RICs—recently industrialized countries

RPA—research problem area

RR—Round-Up-Ready®

SMY—scientist-man-year

SOU—sector-of-use

SRS—Statistical Reporting Service

SYs—scientist years

T—transition

TC—technological capital

TFP—total factor productivity

TM—technology mastery

TRIPS—Trade Related Intellectual Property Issues

UCS—Union of Concerned Scientists

UNDP—United Nations Development Program

UNESCO—United Nations Education, Scientific and Cultural Organization

UNIDO—United Nations Industrial Development Organization

UPOV—International Union for the Protection of New Varieties of Plants

USAID—United States Agency for International Development

USDA-CSRS/CSREES—United States Department of Agriculture Cooperative State Research Service / Cooperative State Research Education and Extension Service

USDA-NASULGC—United States Department of Agriculture and National Association of State Universities and Land Grant Colleges

USDA-SAES—United States Department of Agriculture and State Agricultural Experiment Stations

USPCs—United States Patent Classes

USPTO—United States Patent and Trademark Office

WTC—Wellesley Technology Concordance

WTO—World Trade Organization

YTC—Yale Technology Concordance

Index